I0510625

APPLY SAFETY RISK
AND RELIABILITY ANALYSIS
OF MARINE SYSTEM

APPLY SAFETY RISK AND RELIABILITY ANALYSIS OF MARINE SYSTEM

Dr. Oladokun S. Olanrewaju, CEng, CMarEng

Co-authors:

Ab Saman A. Kader, W.B. Wan Nik,
Sunday W. Peter, Olusegun Oladimeji

Copyright © 2013 by Dr. Oladokun S. Olanrewaju, CEng, CMarEng.

ISBN: Softcover   978-1-4931-0929-6
               Ebook   978-1-4931-0930-2

All rights reserved. No part of this book may be reproduced or transmitted
in any form or by any means, electronic or mechanical, including photocopying,
recording, or by any information storage and retrieval system,
without permission in writing from the copyright owner.

Rev. date: 10/11/2013

To order additional copies of this book, contact:
Xlibris LLC
1-800-455-039
www.Xlibris.com.au
Orders@Xlibris.com.au
504630

CONTENTS

LIST OF FIGURES

LIST OF TABLE

FOREWORD

Risk and Reliability Analysis for safety preservation of environment remain part of sustainable development is important in the marine sector. I am pleased again to publish the book on Practical Marine Safety and Environmental Risk and Reliability Model. The book is written to present case study and practical application of risk and reliability to system design, operation and maintenance towards meeting sustainability and associated safety, efficiency and environmental protection of marine system. The book present application of risk technique for decision support of required solution to engineering problem and failure, the book emphasize the use of emphasize on the use of gap analysis between standard and system functionality to determine high level objective for system engineering risk analysis and solution. The book promote, the use of proactive approach and hybrid use of qualitative and quantitative analysis, the use of engineering and scientific quantitative risk analysis and law physics to define engineering system and related holistic risk characterization.

Risk of new system development that meets sustainability is determined by closeness to shore exposure to uncertain environmental condition, especially, impose risk close to human is prohibitive and, and risk subject risk by environmental is unacceptable. Reliable system design with minimized risk can best be achieved through holistic approach to risk analysis, this include identification of risk through combination of regulation, enforcement, safety awareness expert input use of science content of the system.

Offshore, Nuclear, aviation and Oil and Gas Industry are pioneer of use best practice quantitative risk analysis for all level of their system operation. The international maritime organization deduces formal safety assessment to address issue of risk, and safety in marine industry.

The maritime industry is well aligned to the global pursuit of risk and reliability for sustainable development solution. A commitment to leading practice sustainable development is critical for a maritime company to maintain its social responsibility to the community. Maritime law is continuously being strengthened by the encourage all of level of the industry to employ risk analysis for decision support. Risk analysis provides a valuable central resource for information on the safety that enable applicable of legislation and best practice for the safe design, operation and maintenance of marine sand ocean systems. It is the responsibility of owners, operators, public and regulators to ensure that marine system properly designed, maintained, equipped and operated getting familiar with system functionality and standards and through origination of community participation forum for risk communication in order to maintain safety and protection of the environment.

Risk based design criteria allow better flexibility for innovative designs at optimum maintenance of safety and pollution performance. The use of risk analysis contribute to

- Decision making to tackle recent environmental problem
- Deduction of possible corrective actions and preventive measures to minimise and avoid design flaws.
- Improve management, safe and environmentally friendly sustainable marine system
- Reduction of level of unknown and encouraging community participation components of sustainable system design.

A new age of knowledge and sensitivity about need for system reliability and environmental safeguard shows that the reality of the chemistry of matter in pollutant and associated reaction is full of unknown. Risk analysis represent novel technique that could lead to reduction in workload, cost and time saving, through process of reengineering that can be used to better serve the ecofriendly need of the planet. The use of risk analysis for marine system is blessing to humanity in term of advantage related to economical freight haulage and handling, ecofriendly low air emission and low noise pollution, reasonably efficient transport system, low maintenance after initial building.

The book has 490 pages, where every page is full of useful hints and tips for risk work on marine system to support decision required. I would like to acknowledge the work done in this book to deal with challenges of gap in system standards and performance. I would also like to thank the many individuals and organizations who contribute

—

to the complementation of the book. The book provides standard of operation and compliance that will help to reduce marine accidents and its consequence.

The book on Apply Safety Risk and Reliability for Marine System consider technical environmental, economic and social aspects for all level of marine system design operation maintenance and construction. The concept of risk analysis is simply the best way of doing things for a given sustainable development. Emergence of new challenges require development of new solutions, likewise better solutions are devised for existing issues through application of risk modeling for marine system be flexible and innovative in developing solutions that match site-specific requirements. Apply risk analysis for marine system is as much about approach and attitude as it is about a fixed set of practices or a particular technology.

The requirement for sustainable development include technically appropriate; environmentally sound; financially profitable; and socially responsible. These are address in risk work to strike required balance for decision on system there longitivity. The book present information on a variety of marine system that illustrates and explain risk practice and sustainable development in maritime industry. The books hope to assist all sectors of the maritime industry to reduce the accident and negative impacts on the properties, community and the environment by applying risk for reliable and sustainable marine system sustainable.

PREFACE

All kind of system design aim to deliver product that came last long, especially, especially, the need for system to live through their life cycle. Most past system design are based on proactive and prescriptive approach. Which leave system with more uncertainty? Proactive approach to system design has been used for all system design in sensitive industry like offshore, nuclear, and aerospace. Maritime industry fall under sensitive industry work and shipping activities, considering the importance of the ocean in support of the planetary system. That moves international Maritime Organisation to developed Formal Safety Assessment (FSA) from offshore quantitative analysis (QRA). FSA The use of FSA is limited in maritime industry practice.

This book describes the application of risk and reliability analysis to different marine system. System covered include: dredging, renewable energy, natural gas, navigation channel, inland water transportation, marine offshore, and aquaculture. The book intended to bring awareness and educate practitioner in maritime industry on the use of risk and reliability for reliable marine system design. The reader can hope to learn by reading the book. The inspiration for writing the book begins with, the worldwide need for sustainable development which include proactive methodology to prevent system failure, accident occurrence, as a move to reduce, prevent or control release of affluence to the natural process. The book is written with not much trouble. The formal safety assessment recommended by IMO for design operation, maintenance of marine system represents the main source of the book. The book is unique because, it covers a wide range of application related to marine system. The book is written within the research done in the period year. The reading of the book is straight forward; the chapters are basically application of risk work to relevant area of marine technology.

CHAPTER 1

Safety risk and reliability for inland water transportation

"The significant problems we have cannot be solved at the same
level of thinking with which we created them"
Albert Einstein

Summary

Collision accident remains a big threat to coastal water transportation operation. Occurrence of collision event exposes vessel owners and operators, as well as the public to risk. The nature of the threat can be worrisome; they may lead to loss of life, damage to the environment, disruption of operation and injuries. This makes hybrid analysis of accident frequency and consequence for risk quantification of accident scenarios through stochastic tools very imperative for reliable design and exercise of technocrat stewardship of safety and safeguard of environmental. The study involve predictive model for collision risk and mitigation option for aversion of collision incident. Accident frequency and consequence are obtained using probability tools. Validity of the result is checked with reliability tools. Finding of the study where check with subsystem and uncertainty risk contributing factor in order to arrive at sustainable decision support for collision aversion for inland water transportation. This Chapter discuss implementation Safety and Environmental Risk and Reliability Model (SERM) for aversion of collision accident for vessel navigating for inland waterways and deduced generic risk mitigation option required for operational, societal, limit definition and technological change decision support for development of sustainable inland water transportation system (IWTS).The probability per year and accident consequence predicted is considered acceptable in maritime and offshore industry, but for a channel using less number of expected traffic, it could be considered high. Providing safety facilities like traffic separation, vessel traffic management could restore maximize sustainable use of the channel.

1.1 Introduction

Vessel collision risk in waterway end up with huge consequence. Collision accident scenarios carry heavy consequence, thus its occurrence is infrequence. These accidents represent a risk because they expose vessel owners and operators, as well as the public, to the possibility of losses such as vessel and cargo damage, injuries and loss of life, environmental damage, and obstruction of waterways. They can also lead to instantaneous and point form release of harmful substance to water, air and water and longtime ecological impact (Eftratios Nikolaidis, 2005). Environmental problem and need for system reliability call for innovative methods and tools to assess and analyze extreme operational, accidental and catastrophic scenarios as well as accounting for the human element, and integrate these into a design environments part of design objectives. Risk and reliability based design entails the systematic integration of risk analysis in the design process targeting system risk prevention, reduction that meet high level goal and leave allowance for integrated components of the system including environment that will facilitate and support a holistic approach for reliable and sustainable waterways appropriate and require trade-offs and advance decision-making leading to optimal design solutions (Wallace R. Blischke, 2000).

Collision risk is a product of the probability of occurrence of the physical event and consequential losses. Earlier risk modeling focuses more on either frequency estimation or consequence and leave analysis with remnant uncertainties. Paralleled analysis of frequency and consequence analysis along with translation of other consequence (economic, oil spill, GHG) could leads to quantification of total collision risk. Collision data may be imperfect or inconstant, making it difficult to account for dynamic issues associated with vessels and waterways requirement. Accounting for these lapses necessitated need to base collision analysis on hybrid use of deterministic, probabilistic or simulation methods depending on the availability of a data. Collision accident consequence data are hard to come by like accident frequency data.

However, whatever data that is available should be meaningful processed as much as possible through available tools especially predictive methods for necessary mitigation decision support for sustainable waterways system design. Developing sustainable inland water transportation (IWT) requires transit risk analyses of waterways components and relationship between factors such as environmental conditions, vessel characteristics, operators' information about the waterway, as well as the incidence of groundings and collisions, using available data. Whatever information is available is useful for risk and reliability based decision work of accidents rate of occurrence, consequence and mitigation (Eftratios Nikolaidis, 2005, McGee et. al., 1999). The analysis consider mainly the waterways dimensions and other related variables of risk factors like operator skill, vessel characteristics, traffic characteristics, topographic, environmental difficulty of the transit, and quality of operator's information in transit which are required for decision support related to efficient, reliable and sustainable waterways developments.

This chapter discusses modeling of waterways collision risk. The chapter presents the model of waterways collision risk and associated rate of occurrence and consequence. Relation with other variables risk factors like operator skill, vessel characteristics, traffic characteristics, topographic and environmental difficulty of the transit are taken into consideration [3, 4]. Accident frequency and consequence are determined stochastically. Waterway variable and parameters are compared. The result hopes to contribute to decision support for development and regulation of inland water transportation.

1.2 Safety and environmental risk for IWT

Risk and reliability based model involve innovative approach and tools to assess operational, accidental and consequential catastrophic scenarios. Risk based design entails the systematic integration of risk analysis in the design process. It target safety and environment risk prevention and reduction as part of design objective. It requires accounting for the human element, and integrates them as required into the design environment. Successful outcome of the risk based design require integrated design environment to facilitate and support a holistic risk approach to ship and channel design is needed. Total risk approaches enable appropriate trade-off for advanced sustainable decision making. Integrated risk based system design requires the availability of tools to predict the safety, performance and system components Waterways accident falls under scenario of collision, fire and explosion, flooding, grounding. Other causal factor could fall under Loss of propulsion, Loss of navigation system, Loss of mooring function and Loss of Other accident from the ship or waterways.

Incidents are unwanted events that may or may not apply to accidents. Necessary measures should be taken according to magnitude of event and required speed of response should be given. Accidents are unwanted events that have either immediate or delayed consequences. Immediate consequences variables include injuries, loss of life, property damage, and persons in peril. Point form consequences variables could apply to further loss of life, environmental damage and financial costs. The earlier stage of the risk process involves finding the cause of risk, level of impact, destination and putting a barrier by all mean in the pathway. Risk work process targets the following:

i.   Cause of risk and risk assessment, this involve system description, identifying the risk associated with the system, assessing them and organising them in degree or matrix. IWT risk can be as a result of the following: Root cause, immediate cause, Situation causal factor, Organization causal factor

ii. Risk analysis and reduction process, this involve analytic work through deterministic and probabilistic method that strengthen reliability in system.

iii. Reduction process that targets initial risk reduction at design stage, risk reduction after design in operation and separate analysis for residual risk for uncertainty as well as human reliability factor.

Historical analyses of system performance are important to establish system performance benchmarks that can identify patterns of triggering events, this may require long periods of time to develop and detect. Uncertainty risk in complex systems can have its roots in a number of factors ranging from performance, new technology usage, human error as well as organizational cultures.

They may support risk taking, or fail to sufficiently encourage risk aversion. To deal with difficulties of uncertainty risk migration in marine system dynamic, risk analysis models can be used to capture the system complex issues, as well as the patterns of risk migration. Assessments of the role of human and organizational error, and its impact on levels of risk in the system, are critical in distributed, large scale dynamic systems like

IWT couple with associated limited physical oversight. Table 1.1 shown models that has been used design of system based on risks in marine industry.

IMO and Sirkar et al (1997) methods lack assessment of the likelihood of the event, likewise other model lack employment of stochastic method whose result could cover uncertainties associated with dynamic components of channel and ship failure from causal factors like navigational equipment, training and traffic control. Therefore, combination of stochastic, statistical and reliability method based on combination of probabilistic, goal based, formal safety assessment, deterministic methods and fuzzy method and use of historical data of waterways, vessel environmental, first principle deterministic and traffic data can deliver best outcome for predictive, sustainable, efficient and reliable model for complex and dynamic system like inland water transportation.

Table 1. 1: Previous risk work

Model	Application	Drawback
Brown et al (1996)	Environmental Performance of Tankers	Quantify only oil spill consequence
Sirkar et al (1997)	Consequence if collision and grounding	Difficulties on quantifying consequence metrics
Brown and Amrozowicz	Hybrid used of risk assessment, probability simulation and spill consequence assessment model	Oil spill assessment limited to use of fault tree
Sirkar et al (1997)	Monte Carlo technique to estimate damage and spill cost analysis for environmental damage	Lack of cost data
IMO (IMO 13F, 1995)	Pollution prevention index from probability distribution damage and oil spill	Lack (Sirkar et al, 1007) rational
Research Council committee (1999)	Alternative rational approach to measuring impact of oil spills	Lack employment of stochastic probabilistic method

Prince William Sound, Alaska (PWS, 1996)	The most complete risk assessment	Lack of logical risk assessment framework (NRC, 1998)
Volpe national Transportation Center (1997)	Accident probabilities using statistic and expert opinion	Lack employment of stochastic methods
Puget Sound Area (USCG, 1999)	Simulation or on expert opinion for cost benefit analysis	Clean up cost and environmental damage omission

1.3 Technique of Risk Analysis

Most of the method above used historical data. The novel method in this chapter used limited data of traffic used to model the physics of the system, the transfer function and stochastically project accident frequency. The projection is generic and can be used for any waterways and it consider random collision not which is not considered by previous model. The general hypothesis behind assessing physical risk model is that the probability of an accident on a particular transit depends on a set of risk variables require for analysis of prospective reliable design (DnV, BV, SSPA, 2002, Moderras, 1993). Accident and incident are required to be prevented not to happen at all.

The consequence of no safety is a result of compromise to safety leading to unforgettable loses and environmental catastrophic. Past engineering work has involved dealing with accident issues in reactive manner. System failure and unbearable environmental problem call for new proactive ways that account for equity requirement for human, technology and environment interaction. The whole risk assessment and analysis process starts with system description, functionality and regulatory determination and this is followed by analysis of:

i. Fact gathering for understanding of contribution factor
ii. Fact analysis of check consistency of accident history
iii. Conclusion drawing about causation and contributing factor
iv. Countermeasure and recommendation for prevention of accident

It is very imperative to develop, refine, verify, validate reliable model through effective methods and tools. The risk process begins with definition of risk which stands for the measure of the frequency and severity of consequence of an unwanted event (damage, energy, oil spill). Frequency at which potential undesirable event occurs is expressed as events per unit time, often per year. The frequency can be determined from historical data. However, it is quite inherent that event that don't happen often attract severe consequence and such event are better analyzed through risk based and reliability model. Risk is defined as product of probability of event occurrence and its consequence. Most risk based methods define risk as:

Risk (R) = Probability (Pa) X Consequence (Ca)   1.01

or in a more elaborate expression risk can be defined as:

Risk = Threat x Vulnerability x {direct (short-term)
        consequences + (broad) Consequences}   1.02

In risk analysis, serenity and probability of adverse consequence hazard are dealt with through systematic process that quantitatively measure perceives risk and value of system using input from all concerned waterway users and experts. Risk can also be expressed as:

Risk = Hazard x Exposure   1.03

Where hazard is anything that can cause harm (e.g. chemicals, electricity, Natural disasters), while exposure is an estimate on probability that certain toxicity will be realized. Severity may be measured by No. of people affected, monetary loss, equipment downtime and area affected by nature of credible accident. Risk management is the evaluation of alternative risk reduction measures and the implementation of those that appear cost effective where:

Zero discharge or negative damage = Zero risk   1.04

The risk and reliability model subsystem focus on the following identified four risks assessment and analysis application areas that cover hybrid use of technique ranging from qualitative to qualitative analysis (Wang, 2000):

i.   Failure Modes Identification Qualitative Approaches
ii.   Index Prioritisation Approaches
iii. Portfolio Risk Assessment Approaches, and
iv.   Detailed Quantitative Risk Assessment Approaches.

Conventionally, risk analysis work often deal with accident occurrence, while the consequence is rarely investigated. Parallel addressing of frequency and consequence analyze can give clear reliable result that meet goal based objectives. Risk and reliability based design can be model by conducting the following analysis:

i. Risk identification: this involve establishment of a realistic representation of the navigational activity that answering question relating to how and why accidents occur including consideration for navigation, accident scenarios, probability of occurrence of category of accidents components.

ii. Risk analyses: selection of a design collision scenario (e.g., ship size, speed) to measure collision severity including what happens (structurally) i.e activities during the period of occurrence including when a collision occurs, all of which enter as components of frequency and consequence analysis.

iii. Priotization of risk level: determine the appropriate level of risk for the selected platform and event.

iv. Risk iteration for sustainability: analyses to better define the risk, consequence and cost of mitigation.

v. Damage estimation: what are the consequences of structural damage, this include the impacts associated with property damages, environmental damages, and loss of life.

vi. Mitigation: acceptance criteria, evaluation of preventative and protective measures as well as answering question relating to how can each of the above are addressed. Risk acceptability also involves accident prevention, minimization of structural damage, mitigation of damage consequence, response to damage and loss of life as well as decision on bridge.

vii. Repriotization of exposure category: mitigate risk or consequence of events that meet ALARP principle.

viii. Reassess high risk events for monitoring and control plans.

ix. Recommendation, implementation and continuous monitoring and improvement.

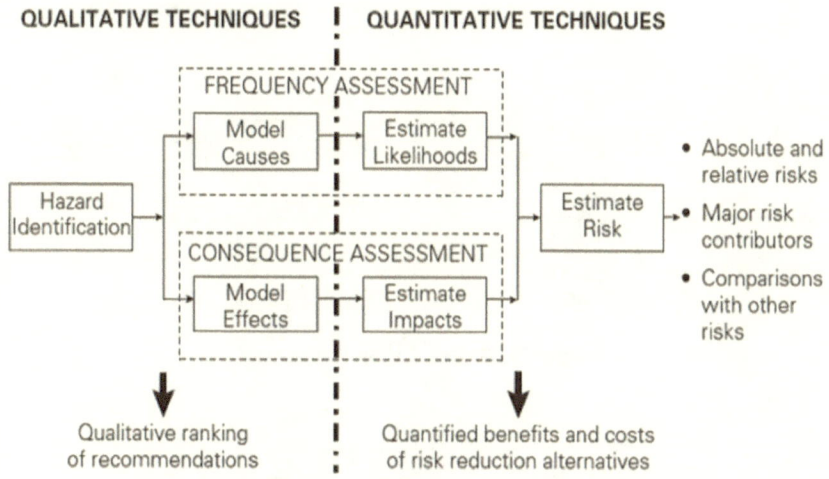

Figure 1. 1: Components of risk and reliability analysis

Figure 1.1 shows components of scientific based risk and reliability analyses. Risk and reliability based design. Effective risk assessments and analysis MODELLING required three elements highlighted in the relation below.

Risk modeling = Framework + Models + Process   1.05

Reliability based verification and validation of system in risk analysis should be followed with creation of database and identification of novel technologies required for implementation of sustainable system.

1.3.1 Risk Framework

Risk framework involves system description, risk identification, criticality, ranking, impact, possible mitigation and high level objective to provide system with what will make it reliable. The framework development involves risk identification which requires developing understanding the manner in which accidents, their initiating events and their consequences occur. This includes assessment of representation of system and all linkage associated risk related to system functionality and regulatory impact (See Figure 1.2)

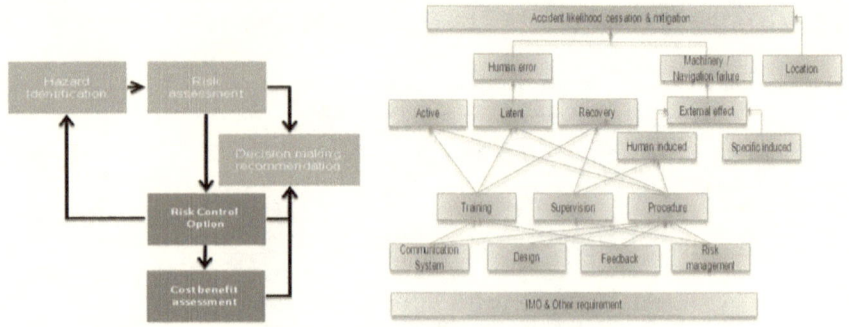

Figure 1. 2: IMO Risk framework

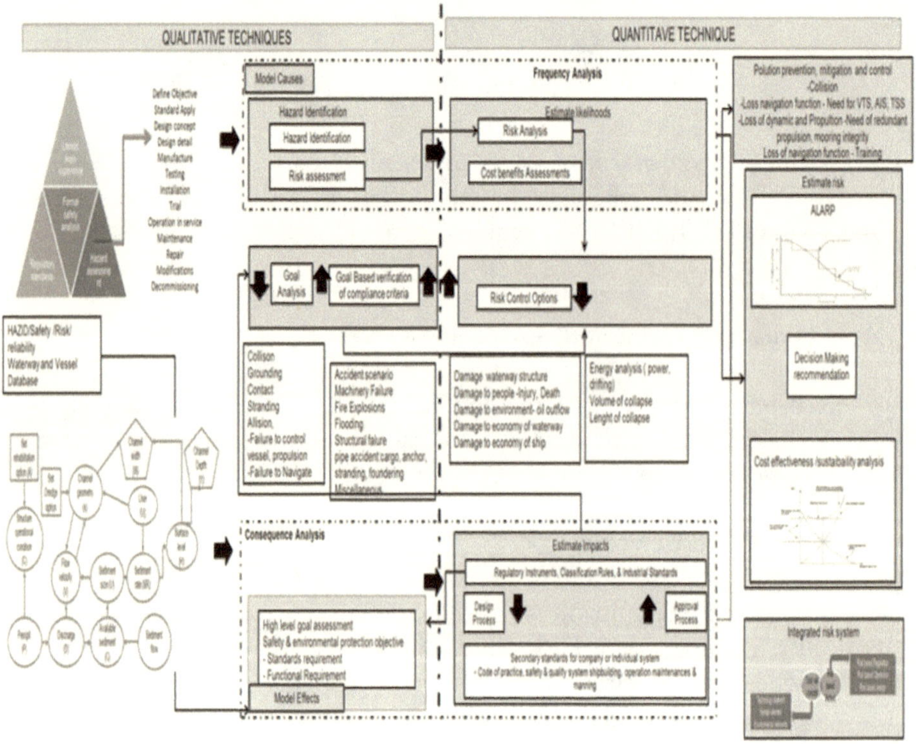

Figure 1. 3: Risk frameworks

Risk framework (Figure 1.3) is developed to provide effective and sound risk assessment and analysis. The process requires accuracy, balance, and information that meet high scientific standards of measurement. The information should meet requirement to get the science right and getting the right science. The process could target interest of stakeholder including members of the port and waterway community, public officials, regulators and scientists. Transparency and community participation helps ask the right questions of the science and remain important input to the risk process, it help checks the plausibility of assumptions and ensures that synthesis is both balanced and informative. Employment of quantitative analysis with required insertion of scientific and natural requirements provide analytical process to estimate risk levels, and evaluating whether various measures for risk are reduction are effective [24,26].

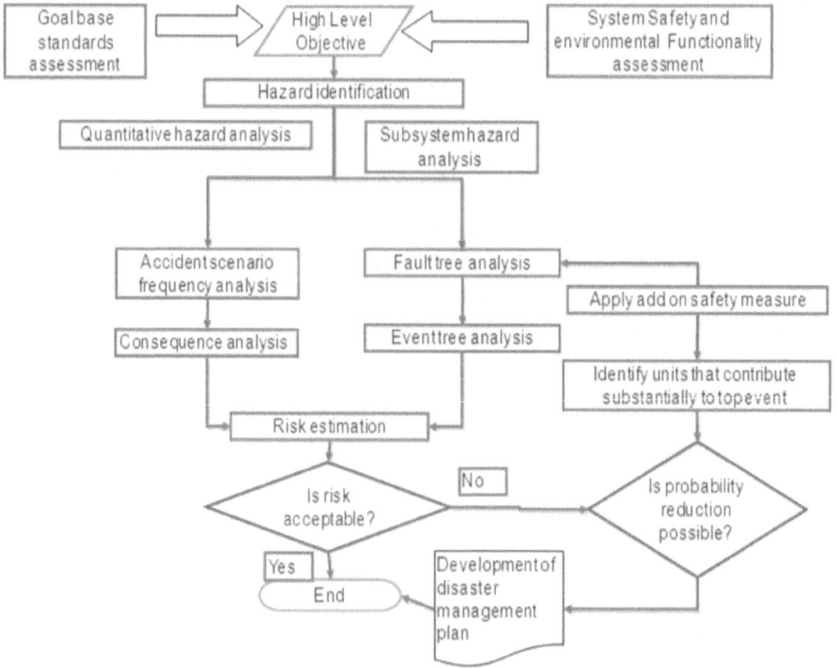

Figure 1. 4: Risk Components of risk and reliability analysis

1.3.2 SERM Process

SERM address risks over the entire life of the complex system like IWT system where the risks are high or the potential for risk reduction is slim. SERM address quantitatively, accident frequency and consequence of IWT. Other risk and reliability components include human reliability assessment which is recommended to be carried out separately as part of integrated risk process. Other waterways and vessel requirement factors that are considered in SERM model are: Construction, Towing operations and abandonment of ship, Installation, Hook-up and commissioning Development and major modifications. Integrated risk based method combined various technique as required in a process. Table 1.2 shows available risk based design for techniques. This can be applied for each level of risk. Each level can be complimented by applying causal analysis (system linkage), expert analysis (expert rating), and organizational analysis (Community participation) in the risk process. From Figure 1.2, the method use is risk analysis that involves frequency analysis where the system is modelled with hybrid of deterministic, probabilistic and stochastic process.

Table 1. 2: Risk based design techniques

Process	Suitable technique
HAZID	HAZOP, What if analysis, FMEA, FMECA
Risk analysis	Frequency, consequence, FTA,ETA
Risk evaluation	Influence diagram. decision analysis
Risk control option	Regulatory, economic, environmental, function elements matching and iteration
Cost benefit analysis	ICAF, Net benefit
Human reliability	
Uncertainty	Simulation/probabilistic
Risk monitoring	

Technically, the process of risk and reliability study involves the following four areas:

i. System definition of high goal objective: This requires defining the waterways by capturing gap between system functionality and standards. The scope of work for safely, environment risk and reliability analysis defines the boundaries for the study. It involves identifying which activities are included and which are excluded, and which phases of the system's life are to deal with.

ii. Qualitative hazard identification and assessment: It involves hazard identification through qualitative review and assessment of possible accidents that may occur, based on previous accident as well as experience or judgment of system users where necessary. Though, using selective and appropriate technique depends on the range, magnitude of hazards and indicates appropriate mitigation measures.

iii. Quantitative hazard frequency and consequence analysis: once the hazards have been identified and assessed qualitatively. Frequency analysis involves estimation of how likely it is for the accidents to occur. The frequencies are usually obtained from analysis of previous accident experience, or by some form of analytic modelling employed in this thesis. In parallel with the frequency analysis and consequence modelling evaluates the resulting effects of the accidents, their impact on personnel, equipment and structures, the environment or business.

iv. Risk acceptability, sustainability and evaluation: Is the yardsticks to indicate whether the risks are acceptable, in order to make some other judgment about their significance. This begins by introducing non-technical issues of risk acceptability and decision making. In order to make the risks acceptable. The benefits from these measures can be evaluated by iterative process of the risk analysis. The economic costs of the measures can be compared with their risk benefits using cost benefit analysis leading to results of risk based analysis. This input necessities to the design or on-going safety management of the installation, to meet goal and objectives of the study.

The process of risk work can further be broken down into the following elements:

i. Definition and problem identification
ii. Hazard and consequence identification
iii. Analysing the likelihood's of occurrence
iv. Analysing consequences
v. Evaluation of uncertainty
vi. Risk control option (RCO) and risk control measure (RCM
vii. Sustainability of (cost safety, environment, injury, fatality, damage to structure, environment) and risk acceptability criteria
viii. Reliability based model verification and validation: statistical software, triangulation, iteration.
ix. Recommendation for implementation: Implement, establishing performance standards to verify that the arrangements are working satisfactorily and continuous monitoring, reviewing and auditing the arrangements

—

Employment of these benefit provide a rational. Formal environmental protection structure and process for decision support guidance and monitoring about safety issues. The scope of sustainable risk based design under consideration involves stochastic, analytic and predictive process work leading to avoidance the harms in waterways. Figure 1.6 shows block diagram of SERM components for IWT. Safety and Environmental Risk and Reliability Model (SERM) for IWT required having clear definition of the following issues:

i.   Personnel, attendance
ii.   Identify activities
iii. Vessel accidents including passing vessel accident, crossing, random
iv.   Vessel location and waterway geography on station and in transit to shore.
v.   Impairment of safety functions through determination of likelihood of loss of key safety functions lifeboats, propulsion temporary refuge being made ineffectiveness by an accident.
vi.   Risk of fatalities, hazard or loss of life through measure of harm to people and sickness.
vii. Property damage through estimation of the cost of clean-up and property replacement.
viii. Business interruption through estimation of cost of delays in production.
ix.   Environmental pollution may be measured as quantities of oil spilled onto the shore, or as likelihood's of defined categories of environmental impact or damage to infrastructures.

Allowance should be made to introduce new issue defining the boundary in the port from time to time. The choice of appropriate types of risk tool required for the model depend on the objectives, criteria and parameter that are to be used. Many offshore risk based design model consider loss of life or impairment of safety functions. There is also much focus on comprehensive evaluation of acceptability and cost benefit that address all the risk components.

Quantitative risk analysis involves pure mix of technical and scientific as well as economic and social risk analysis. When the frequencies and consequences of each modelled event have been estimated, they can be combined to form measures of overall risk including damage, loss of life or propulsion, oil spill. Various forms of risk presentation may be used. Risk to life is often expressed in two complementary forms. The risk experienced by an individual person and societal risk. The risk experienced by the whole group of people exposed to the hazard (damage or oil spill).

1.3.3 Collisions risk Modelling

Collision in waterways is considered low frequency and high consequence events that have associative uncertainty characteristics / component of dynamic and complex physical system. This makes risk and reliability analysis the modest methods to deal with uncertainties that comes with complex systems. Employment of hybrid deterministic, probabilistic and stochastic method can help break the barriers associated with transit numbers data and other limitation. Conventionally, risk analysis work often deal with accident occurrence, while the consequence is rarely investigated, addressing frequency and consequence analyze can give clear cuts for reliable objectives. Risk and reliability based design can be model by conducting the following analysis that includes the following process [13, 16]:

i.   Risk identification
ii.   Risk analyses
iii. Damage estimation
iv.   Priotization of risk level
v.   Mitigation
vi.   Repriotization of exposure category: mitigate risk or consequence of events that meet ALARP principle.
vii. Reassess high risk events for monitoring and control plans.
viii. Recommendation, implementation, continuous monitoring and improvement.

Collision is likely to be caused by the following factors shown in Figure 7 derived from fault three analyses using RELEX software. The relex software is based on fault three analysis where consequence of causal events are add up through logic gate to give minimum cut set probability that trigger the event. It is more effective for subsystem analysis.

$$P \text{ (collision)} = P \text{ (propulsion failure)} + P \text{ (loss of navigation failure)} + P \text{ (Loss of vessel Motion)} \qquad 1.06$$

There is also causes are mostly as a result of causes from external sources like small craft, are cause of cause, cause from other uncertainty including human error may attract separate subsystem analysis. A critical review of risk assessment methodologies applicable to marine systems reiterate that the absence of data should not be used as an excuse for not taking an advantage of the added knowledge that risk assessment can provide on complex systems (McGee et al.,1999. Approximation of the risks associated with the system can provide a definition of data requirements. The treatment of uncertainty in the analysis is important, and the limitations of the analysis must be understood. However, data management system and better approach can always accommodate little data or no data.

1.3.4 Risk and reliability modelling data requirement

Prediction of accident frequency model for navigation channel lead to fundamental model of transit risk that consider factors such as traffic type and density, navigational aid configuration, channel design, configuration and classification. For cases where there are insufficient historical record to support their inclusion, more comprehensive models of transit risk will have to rely on integral use of hybrid of deterministic, probabilistic, stochastic method whose result could further be simulation and expert judgment to optimize deduced result.

1.3.5 Study area and baseline data requirement

Figure 1.5 shows major parameter that required for waterway reliability design. Risk based collision model are derivative for improvement of maritime accident data collection, preservation and limit acceptability using information relating to the following, Kielland et. al. 1994). ports for entering incidents

Wind speed & direction, visibility, water level, current speed & direction, etc. eliminate/correct erroneous and duplicate entries (e.g. location information) record data on actual draft and trim, presence and use of tugs, presence of pilots types of cargo and vessel movements report "barge train" movements as well as individual barges improve temporal resolution (transits by day or hour)

Width data play very important role in possibility of accident and damage cause to physical structures. Therefore beside fatality and societal consequence analysis, extent of damaged can be analysed to determine risk of collision in waterways. Figure 1.7-1.8 and Table 1.3-1.7 show figurative data of channel parameter required for channel risk and reliability analysis, estuarine data: Sungai Langat: 135.7 km => North Estuary 44.2 km, South Estuary: 9.9 km.

The main risk contributing factors can fall under the following: operator vessel characteristics traffic characteristics topographic difficulty of the transit— environmental condition including water level and tide quality of operator's information including uncertainty in Surveys/Charts

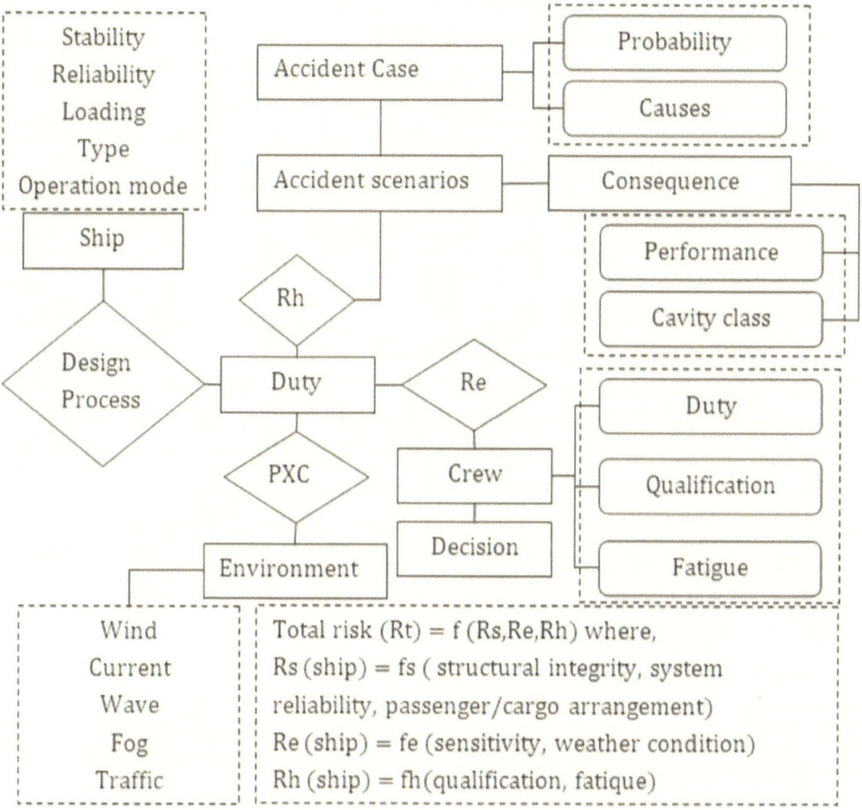

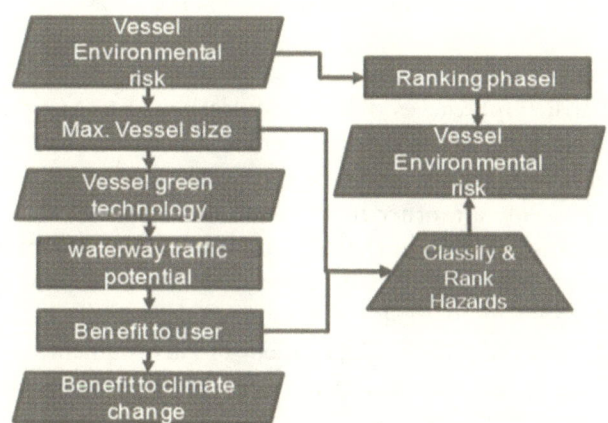

Figure 1. 5: Main components of risk based design for IWT.

Figure 1. 6: IWT safety and environmental model
components block diagram

The study area is Langat River in the west coast of Malaysia, 220m long navigable inland waterways that has been under-utilized. Personal communication and river cruise survey revealed that collision remain the main threat of the waterways despite less traffic in the waterways system. This make the case to examine collision risk and the need to establish risk and reliability based model for sustainable development of the waterways. Data related to historical accidents, transits, and environmental conditions were collected. Accident data for complex system like waterways have inherits few data. Low probability event attract high impact. These make probabilistic methods the best preliminary method to analyze the risk of complex and dynamic system which can be optimized through simulation methods especially visual reality as required. Figure 1.7 shows the map of Langat River tributary. Barges and tugs of capacity 5000T and 2000T are currently plying this waterway at draft of 9m and 15m respectively. Safety associated with small craft is not taken into account.

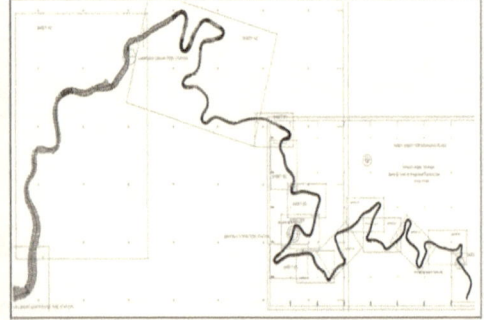

Figure 1. 7: Langat River map [JPS, 2008]

Collisions (including contact between two vessels, between vessel and a fixed structure), causes of collision are linked to navigation system failure, mechanical failure and vessel motion failure are considered in case study of risk and reliability modeling for river transportation presented in This chapter. Below are other information related to channel vessel and environment employed are in the risk process. Lacking information about the distribution of transits during the year, or about the joint distribution of ship size and flag particular, environmental conditions are derivative from probabilistic estimation (Department of Environment, 2000). Figure 1.8 shows Langat River channel parameters, one way traffic, and 98m for straight channel, width and 120m for bends.

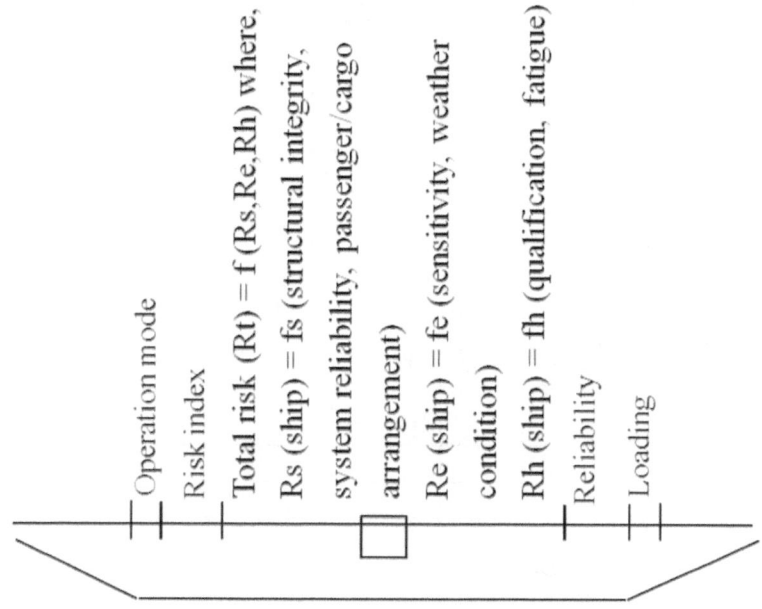

Figure 1. 8: Channel width parameter

Table 1. 3: River Langat tributary

	Approach channel	
	Straight	Bend
Design parameter	98m	120m
	3-6m	3-6m
Side slope	10H:1V	10H:1V
Estuarine 135.7km	North (44.2km)	South (9.9km)

Table 1. 4: River width and depth parameters

Width	**Depth**
Maneuvering lane	Draught
Bank suction	Trim
Wind effect	Squat
Current effect	Exposure allowance
Channel with bends	Freshwater adjustment
Navigationaids	Maneuvering allowance
Pilot	Overdepth allowance
Tugs	Depth transition, tidal allowance

Table 1. 5: Vessel requirement: Barge parameter

Barge parameter	Tonnage	
	2000 tons	**5000 tons**
Length (m)	67.3	76.2
Beam (m)	18.3	21.3
Depth (m)	3.7	4.9
Draft (m)	2.9	4

Table 1. 6: Vessel requirement: Tug parameter

Tugs Parameter	Tonnage	
	2000 tons	**5000 tons**
Length (m)	23.8	23.8
Beam (m)	7.8	7.8
Depth (m)	3.5	3.5
Draft (m)	2.8	2.8
Horse Power (hp)	1200	1200

Vessel movement on Langat River include port call usually consists of two transits in one way traffic, one into and one out of the port. Safe transit data for this study consider the same barge type and size for accidents.

Table 1. 7: Waterway parameter

Type	Channel dimension	
	Basic Maneuvering Lane	1.5B
	Addition for cross wind (less than 15 knots)	0.0B
	Addition for cross current (negligible <0.2 knots)	0.0B
	Addition for bank suction clearance	1.0B
Straight	Addition for aids to navigation (Excellent)	0.0B
	Addition for cargo hazard (medium)	0.0B
	Channel width for Inland Waterways, (B= Beam of the ship)	2B=53 m
Bend	Channel width	3B=64 m

1.3.6 Traffic data

This study analyses accidental system and subsystem accident contributing factors relating to navigational, mechanical failure and human error and exclude those identified as intentional for barge and tugs of 5000T and 2000T are having respective drift of draft greater than 9 to 15m. Table 1.9 shows some of the annual traffic activities data that represent function of accident frequency for Langat River. Seasonal trends can be stochastically modeled from probabilistic result, environmental condition and traffic volume fluctuation is also considered negligible. These are advised to be modeled separately under uncertainty analysis and continue to be incorporated from daily system monitoring and post system analysis. For visibility, navigation is considered to be more risky at night than day time, the analysis follow generic assumption for evenly safe distribution during day and night. Vessel movement for the case under consideration currently has no vessel separation system. However there is traffic movement from both inbound and outbound navigation in the channel. The same type of barge size is considered for the estimation work [29].

Table 1. 8: Tug boat & vessel activities along river for 2008

Vessel movement	Description
Jetty	3 nos
Daily	9 times
Weekly	63 times
Monthly	252 times
Annually	3024 times
Daily no of barge	Total 12 nos
Incoming (every 4 hours)	6 nos
Outgoing (every 4 hours)	6 nos
Speed	2~3 knots
Traffic	Single way
Lay-bya	Propose four location

1.3.7 Damage data

Collision data are drawn from relevant marine administrator. Data gaps can be covered by stochastic and probability estimations. Collision hit that represented function of collision consequence of ship barge and tugs of 5000T and 2000T having respective draft of 9—15m is presented in Table 1.9 where magnitude of hit was described theoretically. Using probabilistic risk method hidden facts can be covered under absolutism principle of discrete system characteristics. Seasonal trends can be

—

49

modelled estimated from result of stochastic outcome. Therefore, it is clear that that limitations of data collection demand use of hybrid of historical and stochastic analysis. Future data collection effort can open opportunity to cover remnant deficiencies gap, improve validation work, improved analysis and understanding of accident risk. The general hypothesis behind assessing physical system risk frequency, consequence and subsystem model is that the probability of an accident, impact and energy release on a particular events and the variables that determine magnitude of losses involved, including short term effect and long term environmental problems and preventive measure that enter into maintaining safety and preserving the environment Emi, H. et al 1997, Cockcroft A. N. Et al., 1989), IMO, 1993).

Table 1. 9: Chronology of accident along Langat River
due to inland navigation activities

No	Date	Time	Case	Impact description
1	12.01.08	8.30pm	Hit	2 nos. of concrete pile have a scratched mark, 5 number of concrete pile was broken.
2	25.02.08	night	Hit	The protection pipe nearby jetty was damaged & missing. 3 numbers. of welded pipe were missing.
3	08.03.08	930pm	Hit / Tug And vessel	The protection pipe nearby jetty was damaged & missing. 3 numbers. of welded pipe missing/bended I—beam
4	07.04.08	3.30pm		Temporary jetty at Pier 9a.
5	07.05.08	5.00pm		Temporary jetty at Pier 9a.
6	09.05.08	8.30am		Temporary jetty at Pier 9a.
7	17.06.08	4.00am		I-Beam that used for Gen-Set was bended. Workers jetty was badly damage.
8	28.07.08	10.30am		Bended I-Beam.
9	23.08.08	8.30am		Protection pipe at Pier 9a.
10	18.09.08	6.30am		Temporary jetty at Pier 9a. Welding set & crane were collapsed & drown into the river.

1.4 System level collisions risk modelling

Collision in waterways is considered low frequency and high consequence events that have associative uncertainty and characteristics of dynamic and complex physical system. This makes risk and reliability analysis the modest methods to deal with uncertainties that comes with complex systems. Employment of hybrid deterministic, probabilistic and stochastic method can help integrate parametric elements of the system required for system behavior analysis. The Langat River involves model systemic analysis procedures for inland waterways that determine the probability of failure or occurrence, risk ranking, damage estimation, sustainability, high risk to life, safety, and cost benefit analyze, sustainability and acceptability criteria. In This chapter, aspect of frequency, consequence model is used to represent total risk estimation of the system. Couple with analyze frequency of occurrence, potential accident causal factor and effect of occurrence is modeling of risk contributing factor of subsystem analysis which is performed using software to model fault three and event tree Brebbia et. al., 2000, DnV.2001). Figure 1.9 shown vessel requirement parameter for the waterway

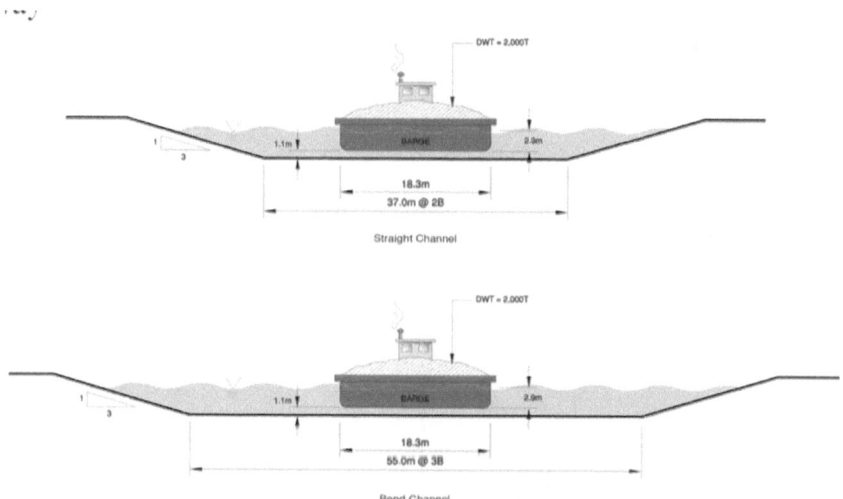

Figure 1. 9: Langat vessel particular

Collision is likely to be caused by the following factors modelled from Relex software as shown in Figure 1.11. Other causes can be as a result of causes from external sources like small craft, under which there are cause of cause, cause from other uncertainty including human error that may attract separate subsystem analysis which are excluded in this analysis. Equation 1 shows the total risk calculation equation for system subsystem and system uncertainty. Figure 1.10 shows vessel and barge movement in Langat River. Figure 1.12 shows waterway system input and output block representation.

P (collision) = P (propulsion failure) + P (loss of navigation failure)
                    + P (Loss of vessel motion)
                    + P (Uncertainty, system complexity)   Eq. 1.07

Figure 1. 10: Collision contributing factors

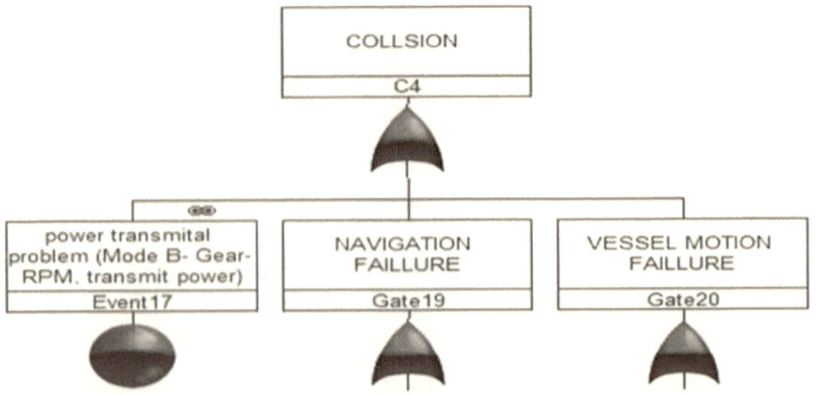

Figure 1. 11: Tugs puling barge in Langat

1.4.1 Traffic frequency estimation modelling

Equation one shows first principle deterministic equation for the physical model of the whole system.

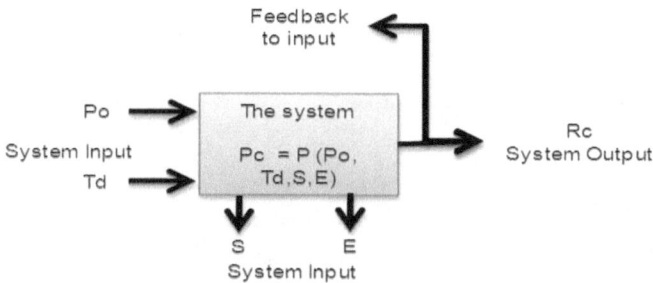

Figure 1. 12: Risk and reliability system analyses system

Traffic density of meeting ship: $\rho = \dfrac{N_m}{v.\tau.W} = \dfrac{N_m}{v1.W} =$ 1.06

1.4.1.1 Analysis of present traffic situation

$$\rho = \frac{N_m}{v.\tau.W} \; Ships/m^2 \qquad\qquad 1.09$$

Expected number of collision

$$Ni = 9.6.B.D, \rho_s = 9.6x\ 21.3\ x\ 64\ x\ 2.2\ x\ 10^{-9} = 2.9\ x\ 10^{-5} \ /passage \qquad 1.10$$

Where: B1 = mean beam of meeting ship (m), V1 = mean speed of meeting ship (knots), B2 = beam of subject ship (m), V2 = speed of subject ship (knots), Nm = arrival frequency of meeting ships (ship/time), D= relative sailing distance.

Figure 1.13 a, b and c show various waterways collision situations, viz : for head—on, overtaking and crossing (angle) collision scenario. Table 1.10, 1.11, and 1.12 shows relevant expression from previous waterways analysis that can be used for required approximation for different collision situation.

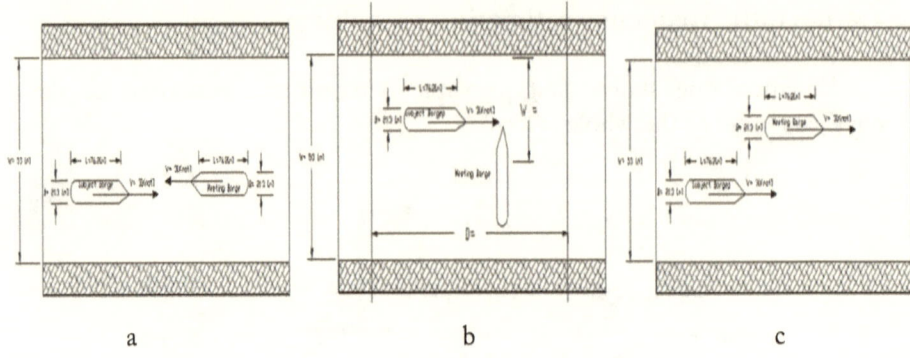

a   b   c

Figure 1. 13: Collision situations, a. overtaking,
b. passing cases, c. crossing

Table 1. 10: Expression for collision situation
(Fujii, 1982, Lewison, 2000).

Expression	Head—on	Overtaking	Crossing	Crossing at specific angle	Circular	Random
Basic	$4xBx$	$\dfrac{(B_1 + B_2)}{W}x$ $\dfrac{(V_1 + V_2)}{V_1 xV_2}x$	$\dfrac{4x(B + L)}{xDx\rho_s}$	$\dfrac{dxwx\rho_l}{V_2}x$ $\dfrac{V_2}{sin\theta}x\dfrac{V_1}{tan\theta}$	$\rho_n xWx\dfrac{8xd_i}{\pi}$	$N_i = \dfrac{N}{\tau xV}x$ $(\dfrac{4}{\pi}xL + 2B)$
Standard	$Dx\rho_s$	DxN_m	$14xBx$ $Dx\rho_s$	xV_i	$48xDxBx\dfrac{\rho_n}{\pi}$	$9.6xDx\rho_n xB$
Relative	1	1	3.5	3.5	3.8	2.4

Approximations; $L = 6B$, $D = W$, $N_t = P_t$ 1.11

Necessary period for ship to pass the fairway
$T=D/v = 3000/3 = 1000$ sec   1.12

Table 1. 11: Failure per nautical mile and failure
per passage for collision situation

Collision Scenario	μ_c (Failure per nautical miles or hours)	Pc (Failure per passage or encounter)
Head On	1.5×10^{-5}	2.7×10^{-5}
Overtaking	1.5×10^{-5}	1.4×10^{-5}
Crossing	3.0×10^{-5}	1.3×10^{-5}

Therefore average Pc and $\mu_c = 2.5 \; x \; 10^{-5}$ 1.13

Probability of loosing navigation control within the fairway

$P_c = \mu_c . T$ *failure / passage*   1.14

Table 1. 12: Failure per nautical mile and failure per passage for different waterways

Collision Scenario	μ_c (Failure per nautical miles or hours)	Pc (Failure per passage or encounter)
Fairway	$2.5 \; x \; 10^{-5}$	$7.0 \; x \; 10^{-5}$
UK	$1.5 \; x \; 10^{-5}$	$1.4 \; x \; 10^{-5}$
US	$3.0 \; x \; 10^{-5}$	$1.3 \; x \; 10^{-5}$

Probability of collision Pa= Pi. Pc (collision / passage) 1.15

Collision per annual (Na) = Pa Collision per year 1.16

The generic models represent any of the above accident situations, which collision be represented by random collision scenarios which has been modelled in this research. Figure 1.14 show a fairway section wide characteristics dimension (W), and randomly distributed traffic. Where: Pi—probability of collision (C) incident, Se—exposed ship to the encounter ship of period (Te), Vm—speed of entering ship exposed to traffic of period ™, Pn—the probability that (Sn) is exposed present in the fairway, τ—Annual operational time, N—Number of ship movement, Ni—The expected number of impacts between incoming traffic Sm and existing traffic (Sn), Pnm—conditional probability per unit for collision event that Sn collides with Sm.

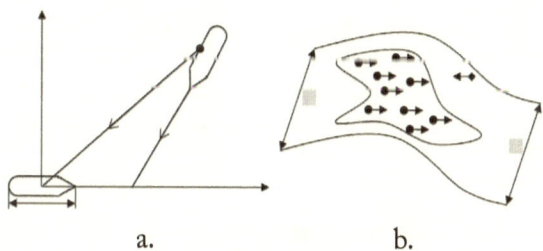

a.   b.

Figure 1. 14: Random collision scenario
a. Waterways, b. Ship movement

We can estimate the probability of collision between entering vessel (Sm) and existing vessel (Sn) where all course are assume to have the same likelihood.

$$P_n = \frac{W \cdot L}{Ve \quad r} =$$

$$\text{1.17}$$

$$N_i = T_M \cdot \Sigma \sum_{E=1}^{E} P_e P_{en}$$

$$\text{1.18}$$

Considering geometry dimensioning we can have where is the final meeting conditioned is transformed by aligning Vm direction along x—axis with basis in the origin. The impact diameter is the dim is normal to the relative speed and is given by vector sum: Pi can be express as the relationship between diameter and characterized width (W)

$$\frac{P_I}{T_E} = \frac{d_i V_r}{W^2}.$$

$$\text{1.19}$$

$$P_{ME} = \frac{P_i V_r}{T_n} = \frac{1}{\pi} \int_0^{\pi} \frac{d_i V_r}{W^2}.d\theta.$$

$$\text{1.20}$$

Model can be approximated as followed:

Number of impact: $N_i = \dfrac{N}{\tau V_r}(\dfrac{4}{\pi}.L + 2.B)$

$$\text{1.21}$$

Expected number of ship in the fairway at any given time:

$$N_m = (\frac{W \cdot L)}{V \quad \tau}.N$$

$$\text{1.22}$$

The traffic density $P_n = (\dfrac{N_m}{W^2} = \dfrac{W}{V} \cdot \dfrac{1}{\tau} \cdot \dfrac{1}{W^2}) = \dfrac{N}{V.\tau.W}$

$$\text{1.23}$$

Considering loss of control potential

$$Pa = P_c.P_i \rightarrow Na = 2\mu_c.D.N_e P_i$$

$$\text{1.24}$$

The number of encounter situation:

$$Ne = \frac{1}{2} N_m . P_i \qquad\qquad 1.25$$

Therefore:

$$Na = \frac{1}{2} \mu_c . D . P_i . N_m \qquad\qquad 1.25$$

In the frequency analysis, the annual frequency of each failure case is estimated. Separate frequencies are estimated for each operating phase as required. In modelling the development, consequences and impact of the events, each failure case is split into various possible outcomes. The outcomes are the end events on an event tree or chain of event trees. Each outcome probability is estimated by combining the probabilities for appropriate branches of the event tree.

The outcome frequency (Fo) is then: $F_o = F_e \prod P_b$ 1.27

Where: F_e is failure frequency, P_b probability of one segment, not all possible outcomes are modelled. Representative scenarios are selected for modelling, and the scenario frequency is taken as:

$$F_s = \sum_{outcomes} F_o \qquad\qquad 1.28$$

Failure per nautical mile and failure per passage can be selected from previous representative work. Necessary period for ship to pass the fairway T=D/v = 3000/3 = 1000 sec. The result of accident frequency (Na) can be compared with acceptability criteria for maritime industry (See Table 1.13). If it is two high the system could be recommended to implement TSS. If the result is high TSS can be modelled to see possible reduction due to its implementation.

Table 1. 13: Accident frequency risk acceptability criteria

Frequency Classes	Quantification
Very likely	Once per 1000 year or more likely
Remote	Once per 100-1000 year
Occasional	Once per 10-100 year
Probable	Once per 1-10 year

1.4.2 Damage estimation modeling

Initial kinetic energy of ship = $E = \frac{1}{2} \frac{M}{1000} V^2 \cdot K$ (MJ)   1.29

Where: E = impact energy (MJ), M= vessel mass (tonnes), V= vessel speed
K= hydrodynamic added mass constant, taken as (Den 1990)
K factor assignment of 1.1 for head on collision (powered) impacts, and
1.4 for broadside (drifting)

Absorbed collision impact $Ei = 47.2.Vc + 32.8$ MJ   1.30

Where Vc = collapse material volume (m³)

Consequence impact acceptability can be deduced as shown in
Table1.14. Severity number can be assigned depending on modality of risk
analysis and findings.
RPN = S x O X D   1.31
RPN calculation of propulsion failure is the highest.

Advantage derived from channel improvement work like channel
widening, deepening and straightening could be quantify into
sustainability equity for determination of risk cost control option for
Langat River sustainable navigation and transportation.

Table 1. 14: Consequence acceptability criteria

Quantification	Serenity	Occurrence	Detection	RPN
current failure that can apply to death failure, performance of mission	Catastrophic (10)	2	10	200
failure leading to degradation beyond accountable limit and causing hazard	Critical (7)	2	5	70
controllable failure leading to degradation beyond acceptable limit	Major (5)	4	2	20
nuisance failure that do not degrade system overall performance beyond acceptable limit	Minor (1)	6	2	12

1.5 Application

The outcome can be compared with risk acceptability criteria for offshore and maritime industry. The acceptability matrix is on ALARP principle. Figure 1.15 shows the result of correlation of combined graphs for accident variables of frequency analysis with changing channel variables (V, W, Nm, and B). For accident per 10,000 years, speed and number of ship is critical. Maximum speed of 4 knot is considered best for the channel. Following precision theory, impact is more like for average excess speed of 12 knot at angle around 100 to 150 degree.

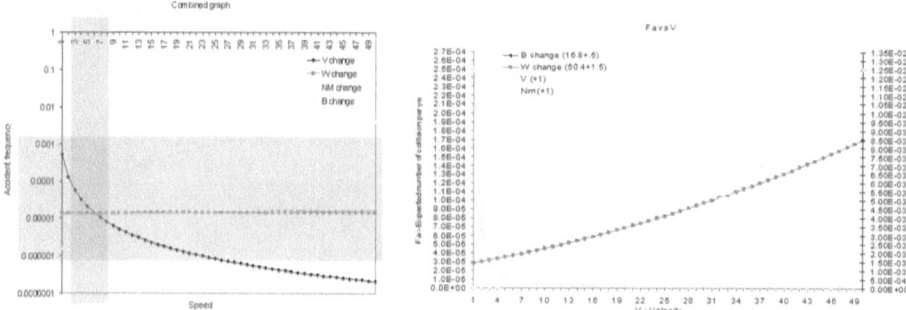

Figure 1. 15: Combined graphs for frequency analysis
with changing channel variables

Figure 1.16 shows accident impact probability correlation with speed and angle of impact. Impact probability is likely at angle between 105 to115 degrees. Average speed of 10 knot gives minimum impact.

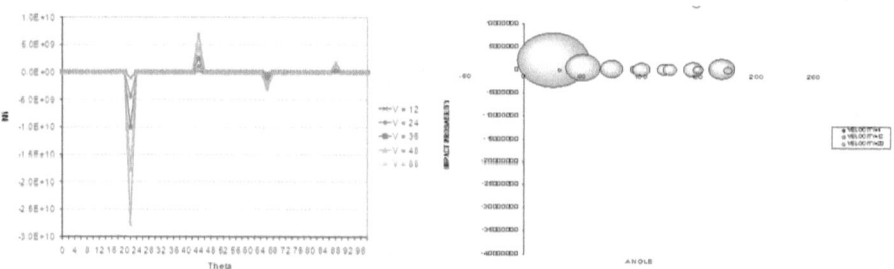

Figure 1. 16: Impact probability Vs angle of impact theta

Figure 1.16 shows accident is intolerable for vessel of 12 knot. The amount of energy is unacceptable at from vessel of 20,000 and speed of 40 knot.

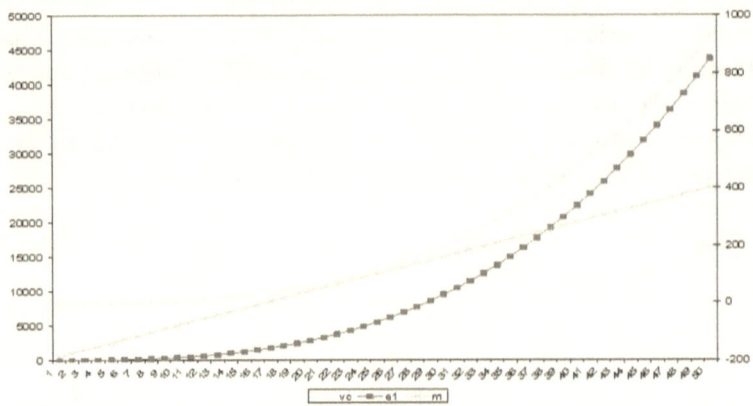

Figure 1. 17: Correlation between accident energy,
volume of collapse and mass

Figure 1.17 shows acceptable damage under different mass and energy, acceptable damage is at 12, 000KJ. Figure 1.18 shows correlation for drifting collision damage, damage tolerable between 5000KJ –20, 000KJ. The consequence could further be broken down into effect for ship, human safety, oil spill and ecology.

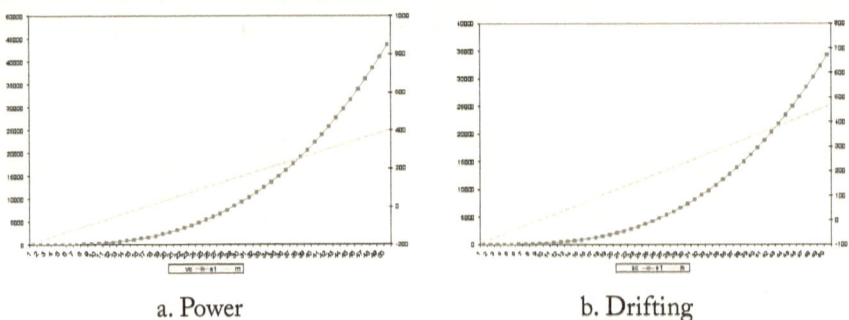

a. Power   b. Drifting

Figure 1. 18: Correlation collision

1.6 ALARP principal, risk acceptability criteria and risk control option

The analysis is followed by influence diagram risk control option and sustainability balancing of cost benefit towards recommendation for decision support for efficient, reliable and effective. Historical observation reveals that the collision in Langat River has significantly increased. The ALARP (As Low s Reasonable Possible) graph show that the risk is low, but this conservatively unacceptable for a channel with less traffic density. The ALARP work is followed is followed by risk acceptability criteria whose analysis is followed with risk cost control option(RCO) using cost of averting fatality index (ICAF). Advantage derived from channel use and improvement work like channel widening, deepening and straightening could be quantify into sustainability equity for determination of RCO for efficient and sustainable transportation on Langat River. Table 1.15 shows typical risk matrix diagram to match ALARP region.

Table 1. 15: Risk priority matrix

L = low risk; M moderate risk; H = high risk; V =very high risk

Frequency of Occurrence (or Likelihood)	Consequences (Severity of Accident)				
	Incidental (1)	Minor (2)	Serious (3)	Major (4)	Catastrophic (5)
Frequent (5)	M	H	VH	VH	VH
Occasional (4)	M	M	H	Risk without measure	VH
Seldom (3)	L	M	H	H	VH
Remote (2)	L	L	Risk after measure	H	H
Unlikely (1)	L	L	M	M	H

Risk acceptability criteria established in many industries and regulations are for limit definition of the risk. Risk is never acceptable, but the activity implying the risk may be acceptable due to benefits safety, fatality, injury, individual and societal risk, environment, economy. The rationality may be debated, societal risk criteria are used by increasing number of regulators. The consequence analysis is followed by influence diagram that is based on ALARP principle. ALARP here the two

—

61

main component (frequency Vs consequence) of risk whose combined plot against each other is measure of overall risk. This is followed by comparison with risk acceptability criteria. (See table 1.14). Risk is never acceptable, but the activity implying the risk may be acceptable due to benefits safety, fatality, injury, individual and societal risk, environment, economy.

The rationality may be debated, societal risk criteria are used by increasing number of regulators. After other channel complexity, reliability and sustainability analysis, decision can be made for waterways requirements. ALARP. Comparison shows that Langat River risk is not appalling. The risk curve is at the lower portion of ALARP.

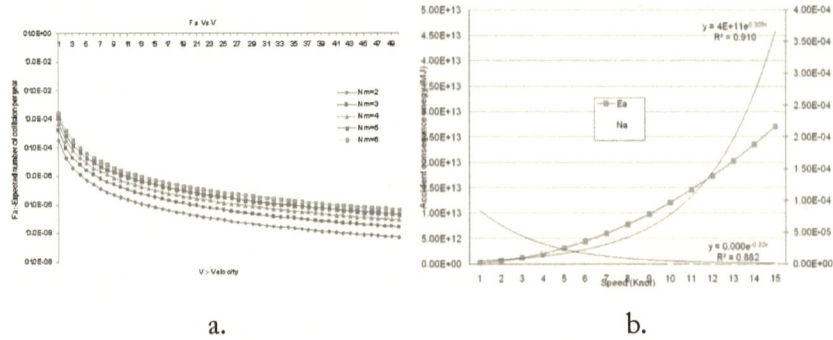

a.   b.

Figure 1. 19: Accident energy Vs Accident Frequency
@ Changing beam of ship

This curve lies in the safe zone of ALARP graph. It indicates that the collision risk in Langat River is acceptable. But in the context of channel with less traffic density, the figure does not look good. If various channel improvement plan are implemented then it is good, the level of risk will further come down. Figure 1.19 shows the relationship between accident frequency and energy. At accident energy of 20MJ and frequency of 10e4, mass of approximately 10, 000 is tolerable to navigate in Langat waterway for vessel mass of 10000, there is highest risk of potential for collision. Current beam of ship should be maintain, increase in beam of ship could lead to rise in risk. Ship risk becomes higher at large beam of ship.

—

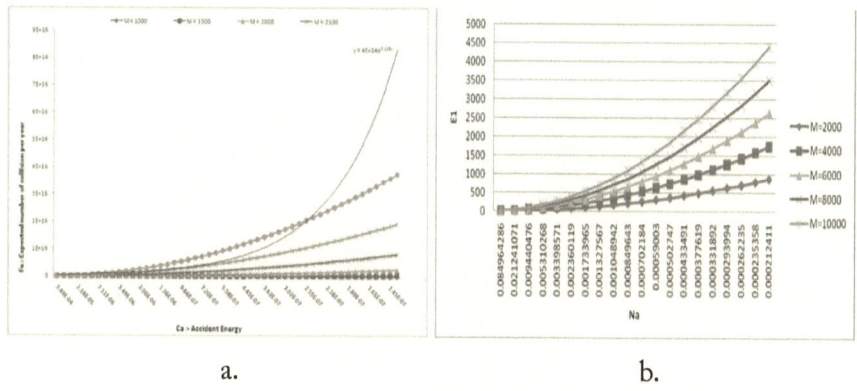

a.   b.

Figure 1. 20: Accident energy Vs Accident Frequency @ Changing mass

Figure 1.21 shows width changes and accident frequency, current width of the channel is only good for the next 100 years; it is unacceptable for maritime industry risk acceptability criteria of per 10, 000 years. With this, risk definition for width of the channel can be provided.

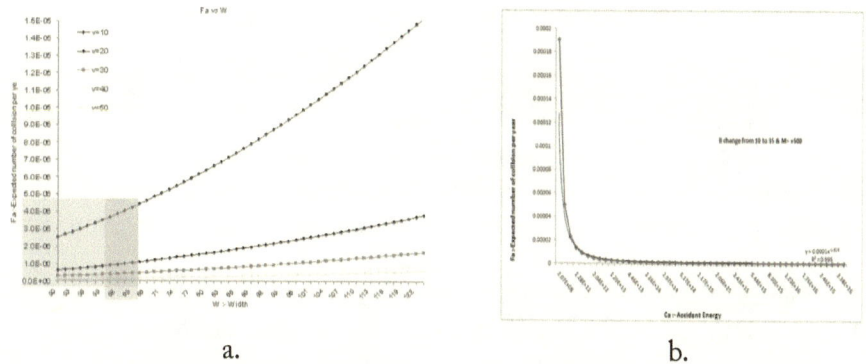

a.   b.

Figure 1. 21: Accident energy Vs Accident Frequency @ Width

Figure 1.22 shows that the current number of ship should be maintain on Langat River, increase could give a sharp rise of risk due to exponential correlation, risk become higher with more number of ships.

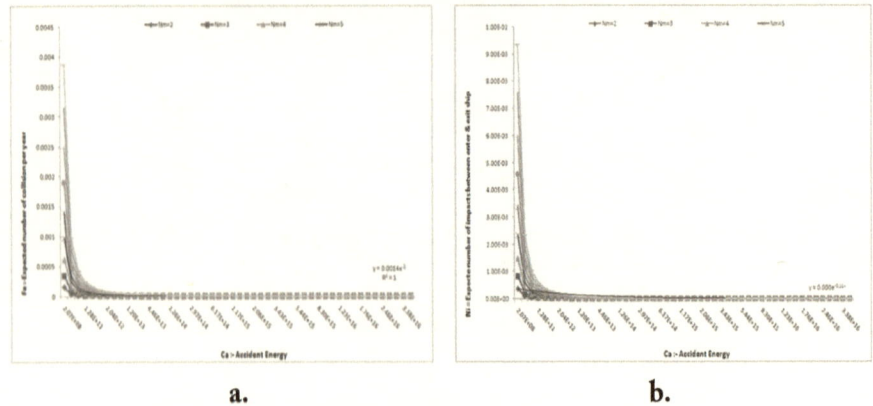

a.   b.

Figure 1. 22: Accident frequencies per year Vs Accident
Consequence@ Changing number of ships wide of channel.

1.6.1 Other Consequence Quantification

Consequence deduce from energy calculation can be translated to
loss of fatality and Green House Gas (COx) release. Figure 16 a, b and
c shows expected fatality, Cox and cost respectively. Two dimensional
cross plotting of frequency and consequence in Figure 16a shows risk is
acceptable at least no fatality (0.11), 0 Fatality at 50MJ release of energy
for current 12 numbers of ships using the channel. Two dimensional plots
for prediction COx in Figure 16b shows that for the current number
of ships on Langat River, the impact is acceptable, but at long run, it is
unacceptable.

Figure 1.23 shows total cost of averting fatality (CAF) for the system.
Where G (Gross), N (Net), I(imply) and CURR is cost of unit risk
reduction. There is maximum cost agreement between NCAF and CURR
at speed 20 knot which correspond to 3 mil acceptability cost for maritime
industry.

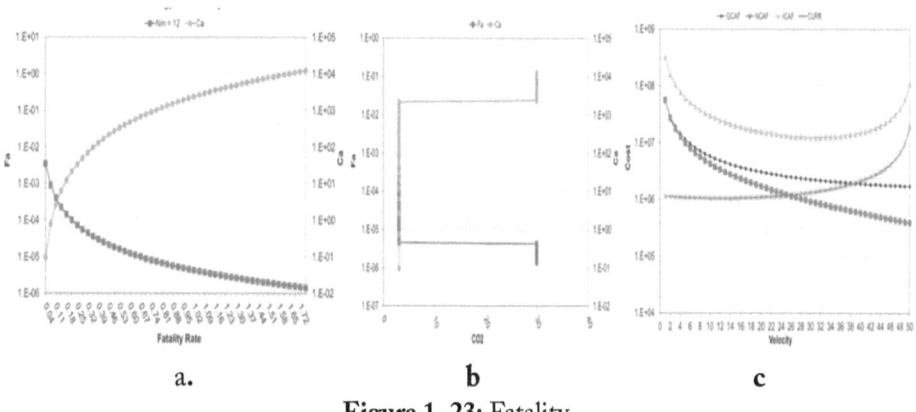

a.   **b**   c

Figure 1. 23: Fatality

1.7 Additional system risk and reliability analysis

1.7.1 Subsystem level analysis

By studying the risk models associated with collision scenarios, four sub models in particular stands out where further risk reduction could be effective : Power transmission, navigation, vessel motion, and human reliability. Frequency calculation through Fault Tree Analysis (FTA) modelling involve top down differentiation of event to branches of member that cause them or participated in the causal chain action and reaction. It is a systematic inclusion of all components involved in particular events or failure say propulsion failure. Cumulative of FTA analyses leads to scenario generation by using minimum cut set of accident cause. Highly probable cut set will satisfy the following condition.

1.7.2 Channel Complexity analysis

Various channel components enter channel complexity regime, these can be visibility weather, squat, bridge, river bent and human reliability. It is important to account for each of them in channel design work. Figure 1.24 show channel complexity for Langat. Poor visibility might be expected to increase the risk of groundings and collisions. The increase in accident risk due to poor visibility is more consistent and more significant than the change associated with high wind. A model extracted from Dover waterway studies concluded with the following [28]:

Fog Collision Risk Index (FCRI) = $(P_1 + VI_1 + P_2 + VI_2 + P_3 . VI_3)$ 1.32

—

Where: P_k = Probability of collision per million encounters, VI_k = Fraction of time that the visibility is in the range k, K = Visibility range: clear (>4km), Mist/Fog (200m—4km), Tick/dense (less than 200m).

Figure 1.24 shows channel complexity for Langat River, Channel straightening and alignment, required radius of curvature at bends for 5000 DWT,

Towed barge Length = Barge Length + Tug Length + Tow Line, R> (4-6) length of barge train (to meet the navigation requirement): PIANC

Empirically derived means to determine the relationship between accident risk, channel complexity parameters and VTS is given by equation

$$R = -0.37231 - 3.5297C + 16.3277N + 0.2285L - 0.0004W + 0.01212H + 0.004M \quad 1.33$$

For predicted VTS consequence of 100000 transit, C = 1 for an open approach area and 0 otherwise, N = 1 for a constricted waterway and 0 otherwise, L = length of the traffic route in statute miles, W = average waterway/channel width in yards, H = sum of total degrees of course changes along the traffic route, M = number of vessels in the waterway divided by L. Barge movement creates very low wave height and thus will have insignificant impact on river bank erosion. Speed limit can be imposed by authorities for wave height and loading complexity. Other important analysis is reliability analysis to cater for uncertainty estimation. It is important for this to be carried out separately. Reliability work could include projection for accident rate for certain number of year, using poison distribution or determination of exact period for next accident using binomial function. Ship collisions are rare and independent random event in time. The event can be considered as poison events where time to first occurrence is exponentially distributed (Parry, 1996, Richards et al., 2001. Another critical analysis is stochastic estimation for human reliability assessment which can be done using questionnaire analysis or the technique of human error rate prediction THERP probabilistic relation.

$$P_{EA} = HEP_{EA} . k \sum_{k=1}^{m} PSF_K . W_K + C \qquad 1.34$$

Where: P_{EA} = Probability of error for specific action, HEP_{EA} = Nominal operator error probability for specific error, PSF_k = numerical value of kth performance sapping factor, W_k = weight of PSF_k (constant), m=number of PSF, C= Constant.

Navigation of vessel in shallow water at a hull displacement cause vertical sink age, or "squat," as a result of a pressure drop beneath its hull, to avoid ship groundings with possible severe economic and environmental consequences, the relevant governmental, port, and maritime agencies and organizations need a reliable method of predicting ship squat. Squat analysis and channel clearance based on the physical characteristics of a channel and the ships that travel through it can be used to issue appropriate regulations regarding vessel size, speed and to plan channel dredging operations. Model on weather and human reliability assessment, and expert judgment as well as simulation could help perfect the reliability on safety and environmental risk study for inland water ways (Jonh, 1994), (Amrozowicz, 1996), (Millward, A. 1990).

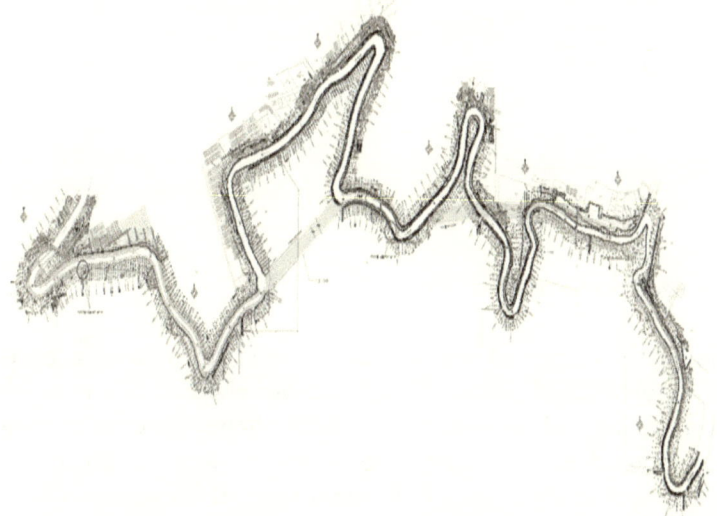

Figure 1. 24: Langat channel complexity

1.8 Between reliability and validation

Accident means, variance and standard deviation from normal distribution

For 10 years =>Mean (μ = 10 x Na   1.35

—

Variance (σ) = 10 x Na x (1-Na)

Standard Deviation $= \sqrt{\sigma}$, Z = $(X-\mu)/ \sigma$ 1.36

Year for system to fail from binomial, mean time to failure and poison distribution.

$$F_r \ (N/\gamma, T) = e^{\gamma.T}( )\gamma, (\gamma.T)^N) . N! \qquad 1.37$$

Comparing the model behaviour apply to other rivers of relative profile and vessel particular. Triangulating analysis of sum of probability of failure from subsystem level failure analysis. System improvement, for example Traffic Separation Scheme (TSS) Implementation effectiveness, could achieve reduction in head collision. This can be done through integration of normal distribution along width of the waterways and subsequent implementation frequency model. And the differences in the result can reflect improvement derived from implementation of TSS.

$$f_{south} \ (x) = \frac{1}{\mu\sqrt{2\pi}} e^{-\frac{1}{2}}(x - \frac{12}{\mu}) \qquad 1.38$$

$$f_{north} \ (x) = \frac{1}{\mu\sqrt{2\pi}} e^{-\frac{1}{2}}(x - \frac{12}{\mu}) \qquad 1.39$$

1.8.1 Validation Analysis

Minitab is employed for validation and the following result was yielded. Figure 1.25 shows accident frequency residual plot from Minitab is shown with good fitness. Figure 1.25 shows accident consequence validations, accident consequence good to fit, residual graph of CDF profile tracing infinity. Figure 1.26 shows residual histograms distribution diagram for accident frequency, skewed to low risk area, outlier can be removed. Figure 1.26 shows Log normal plots Accident frequency (Na), distribution shows a good to fit.

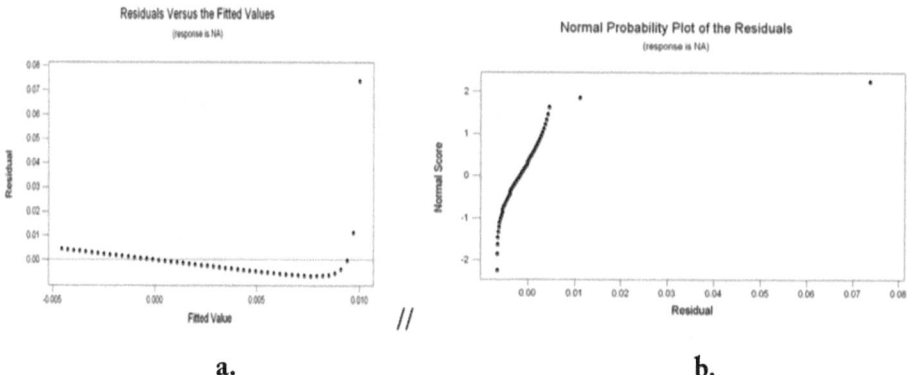

a.   **b.**

Figure 1. 25: a. Accident frequency residual plot
b. Accident consequence validation

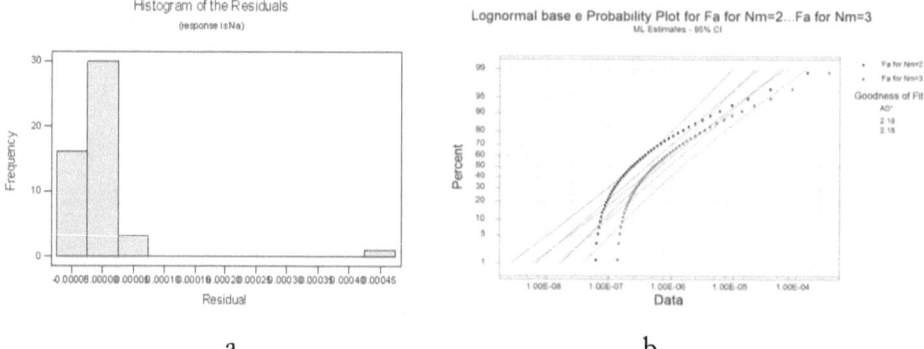

a.   b.

Figure 1. 26: a. Residual histograms distribution diagram,
b. Log normal plot

Figure 1.27 shows the matrix plot of Fa, operating region of channel variables shown in the plot is in agreement with operational risk.

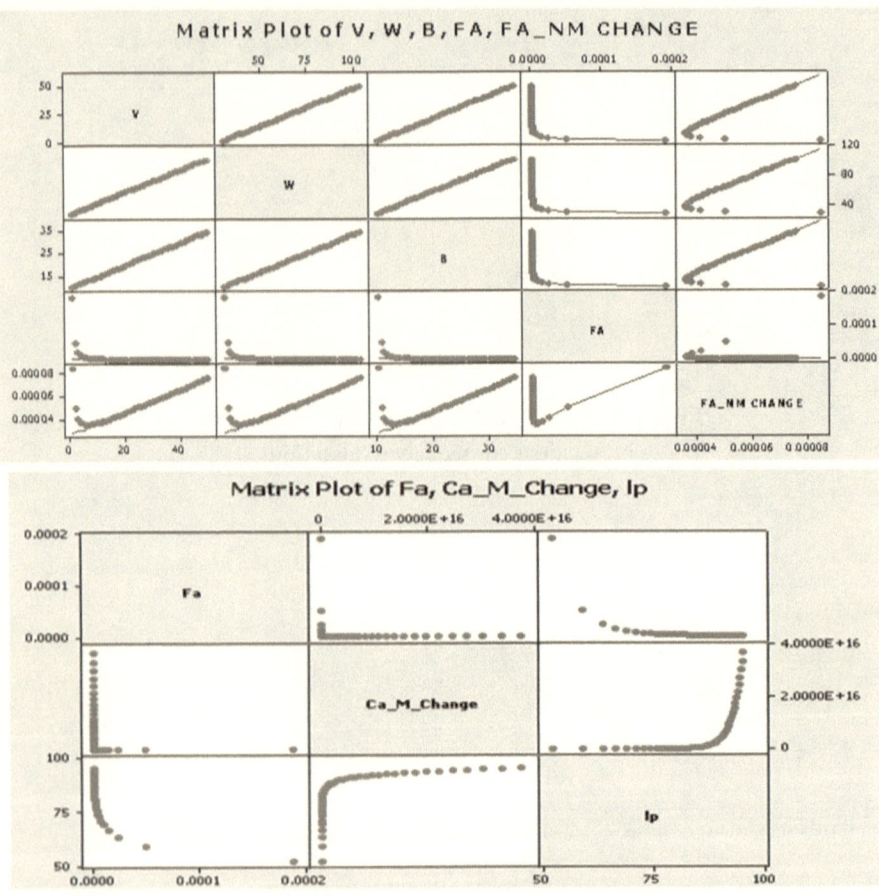

Figure 1. 27: Matrix plot for Fa and Ca

Operating speed of vessel is between 0-40, while operating Width of the channel is between 50-100 and the beam is between 10-30. The matrix observation also show that B and W parameter risk are tolerable for long-term. The matrix plot for accident frequency and consequence shows that curve fit quite well in low risk areas.

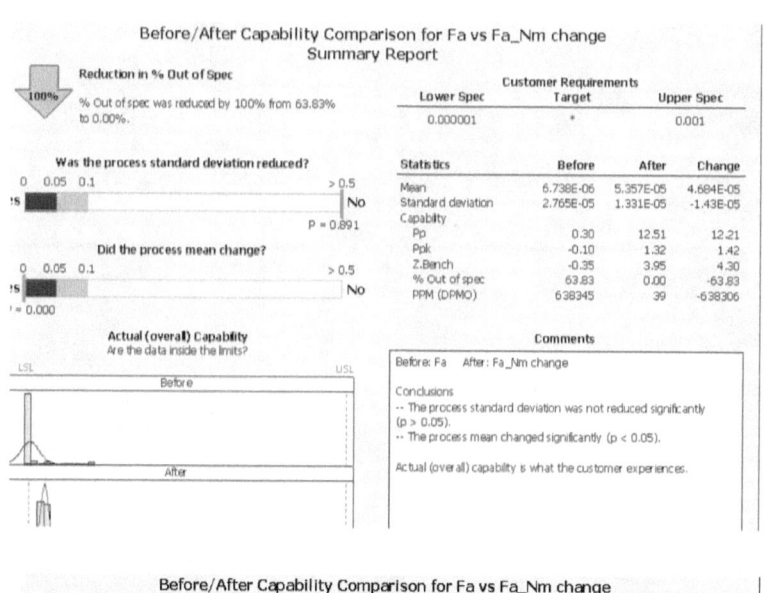

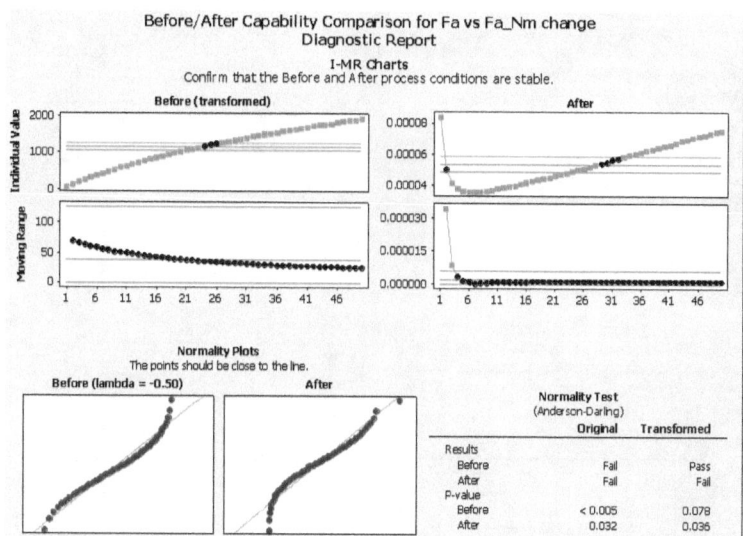

Figure 1. 28: Performance of Fa

Figure 1.28 a & b shows the overall capability for Fa at 95% confident, only 13% of the process is out of spec, the process is stable as less than 0.05 and the bell shape distribution fit with only 1 out layer. All the representation of the data is within limit. The capability plot for Ca confirm acceptable result for stable process. Response to energy distribution is well distributed.

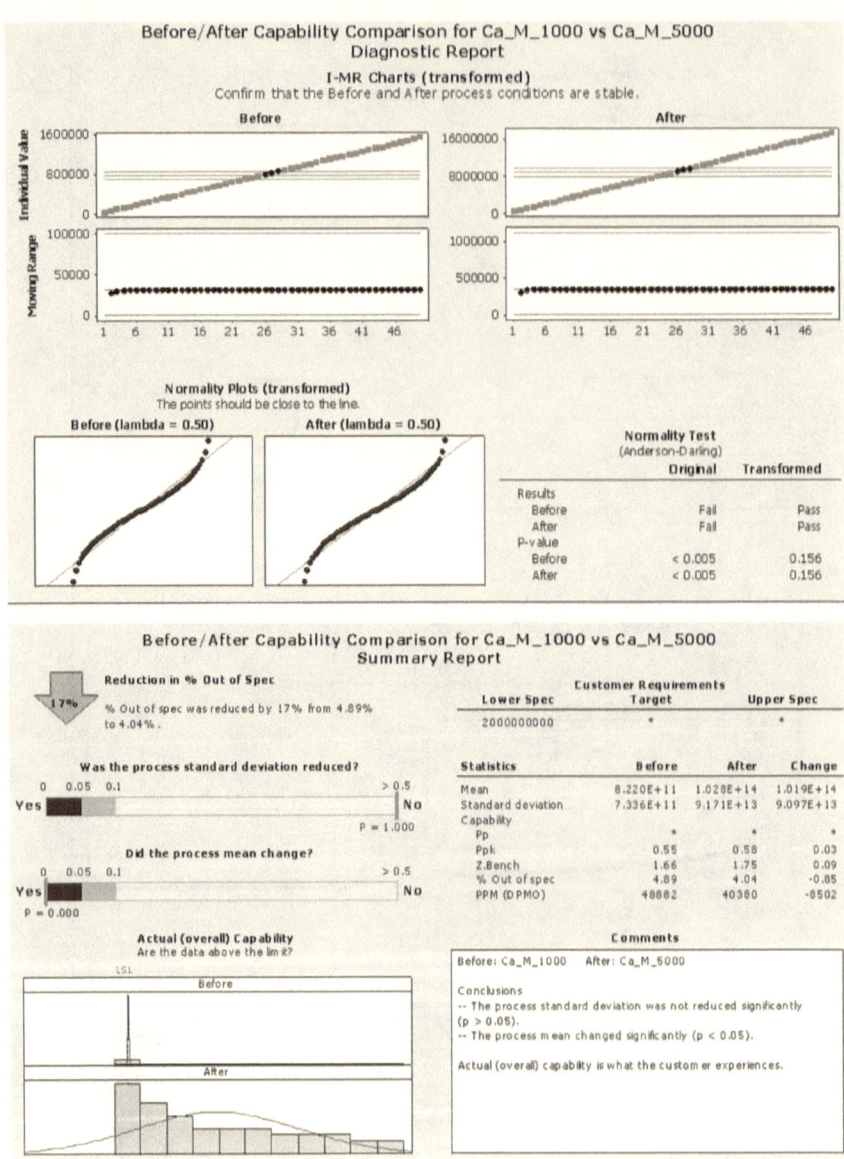

Figure 1. 29: Performance of Ca

Figure 1.29 shows residual histograms distribution graph for accident consequence for powered collision energy (E1), distribution fall around low risk area for drifting collision case, the profile trace CDF infinity, showing indication of good damage data set for the prediction. Figure 1.30 a and b show collision impact variation for powered collision energy for drifting or anchored collision (E1), normal distribution and histogram curve.

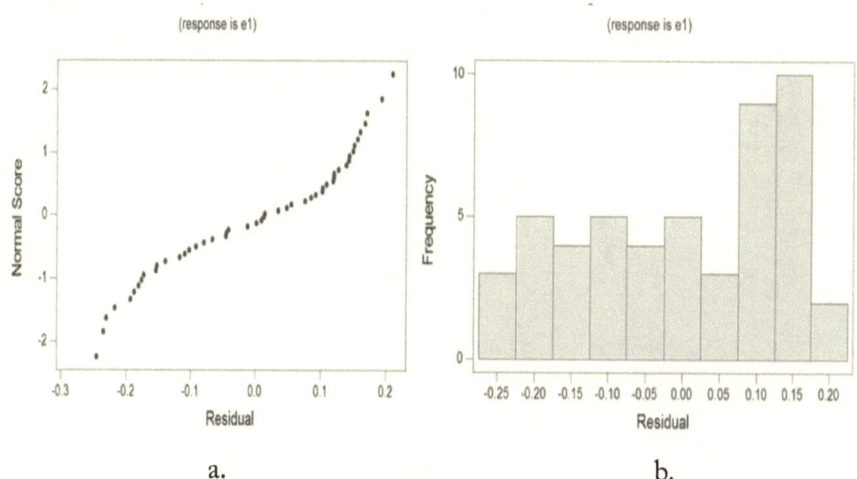

a.   b.

Figure 1. 30: a. Normal distribution plot for E1 residual,
b. Histogram plot of the E1 residual

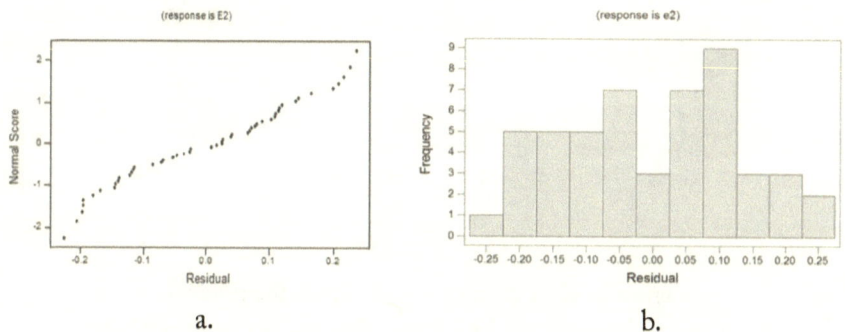

a.   b.

Figure 1. 31: a. Normal Probability plot of the residual,
b. Histogram plot of the residual

Figure 1.32a shows normal plot for interpolated accident energy. Figure 1.32b shows regression plot for E1 and Vc1. Figure 1.32c shows Lp trend reliability. The plot show good trend agreement with prediction for Lp and, Vc.

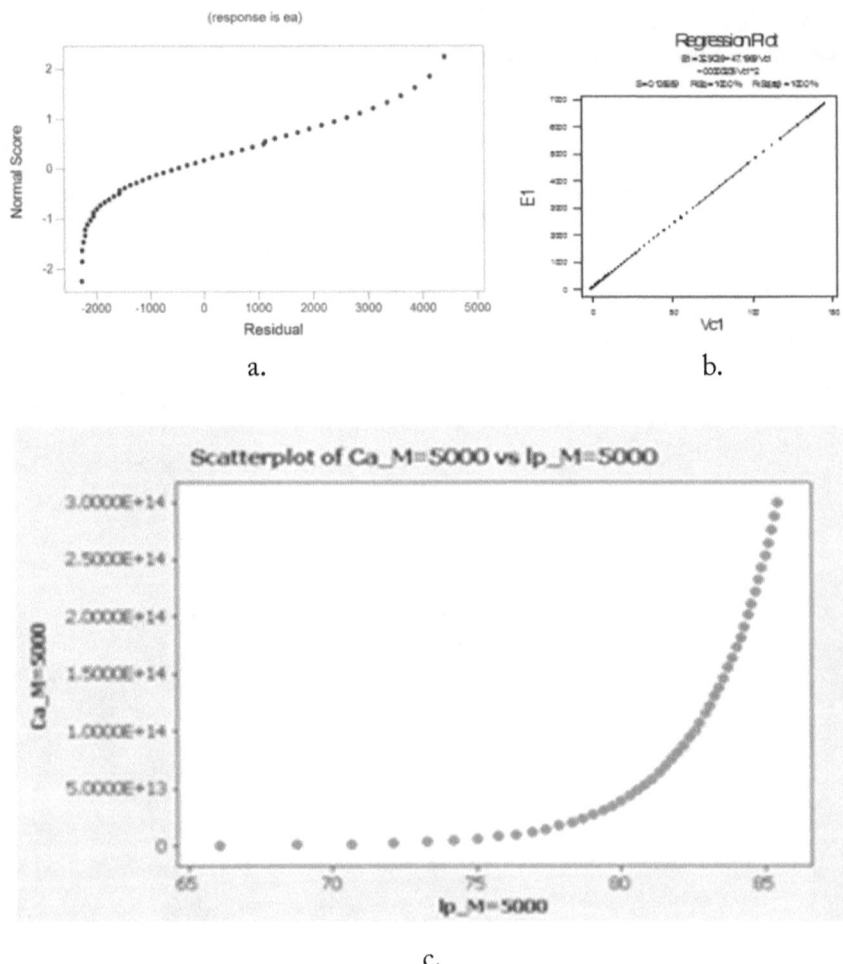

Figure 1. 32: a. Normal probability of the residual,
b. regression plot for E1 & Vc1, and c. Regression plot for E1& Vc1

Figure 1.33 present matrix curve fit quite well, and the graph show V, Nm, W, and B risk acceptability limit. The matrix show risk acceptability state line and distribution of parameters within acceptability and non-acceptability region. Table 1.16 shows regression equation deduced from the model.

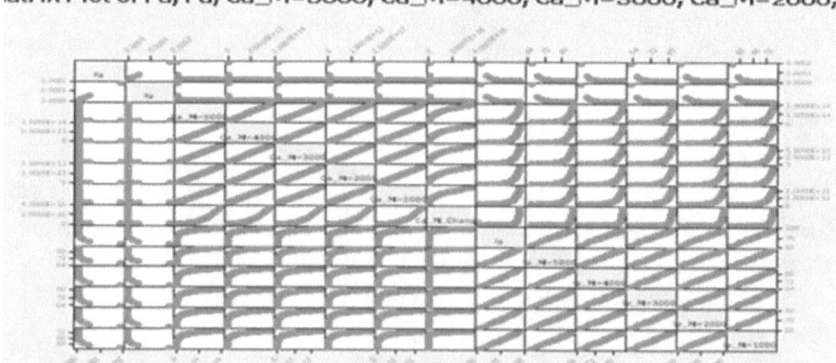

Figure 1. 33: Combine matrix risk curve for Ca AND Fa

Table 1. 16: Regression equation

Plot	Variables	Equation	Correlation	Trend
Frequency model				
Fa	@Nm changing Speed	$y = 2E\text{-}05e^{-0.11x}$	$R^2 = 0.826$	Exponential
Fa	@V	$y = 2E\text{-}05e^{-0.11x}R^2 = 0.826$	$R^2 = 1$	Square
Fa	W	$y = 2E\text{-}08x + 1E\text{-}05$	$R^2 = 1$	Square
Na	B	$y = 9E\text{-}07x + 0.000$	$R^2 = 0.999$	Linear
ALARP Risk curve				
Fa VCa	V	$y = 0.0003x^{-2}$	$R^2 = 1$	Power
Fa VCa	B	$y = 0.0001x^{-1.616}$	$R^2 = 0.995$	Power
Fa VCa	B =18.3, M=2000	$y = 0.0003x^{-2}$	$R^2 = 1$	Power
Fa VCa	B =21.3, M=2000	$y = 0.0004x^{-2}$	$R^2 = 1$	Power
Fa VCa	Nm	$y = 0.0014x^{-2}$	$R^2 = 1R^2 = 0.826$	Power
	Nm=3	$y = 2E\text{-}05e^{-0.111x}$		Exponential
Fa VCa	W=30-2480, M=5000	$y = 0.0003x^{-2}$	$R^2 = 1$	Power
Fa VCa	W=46	$y = 0.0003x^{-7}$	$R^2 = 1$	Power
Damage				
Fa Vs Lp	Speed	$y = 0.0003x^{-2}$	$R^2 = 1$	Power
Lp Vs Ca	Lp	$y = 11.219\ln(x) + 51.188$	$R^2 = 0.9997$	Logarithm
Lp vs Fa	M	$y = 5.1793\ln(x) + 63.748R^2 = 0.9988$		Logarithm
Fa Vs Vc1	Vs	$y = 0.0003x^{-2}$	$R^2 = 1$	Power

Fa Vs Vc2	Vs	$y = 0.0003x^{-2}$	$R^2 = 1$	Power
Ca vs Fa	Vc1(Drifting collision)	$y = 0.4457x^2 - 9.1828x + 40.251$	$R^2 = 0.9958$	Polynomial
Ca vs Fa	Vc2(Powered collision)	$y = 0.5673x^2 - 11.687x + 51.418$	$R^2 = 0.9958$	Polynomial

System level damage estimation modeling under risk of consequence and its important in risk analysis process using hybrid of first level approximation and stochastic associated with risks has been modeled. Frequency and consequence graph has been generated and risk components variables have been simulated to facilitate choice of decision support for sustainable waterways transportation system design. Implications of quantifying cost and benefit requirement as part of decision support under RCO, ICAF and ALARP principle has been discussed. Total risk has been for Langat River has been model from SERM. Accident collision per year has been determined, large barge have higher accident rates than small ships. Variables that affect accident rates have been simulated for necessary limit acceptability purpose for the channel. Accident rate has increased compare to previous year, a situation that required attention for solution. Damage estimation does have some degree of linear relationship with the risks to people and the environment.

Risk measurement can further be reliable by determining risk control option through sustainability balance between environmental, economic and safety. Quality of operator's, real-time environmental information about environmental conditions, including currents, tide levels, and winds in transit can be useful in sustainability balance work for waterways developments. It is important to use caution when comparing accident rates across ports and over time because of differences in reporting criteria. Other channel complexity analysis towards reliability and covering uncertainty has been proposed. Advantage of Implementing of TSS is discussed. Collision risk is much common and propulsion failure is one of the contributing factors. Preceding system description and hazard identification, Fault tree and event three is utilized to determine risk index of propulsion failure and interpolation of the index into ALARP influence diagram provide decision support for cost control option towards sustainable, reliable, efficient propulsion technology design and operability signature. However the annual accident data collected is good for preliminary analysis. Probabilistic method can give information about possible temporal factor changes.

—

References

1. Amrozowicz, M.D. (1996). The Quantitative Risk of oil Tanker Groundings. Master's degree thesis, Ocean Engineering Department. Massachusetts Institute of Technology, Cambridge,Massachusetts.
2. Brebbia C.A, Olivella. J. Gallor. W., 2000). The safety of ship movement in port water area. Maritime Engineering and port. WIT press, pg 80-89.
3. Cockrofft A.N and Lameijer, (1989). Collision avoidance rule, International Regulations for preventing collisions at sea. Newnes. Department of Environment. (2000) "Modelling and Data integration in the study of sediment". Kuala Lumpur. Malaysia.
4. DnV. (2001). Marine Risk Assessment. Her majesty stationary office. United Kingdom.
5. DnV, BV, SSPA (2002). Thematic Network for Safety Assessment of Waterborne Transportation. Denmark.
6. David Vose. 1996. Risk Analysis—A Quantitative Guide. John Wiley & Sons, INC. Canada pp. 67-87.
7. Eftratios Nikolaidis. (2005). Probabilistic analysis of dynamic systems. Engineering design reliability.CRC press. Pg 17-1.
8. Emi, H. et al (1997). An Approach to Safety Assessment for Ship Structures. ETA analysis of engine room fire, Proceedings of ESREL '97, International Conference on Safety and Reliability. Vol. 2.
9. Fujii Y. (1982). Recent Trends in Traffic Accidents in Japanese Waters. Journal of Navigation. Vol35 (1), pp. 88—102
10. IMO. (1993). Regulation on ship subdivision and damage stability of cargo ship, above 100m. Resolution a 684(17). Sales number IMO 871E. International Maritime Organization, London.
11. Jonh R. Dudley. (1994). towards safer seas & cleaner seas. Pg139-151
12. Kielland, P., Tuhman, T., (1994), On estimating map model errors and GPS position errors". Ottawa, Canada: Canadian Hydrographic Service.
13. Lewison, Gr. G. 2000. The Risk Encounter Leading to Collision. Journal of Navigation. Vol 31 (3). pp. 288—109.Wang J. (1999) A subjective modelling tool applied to formal ship safety assessment, Ocean Engineering 27 2000. pp. 1019–1035.

14. McGee, s.p., Troesch, A. AND Vlahopulos, N. (1999). Damage length predictor for high speed craft. Marine Technology, Vol. 36(4), 201-214.
15. McGowan J. (2001). M/v Santa Clara I. Why the Incident is so Unique. Proceeding of U.S. Marine
16. Safety Council DnV., Marine Risk Assessment, Her majesty stationary office United Kingdom.
17. Minorsky, V.U., (1959). "An analysis of collision with reference to protection of nuclear power plant". Journal of ship research. Vol. 3(2), 1-4.
18. Parry, G. (1996). The Characterization of Uncertainty in Probabilistic Risk Assessments of Complex Systems. Reliability Engineering and System Safety. 54:2-3, 119-126.
19. Richards C. Dorf. Eban Goodestein. (2001). How much environmental damage is too much? Technology. Human and society. Academics press pg 53-59
20. Millward, A. (1990). A Preliminary Design Method for the Prediction of Squat in Shallow Water". Marine Technology. 27(1):10-19.
21. Moderras M. (1993). What Every Engineer Should Know about Reliability and Risk Analysis. MarkelDeker Inc, Switzerland, pp. 299-314.
22. Sirkar, J., Ameer, P., Brown, A., Goss, P., Michel, K., Nicastro, F. and Willis, W. (1997). A Frameworkfor Assessing the Environmental Performance of Tankers in Accidental Groundings and Collisions. *SNAME Transactions*
23. Wallace R. Blischke. (2000). Reliability modeling, prediction, and optimization. pg 119-123
24. Wang X. John X, 2000). "What Every Engineer Should Know about Risk Engineering and Management". Markel Deker Inc. Switzerland, pp. 112-128.

CHAPTER 2

Safety & Environmental Risk Compliance for Nature Gas Ship Design & Operation

"Imagination is more important than knowledge"
Albert Einstein

Summary

The quests for an efficient fuel friendly to the environment have been recognized in maritime industry for a long time through improvements of gasoline and diesel by chemical reformulation. Inconvenience posed by these reformulation chemicals is performance problems; cold-start ability, smooth operation and avoidance of vapor lock. Climate change problem has further aggravated need to use fuel that could contribute to decrease in greenhouse gases and ozone-forming pollutants. Alternative fuels to petroleum have been identified to include, compressed natural gas (CNG); liquefied petroleum gas (LPG); methanol from natural gas LNG. Selection of this towards centralized reduction of Green House Gases (GHGs) will depend on ease of use, performance and cost. LNG cargo is conditioned for long distance transfer while CNG and LPG cargo are conditioned for end user consumption and short distance transfer. It is therefore, clear that promoting the use of CNG will catalyze boosting of economy of coastal ship building and transportation, including environmental friendly utility fuel, and new generation of intermodal transportation and supply chain. Since the danger behind use of this gas could not be either underestimated by virtue regarding coastal operation proximity and consequence. The chapter discusses risk and potential regulation that can formulate beyond compliance, decision towards use of top—down risk based design and operations that will reinforce new integrative, efficient, environmental friendly, reliable multimodal and intermodal links advanced concepts for LPG ship operating in coastal and restricted waters.

2.1 Introduction

Fuel technology has been dominated with ways to improve gasoline and diesel by chemical reformulation that can lead increase efficiency and additional inconvenience leading to ozone depletion, green house and acid rain forming pollutants. Likewise, side effects problems posed to transportation vehicles have been dominated by condition, other performance issues. Time has shown that the global trend in de-Carbonization of the energy system follow the following path: COAL > OIL> NATURAL GAS > HYDROGEN

The drive towards environmentally friendlier fuels points next to Natural Gas (NG) and the infrastructures to support that trend are being pre-positioned by corporate mechanisms as well as governmental bodies worldwide. NG is cheap and its reserve is plentiful. Natural Gas as fuel is becoming more and more established in urban transport and Power Generation sectors. Its use will also take aggressive approach for all inland vessel including ferries in the eyes of potential environmental compliance new regulations. Internationally its operational record and GHG gas score is rated as GOOD. However, CNG, LPG and ethanol has been proven to be environmental friendly and has fuel economy of 50 percent. This shows that, CNG and LPG have potential for large market for use in niche markets in both developed and developing countries. Other gains from CNG and LPG depend on the amount of associated methane emissions from gas recovery, transmission, distribution, and use. On a full-cycle basis, use of LPG can result in 20-25% reduction in GHG emissions as compared to petrol, while emission benefits from CNG are smaller—about 15%. [1]

Furthermore, it is clear that promoting the use of CNG and LPG will be a catalyst to boost economy of coastal ship building, environmental friendly intermodal transportation for supply chain. Efficient and reliable operation can be made afforded by LPG, transportation, supply vessel, tugs to support this potential development. On the regulatory regime, IMO focus more on operational issues relating to carriage of gas with no specification for CNG and LPG, while the ICG code and class society guidelines elaborate on the design as well as operational consideration. Local administration imposes additional regulation as required for their respective implementation. Time has revealed that there will be large demands for these gases.

This Chapter focus on integrative use of IMO prescriptive goal and risk based standards with holistic consideration of factors require for safe design and operation of LPG ships in inland water. Including hybrid use of elements of Formal Safety Assessment (FSA) and Goal Based Standards (GBS) to prevent, minimize control and guarantee the life span of LPG ships and protection of environment. The chapter will discussed top down environmental risk generic risk model and operations of LPG ship. It will describe the characteristics of LPG, regulatory issues and environmental issues driving today's beyond compliance and selection of new technology policy. Since it is the consequence of accident and incident that leads to environment disaster, the chapter will discussed issues that allow prevention and control of accident. Since issues relating to global warming, GHG releases is strictly linked to ship energy source, the chapter will also discuss impact areas and potential new technology driving beyond compliance policy adoption for LPG design and operation.

2.2 Natural Gas and its Products

Natural gas in its liquid state (LNG) or liquid natural gas that comprise of liquid hydrocarbons that are recovered from natural gases in gas processing plants, and in some cases, from field processing facilities. These hydrocarbons involve propane, pentanes, ethane, butane and some other heavy elements. LNG accounts for about 4% of natural gas consumption worldwide, and is produced in dozens of large-scale liquefaction plants. Natural gas contains less carbon than any other fossil fuel and, therefore produces less carbon Dioxide (CO_2) when compared to any conventional vehicles. Its usage also results in significantly less carbon monoxide (CO), as well as less combustive organic compounds than their gasoline counterparts. It is produced by cooling natural gas to a temperature of minus 260 degrees F (minus 160 Celsius). At this temperature, natural gas becomes liquid and its volume reduces 615 times. LNG has high energy density, which makes it useful for energy storage in double-walled, vacuum-insulated tanks as well as transoceanic transportation.

–

The production process of LNG starts with Natural Gas, being transported to the LNG Plant site as feedstock, after filtration and metering in the feedstock reception facility, the feedstock gas enters the LNG plant and is distributed among the identical liquefaction systems. Each LNG process plant consists of reception, acid gas removal, dehydration removal, mercury removal, gas chilling and liquefaction, refrigeration, fractionation, nitrogen rejection and sulfur recovery units. LPG and CNG are made by compressing purified natural gas, then stored and distributed in hard containers. Mostly, LPG station is created by connecting a fuel compressor to the nearest natural gas pipeline distribution system. The process through which Liquefied Natural Gas is produced consists of three main steps, namely:-

a) **Transportation of Gas** –The best place to install the plant is near the gas source. The gas is basically transported through pipelines or by truck and barge.

b) **Pretreatment of Gas**-The liquefaction process requires that all components that solidify at liquefaction temperatures must be removed prior to liquefaction. This step refers of treatment the gas requires to make it liquefiable including compression, filtering of solids, removal of liquids and gases that would solidify under liquefaction, and purification which is removal of non-methane gases.

c) **Liquefaction of Gas** –Today, alternative fuels to petroleum has been identified to include Compressed Natural Gas (CNG); Liquefied Petroleum Gas (LPG); methanol from natural gas, coal or biomass; ethanol from biomass; electricity and hydrogen. However NG quality may be expressed with the Wobbe Index— Methane Number MN80 (Volume percent hydrogen atoms / carbon atoms) or Methane >=88%

Since 1960s, CNG and LPG are recognized as vehicle fuel alternative to oil-based gasoline and diesel fuel that reduces pollution of the air. It is a natural gas compressed to a volume and density that is practical as a portable fuel supply. Compressed natural gas (CNG) and Liquefies petroleum gas (LPG) are use as consumer fuel for vehicles, cooking food and heat homes. There exist a vast number of natural gas liquefaction plants designs, but, all are based on the combination of heat exchanger and refrigeration. The gas being liquefied, however, takes the same liquefaction path. The dry, clean gas enters a heat exchanger and exits as LNG. The capital invested in a plant and the operating cost of any liquefaction plant is based on the refrigeration techniques.

Natural gas is transported through pipelines to refuelling stations then compressed at a pressure of 3,000 psi with the help of specially installed compressors that enables it to be loaded as gas cylinders for vehicles. The process consists of drawing the natural gas from underground pipelines by the compressor. The composition of pipeline natural gas varies considerably depending on the time of year, pipeline demand, and pipeline system. It may contain impurities, like oil, particulates, hydrogen sulphide, oxygen or water. Hence, the modern day, quality LPG plant system consists of facilities to address these problems. Using LNG as the feedstock to make CNG and LPG eliminates or mitigates each of the above stated concerns as contains no water or any such impurity.

This eliminates the concerns for corrosion, plugging of fuel lines, and the formation of hydrates. Significant design innovation will involve development of liquefied gas technology that promises lower costs and shorter scheduling time than either Liquefied Natural Gas technology or a pipeline transport as well as provision of unique solution to the development of distressed or stranded gas reserves and alternative to associated gas re-injection. Liquefied Petroleum Gas (LPG) can also be produced either as a by-product when refining crude oil or direct from the gas wells. The two most common LPG gases are known as Commercial Propane and Commercial Butane as defined in BS 4250 [2]

Up to 15kg and generally used for leisure applications and mobile heaters. Commercial Propane is predominately stored in red cylinders and bulk storage vessels and especially used for heating, cooking and numerous commercial and industrial applications. LPG has one key characteristic that distinguishes it from Natural Gas. Under modest pressure LPG gas vapor becomes a liquid. This makes it easy to be stored and transported in specially constructed vessels and cylinders. The combustion of LPG produces Carbon Dioxide (CO^2) and water vapor therefore sufficient air must be available for appliances to burn efficiently. Inadequate appliance and ventilation can result in the production of toxic Carbon Monoxide (CO). All things being equal, it produces much less hydrocarbon compare to diesel. Hazards associated with LPG ships are linked to the gas characteristics that attract beyond compliances operability and design policy. Selection of this towards centralized reduction of GHGs will depend on ease of use, performance and cost.

2.3 Natural Gas Properties

Everyone dealing with the storage and handling of LPG should be familiar with the key characteristics and potential hazards. Matter either in their solid, a liquid or a gaseous form is made from atoms which combine with other atoms to form molecules. Air is a gas, in any gas, large numbers of molecules are weakly attracted to each other and are free to move about in space. A gas does not have a fixed shape or size. Each gas that the air is composed of consists of various different properties that add to the overall characteristics of a particular gas.[3] Gases have certain physical and chemical properties that help to differentiate a particular gas in the atmosphere.

Depending on different properties the gases are used widely in several applications. Below are some of the gases properties—Natural gas may consist of:

i.   Methane CH_4—> .80%
ii. Ethane C_2H_6—>.20%
iii. Propane C_3H_8—>20%
iv. Butane C_4H_{10}—>20%
v.   Carbon Dioxide CO_2->.8%
vi. Oxygen O_2—>0.2%
vii. Nitrogen N_2—>5%
viii. Hydrogen sulphide H_2S—>5%
ix. Rare gases-> A, He, Ne, Xe trace

Hazards associated with LPG ships are linked to the gas characteristics and beyond compliances operability and design. CNG are a non toxic gas liquid at—259 °F /—162 °C which ignites at 1350°F / 732°C. The octane number is 120; it can inflame having a share of 5.3 to 15% in air. Methane has only 42.4% of the density of air and thus is lighter and may disappear in case of leakages.

2.4 Natural gas and LPG

LG carriers has proven considerable good safe ship in term of designed, constructed, maintained, manned and operated of all the merchant fleet of today. So far they have low accident record and non major has lead to release of large amounts of LG have ever occurred in the history of LG shipping. Nevertheless, there have been major concerns regarding safety of LG shipping and vivid that one catastrophic accident has the potential for serious consequential fatal and environmental damage. Therefore it became imperative to use IMO Goal—based and risk based instruments to quantify a baseline risk level to identify and evaluate alternative risk control options for improved safety.

Toward zero accident and zero, incident, apart from normal SOLAS standards for all ships, there are additional international regulation / Code for the Construction and Equipment of Ships Carrying Liquefied Gases in Bulk this include—The IGC Code. This Code is applicable to Liquefy gas carriers which are made mandatory under the SOLAS Convention. Thus, Risks associated with LPG ships encompass the following areas:

i. Loading
ii. Shipping in special purpose vessels
iii. Unloading at the receiving terminal.
iv. Third party risks to people onshore or onboard

NG shipping industry is undergoing considerable changes, e.g. an expected doubling of the fleet over a 10-year period, emergence of considerable larger vessels, alternative propulsion systems, new operators with less experience new trading route, offshore operations and an anticipated shortage of qualified and well trained crew to man Liquefies gas carriers in the near future. With this development, there is tendency for gas shipping to experience an increasing risk level in the time to come. Most IMO previous rules were made on reaction basis, in this age of knowledge employment of the new philosophy to design construct and operate based on risk and considering holistic factors of concern for sustainability and reliability remain a great invention of our time to save LPG ship and shipping.

2.5 Maritime Regulation

The International convention for the Safety of Life at Sea (SOLAS) is the fundamental IMO instrument that deal with regulation requirement for basic construction and management for all types of ships. It covers areas like are stability, machinery, electrical installations, fire protection, detection and extinction systems, life-saving appliances, Surveys and inspections, SOLAS also contains a number of other codes related to safety and security that applies to shipping in general. Examples of these are the Fire Safety Systems Code (FSS Code), the International Management Code for the Safe Operations of Ships and for Pollution Prevention (ISM Code) and the International Ship and Port Facility

—

security Code (ISPS Code). These codes imply requirements aiming at enhancing the safety on Liquefy Gas (LG) shipping activities as well as shipping in general [4, 5].

Classification society rules apply for structural strength while special code for ships carrying liquefied gas included in the SOLAS regulations—the IGC code. Other IMO regulations pertaining to safety are contained in the International convention on Load Lines which addresses the limits to which a ship may be loaded, the International Convention for the Prevention of Collisions at Sea (COLREG) addressing issues related to steering, lights and signals and the International Convention on Standards of Training, Certification and Watch keeping for Seafarers (STCW Convention) which addresses issues related to the training of crew. The International Convention for the Prevention of Pollution from Ships (MARPOL) addresses issues related to marine and air pollution from ships. These regulations are applicable to all ships as well as LPG ships. The issue of global warming has initiated MARPOL annex VI, was given preferential acceptance beyond tacit procedure and there is indication that more will follow [6].

2.6 Maritime Regulations for Liquify Gas Regulation

IMO regulation for safety regarding carriage of gas was never specifically for LNG, CNG or LPG carriers. However safety regulations exist in order to ensure the LPG ships are safe. Thus Gas carriers need to comply with a number of different rules that are common to all ship types, as well as a set of safety regulations particularly developed for ships carrying liquefied gas and the their crew as well as site selection and design of LG terminals. This include issues relating to control of traffic near ports, local topology, weather conditions, safe mooring possibility, tug capability, safe distances and surrounding industry, population and training of terminal staff. These considerations contribute to enhance the safety of LPG shipping in its most critical phase, i.e. sailing in restricted waters or around terminal and port areas.

The IGC code prescribes a set of requirements pertaining to safety related to the design, construction, equipment and operation of ships involved in carriage of liquefied gases in bulk. The IACS unified requirements for gas tankers were partly derived from the IGC code. The code specifies the ship survival capability and the location of cargo tanks. According to the type of cargo, a minimum distance of the cargo tanks from the ship's shell plating is stipulated in order to protect the cargo in case of contact, collision or grounding events. Thus the code prescribes

—

requirements for ships carrying different types of liquefied gas, and defines four different standards of ships, as described in Table 2.1. LNG carriers are required to be ships of type 2G and all LNG carriers should be designed with double hull and double bottom, while 2PG type is for LPG Ships.

The IGC code requires segregation of cargo tanks and cargo vapor piping systems from other areas of the ship such as machinery spaces, accommodation spaces, control stations; it also prescribes standards for such segregation. It provides standards for cargo control rooms and cargo pump-rooms are as well as standards for access to cargo spaces and airlocks. It defines requirements for leakage detection systems, as well as loading and unloading arrangements. Different types of cargo containment systems are permitted by the IGC code, and the two main types of containment systems in use in the world liquefied tanker fleet are membrane tanks and independent tanks. Membrane tanks are tanks which consist of a thin layer or membrane, supported through insulation by the adjacent hull structure. The membrane should be designed in such a way that thermal expansion or contraction does not cause undue stress to the membrane. The independent tanks are self-supporting in that they do not form a part of the ship's hull [10].

Table 2. 1: Required for ship carrying liquefied gas

Ship type	Cargo
3G	Require moderate prevention method
2G	Ship less than 150m Require significant preventive measure
2PG	Require significant preventive measure cargo are carried in C tanks
1G	Require significant maximum preventive measure

The IGC code defines three categories of independent tanks: Type A, B and C. Type C tanks are pressure tanks for LPG and will not be required for LNG vessels since LNG are transported at ambient pressure. Regardless of what containment system is used, the tanks should be design taking factors such as internal and external pressure, dynamic loads due to the motions of the ship, thermal loads, sloshing loads into account, and structural analyses should be carried out. A separate secondary barrier is normally required for the gas liquefied gas containment systems to act as a temporary containment of any leakage of LNG through the primary barrier. For membrane tanks and independent type a tanks, a complete secondary barrier is required. For

—

independent type B tanks, a partial secondary barrier is required, whereas no secondary barrier is required for independent type C tanks. The secondary barrier should prevent lowering of the temperature of the ship structure in case of leakage of the primary barrier and should be capable of containing any leakage for a period of 15 days.

The code contains operational requirements related to i.e. cargo transfer methods, filling limits for tanks and the use of cargo boil-offs as fuel as well as requirements on surveys and certification. Equivalents to the various requirements in the code are accepted if it can be proven, e.g. by trials, to be as effective as what is required by the code. This applies to fittings, materials, appliances, apparatuses, Equipments, arrangements, procedures. Additional requirements regarding insulation and materials used for the cargo containment systems as well as construction and testing, piping and valving etc. are included in the IGC code. The IGC code also requires certain safety equipments to be carried onboard LPG carriers. These include ship handling systems such as positioning systems, approach velocity meters, automatic mooring line monitoring and cargo handling systems such as emergency shutdown systems (ESD) and emergency release system (ERS). In addition, systems for vapor, fire detection, fire extinguishing (dry chemical powder) and temperature control are required.

In addition to the numerous regulations, codes, recommendations and guidelines regarding gas carriers issued by IMO, there are extensive regulations, recommendation and guidelines under international and local umbrella related to safety LPG shipping exist that undoubtedly contributing to the high safety standard and the good safety record that has been experienced for the fleet of LG carriers, e.g. standards of best practice issued by SIGTTO (The Society of International Gas Tanker & Terminal Operators) [4, 11].

2.7 Training requirement

Any person responsible for, or involved with, the operation and dispensing of LPG should have an understanding of the physical characteristics of the product and be trained in the operation of all ancillary equipment. Thus acquiring sufficient crew with the required level of experience, training and knowledge of LG are believed to be one of the major safety-related challenges to the maritime LG industry in the years to come. In addition to strict regulations on the ship itself, there are also extensive international regulations specifying the necessary training and experience of crew that operate LPG carriers. These include the international rules on training requirements are contained in regulations such as the STCW 95, ISM code, tanker familiarization training, as well as flag state or company specific training requirements that go beyond these international regulations [12, 13].

The competence level of Liquefied gas crew has generally been regarded as quite high compared to that of other ship types. A study presented in demonstrates that the performance score of crew onboard gas and chemical tankers are the best among cargo carrying ships, second only to that of passenger vessels. STCW 95 contains minimum training requirements for crew engaged in international maritime trade. In particular, chapter V of the STCW code contains standards regarding special training requirements for personnel on certain types of ships, among them liquefied gas carriers. One requirement for masters, Officers and ratings assigned specific duties and responsibilities related to cargo or cargo equipment on all types of tankers, e.g. LNG tankers, is that they shall have completed an approved tanker familiarization course. Such a course should have minimum cover the following topics:

i. Characteristics of cargoes and cargo toxicity
ii. Hazards and Hazard control
iii. Safety equipment and protection of personnel
iv. Pollution prevention

The course must provide the theoretical and practical knowledge of subjects required in further specialized tanker training. Specialized training for liquefied gas tankers should as a minimum include the following syllabus:

i. Regulations and codes of practice
ii. Advanced firefighting techniques and tactics
iii. Basic chemistry and physics related to the safe carriage of liquefied gases in bulk
iv. Health hazards relevant to the carriage of liquefied gas
v. Principles of cargo containment systems and Cargo-handling systems
vi. Ship operating procedures including loading and discharging preparation and procedures
vii. Safety practices and equipment
viii. Emergency procedures and environmental protection

In addition to these training requirements, masters, chief engineering officers, chief mates, second engineering officers and any persons with immediate responsibilities for loading, discharging and care in transit of handling of cargo in a LG tanker are required to have at least 3 months sea service on a liquefied gas tanker. Due to the extensive training requirements and experience level of their personnel, the maritime LNG industry claims that the crew sailing the LNG fleet are among the best in the world. However, a shortage of experienced LG crew is foreseen in the near future especially with the expected growth of the LPG fleet.

2.8 Transportation of LPG Inland Water

LPG and CNG and LNG are next in line of alterative for transportation to gasoline because of their associated environmental benefits including reduction of GHGs. Thus, it is more useful for countries with natural gas resources and a relatively good gas distribution system. LPG has been explored in the 1930s but it's used has been slowed because of favorable economy of petroleum. However, the current threat of climate change has increased the focus on alternative transport fuels which include. Countries with programmed on the use of CNG and LPG as a transport fuel include the USA, Canada, UK, Thailand, New Zealand, Argentina and Pakistan [1,2] CNG and LPG are used in both private vehicles and transport fleets. It is estimated that about 250 million vehicles are using this fuel worldwide, and its use is on the increase, representing 2% of total global transport fuel use.

The advantages of using LPG are:

   i.   Environmental friendliness reduced engine maintenance cost
   ii. Improved engine and fuel efficiency

However limitations are the following:

   i.   Storage containment
   ii. High cost of conversion
   iii. Need for high skill operator

Each category of this required thorough, holistic risk, goal based design and operability assessment for safety, reliability and protection of environment

2.9 Environmental concern—a driving force for beyond compliance policy

Over the last decade, each passing years has been augmented concerned about issue of environment importance in design, construction, operation and beneficial disposal of marine articraft .the overriding force is increasing the resources of the planet that we live and that only a few are renewable. This accumulated to production that has elements of long-term sustainability of the earth. Precipitated effect over the year has call for public awareness and translated into impact through these the following manners:

a) **Regulations:** public pressure on governmental and non-governmental organization regulation due to untold stories of disaster and impact, the public is very concerned and in need of fact that if the quality of life of people enjoy is to be sustained, for them and the future generation then the environment must be protected. conspicuous issue, expertise and finding of regulations make them to go extra length on unseen issue, contrasting between the two, while commercial force act on hat will be forth problems.

b) **Ship Concept design**—is very important in shipping and it account for 80 percent of failure, therefore compliance and making of optimal design has a great impact in ship whole life cycle. The impact of environment in ship design is very difficult because of large numbers of uncertainties. Environmental impact hat need to be taken into considerations in concept design can be classified into the following:

c) **Operations:** considering limiting life cycle of ships at estimate of 25 years, issues relating to the following are equally not easy to quantify in design work, even thus a lot of research effort has been set on move on this, but the call of the day require allowable clearance and solution to be given to the following: Known emission, Accidental, Ballast waste, Coating

d) **Commercial forces:** where company that or product that operate in non environmental friendly way, people are prone to spurn the company's products and service, therefore having impact on company return on investment.

a) **Construction** and **Disposal**—use of meticulous scantling and factors worth consideration with the ship at the end of her life cycle

—

Shipboard environmental protection should **Pollution Prevention (P2)** or **Pollution Control. Pollution Prevention** Use fewer environmentally harmful substances and generate less waste on board. **Pollution Control:** Increase treatment, processing, or destruction of wastes on board. The basic P^2 principles follow: *Eliminating* the use of environmentally harmful chemicals and *reducing* the amount of waste we generate on board is often better that treating it on board. Typical environmental greenhouse gas release from different prime movers is shown in Table 2.2. Emission is inherent consequence of powered shipping, Fuel oil burning as main source, Continuous combustion machineries—boilers, gas turbines and incinerators. And this made the following issue very important:

i. Worldwide focus of fuel-> Exhaust gas emission law by IMO and introduction of local rules
ii. Emission limits driving evolution to development and adaptation to new technology
iii. Solution anticipated to maintenance of ship life cycle at average of 25 years
iv. Focus is currently more on, NOx and SOx—HC, COx and particulate will soon join
v. Consideration involves not only fuel use and design but also operational issues.

Table 2.2: Environmental performance

Emission	LPG	Gasoline	Diesel
Cox	1	10.4	1.2
HC	1	2.0	1.2
NO	1	1.2	1.1
PM	neg	Present	Very high
Sox	neg	Neg	Very high

Table 2.3: Environment demand for ships

Environmental parameters	**Environmental demand**
Ship design	Need for longer safe life cycle
Construction	High worker safety standards, low energy input
Emission	Minimum pollution and emission, Minimum Sox, Nox and Cox, PMs-Zero discharge

—

Scrapping	Zero harmful emission
Operation waste	Efficient maneuverability
Energy	Maximum fuel efficiency
Antifouling	Harmless
Ballast water	Zero biological inversion or transfer of alien species
Sea mammal interaction	Maneuverability capability
Accident	Able officer, ship structure, integrity
Fire	Harmless
Wave wash of high speed marine	Zero inundation and spray a shore

Table 2.3 shows the environmental regulatory demand of our time for ships that need to be considered in design and operation of LPG ships.

2.10 Hybrid use of High Level Objective Based and Safety Risk Based Design towards Beyond Compliance

It is clear that the shipping industry is over killed with rules and recent environmental issues are have potential to initiate new rules, this made firms to selectively adopt "beyond compliance" policy that are more stringent than the required extant law due to . Beyond compliance policy are mostly intra—firm process—which could be power based or leadership based. it draw insight from institutional theory, cooperate social performance perspective, and stakeholder theory that relate to internal dynamic process. While external forces create expectation and incentive for manager, intra firm politics influence how managers perceive, interpret external pressure and act on them [7, 14] Policy towards beyond compliance fall into 2 categories:

i. Whether they are now required by law but they are consistent with profit maximization.
ii. Requirement by law and firm are expected to comply by them.

Towards sustainable reliability, it is preferable to use stochastic and probabilistic methods that could help improve in the existing methodology this method involve absolutism that will cover all uncertainty complimented by historical and holistic matrix investigation. Hybridizing models is also a plus for the best solution of sustainable maintenance of navigation channel.

Beyond compliance towards meeting required safety level and life cycle and environmental protection required systematic employment of hybrid of OBS and RBS systems. Below is the general step of RBS and OBS which can be apply for above described characteristic of LPG Ships.

2.11 Components of Goal Based Standards

Objective—based standards (OBS) are ship safety standards comprising five tiers (see Figure 2.1):

Level I	:	Consists of goals expressed in terms of safety objectives defined by risk level.
Level II	:	Consists of requirements for ship features/capabilities, defined by risk level, that assure achievement of ship's safety objectives.
Level III	:	Here Tier IV and V are to be verified for compliance with Tier II.
Level IV	:	Consists of rules, guidelines, technical procedures and programs, and other regulations for ship designing and ship operation needs, fulfillment of which satisfies ship's feature/capability requirements.
Level V	:	Consists of the code of practice, safety and quality systems that are

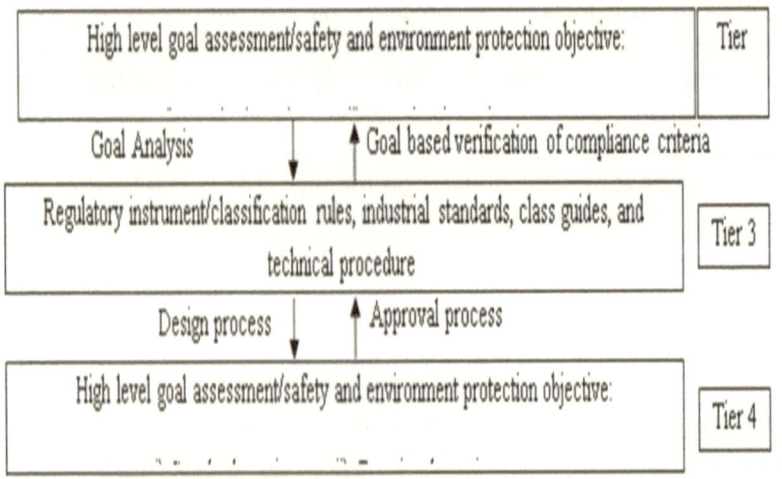

Figure 2. 1: High level Goal Based Assessment

—

2.12 Sustainable Risk Assessment

Sustainability remain a substantial part of assessing risk and life cycle of ships– however, they are very complex and require long time data for accurate. Environmental risk and Environmental impact assessment (EIA) procedure is laid out by various environmental departments and will continue to remain similar except that the components of risk area cover different uncertainty to sustain a particular system are different. EIA has been a conventional process to identify, predict, assess, estimate and communicate the future state of the environment, with and without the development in order to advise the decision makers the potential environmental effects of the proposed course of action before a decision is made.

RBS is improvised version of EIA where holistic consideration, community participation, expert rating, cost benefit analysis and regulatory concerned are core part of the philosophy leading to reliable decision making and sustainable system design and operation. In risk assessment, serenity and probability of adverse consequence (HAZARD) are deal with through systematic process that quantitatively measure, perceive risk and value of ship using input from all concerned—waterway users and experts [8, 14]. RISK = Hazard x Exposure (an estimate on probability that certain toxicity will be realized). While HAZARD: Anything that can cause harm (e.g. chemicals, electricity, natural disasters)

Severity may be measured by:

i.   No. of people affected
ii. Monetary loss
iii. Equipment downtime
iv. Area affected
v.   Nature of credible accident

Risk ranking index according to level of risk the tables bellow show an example of risk matrix (Table 2.3) with assignments of risk level identifies by number index.

Table 2. 3: Risk level matrix

			Consequence Criteria				
			1—Insignificant	2—Minor	3—Moderate	4—Major	5—Catastrophic
Likelihood	A	Consequence certain to occur	Medium (M)	High (H)	High (H)	Very High (VH)	Very High (VH)
	B	Consequence likely to occur	Medium (M)	Medium (M)	High (H)	High (H)	Very High (VH)
	C	Consequent possibly likely to occur some time	Low (L)	Medium (M)	High (H)	High (H)	High (H)
	D	consequence unlikely to occur but could happen	Low (L)	Low (L)	Medium (M)	Medium (M)	High (H)
	E	consequence may occur only in exceptional circumstances	Low (L)	Low (L)	Medium (M)	Medium (M)	High (H)

Risk management is the evaluation of alternative risk reduction measures and the implementation of those that appear cost effective where Zero discharge = zero risk, but the challenge is to bring the risk to acceptable level and at the same time, derive the max Benefit [10]

2.13 Components OF SYSTEM Based SAFETY RISK Assessment

System based safety assessment targets:

i. Identifications of potential hazard scenarios and major impact to ship Shipping and ship design which could lead to significant safety or operability consequences as well recent call for policies chance and procedural major effects
ii. Verification of current design, construction and operations ensure that risk from identified scenarios meet risk acceptability criteria

iii. If not, to recommend additional RBA process and available technology for control and protection that can reduce risk to suitable level.

Below are the general RBA steps:

STEP 1—HAZID:

The HAZID (step 1) should be conducted a in a technical meeting including brainstorming sessions, from various sectors within the LPG industry, i.e. ship owner/operator, shipyard, ship design office/maritime engineering consultancy, equipment manufacturer, classification society and research centre/university. Common identifiable hazards are:

i.   Emission to air, water and soil
ii.   Shipboard cargo tank and cargo handling equipment
iii. Storage of tanks and Piping
iv.   Safety Equipments and Instruments
v.   Ruder failure in inland water
vi.   Crew fall or slip on board
vii. Fault of navigation equipments in inland water
viii. Steering and propulsion failure
ix.   Collision with ship including Passing vessel hydro dynamic effects
x.   Terrorist attack or intentional incident
xi.   Potential Shortage of crew
xii. Navigation and berthing procedure

The results from the HAZID should be recorded in a risk register stating total number of hazards, different operational categories. The top ranked hazards according to the outcome of the HAZID can be selected and given respective risk index based on qualitative judgment by the HAZID participants from diverse field of expert. It should emphasize on the study of existing situations and regulations including policies in place, present performance, flaws and survey on parties feeling on acceptability and procedures.

STEP 2—Hazard analysis

The risk analysis (step 2) comprises a thorough investigation of accident statistics for liquefy gas carriers as well as risk modeling utilizing event tree methodologies for the most important accident scenarios,

—

based on the survey of accident statistics and the outcome of the HAZID leading to generic accident scenarios recommendation for further risk analysis. Figure 2.2 shows formal safety assessment steps.

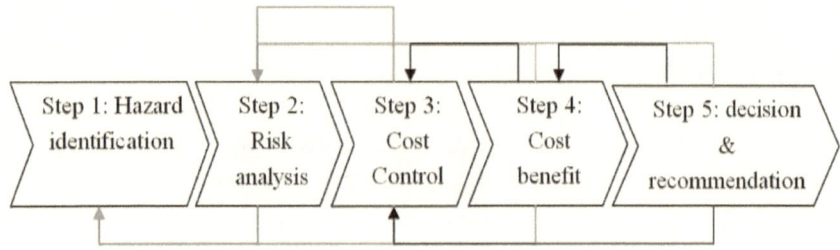

Figure 2. 2: Risk assessment and analysis steps

The risk analysis essentially contains two parts, i.e. a frequency assessment and a consequence assessment. The frequency assessment, involve estimation of frequency of generic incidents using reasonable accident statistics derived from the selected accident scenarios which should also be compared with similar studies for liquefy gas carriers as well as other ship. The consequence assessment should be performed using event tree methodologies. Risk models can be developed for each accident scenario and event trees constructed according to these risk models utilizing accident statistics, damage statistics, fleet statistics, simple calculations modeling and expert opinion elicitation [15]. The frequency and consequence assessments provide the risk associated with the different generic accident scenarios which can be summarized in order to estimate the individual and societal risks pertaining to liquefy gas carrier operations and design. Based on available accident statistics and results from the HAZID, eight generic accident scenario umbrellas that required deep analysis are:

    i.   Collision
    ii.   Fire or explosion
    iii. Grounding
    iv. Contacts
    v.   Heavy weather/loss of intact stability
    vi. Failure/leakage of the cargo containment system
    vii. Incidents while loading or unloading cargo LPG
    viii. Emission ship power sources

—

The first five generic accident scenarios are general in the sense that they involve all types of ships; wile 6 and 7 accident scenarios are specific to gas carriers and 8 concerned new environmental issue driving compliance and technology for all ships. Selected accident scenarios to investigate frequency assessment could provide a sufficiently accurate estimate of initiating frequencies for the eight selected accident scenarios. Figure 2.3 shows risk model for explosion case.

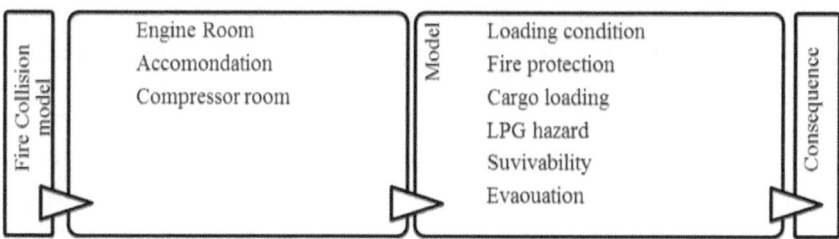

Figure 2. 3: Risk model for explosion scenario

Identification of accident scenario that is significant to risk contribution should consider use of:

i.   Holistic risk assessment of major treat using RBA and OBS model including application of stochastic.
ii. Probabilistic and deterministic methods to increase reliability and reduce uncertainties as much as possible this including using tool comprising foreseeable scenarios and scenario event, such tolls are:
   a. Accident modeling model
   b. Estimation of risk, accident frequency and consequences

STEP 3—RISK CONTROL

Risk control measures are used to group risk into a limited number of well thought out practical regulatory options. Consideration should focus on:

i.   Specification of risk control measures for identified scenarios
ii. Grouping of the measures into possible risk control options using
   a. General approach—which provides risk control by controlling the likelihood of initiation of accidents, and may be effective in preventing several different accident sequences; and

b. Distributed approach—which provides control of escalation of accidents, together with the possibility of influencing the later stages of escalation of other, perhaps unrelated, accidents. And this followed by assessment of the control options as a function of their effectiveness against risk reduction.

STEP 4: COST BENEFIT ASSESSMENT (CBA)

Risk—Cost—Benefit analysis to deduce mitigation and options selection Proposed need for new regulations based on mitigation and options

i. CBA quantification of cost effectiveness that provide basis for decision making about RCO identified, this include the net or gross and discounting values.
ii. Cost of equipment, redesign and construction, documentation, training, inspection maintenance and drills, auditing, regulation, reduced commercial used and operational limitation (speed, loads)
iii. Benefit could include, reduced probability of fatality, injuries, serenity and negative effects as well as on health, severity of pollution and economic losses

STEP 5: DECISION MAKING

This step involves:

i. Discussion of hazard and associated risks
ii. Review of RCO that keep ALARP
iii. Comparison and rank RCO based on associated cost and benefit

Specification of recommendation for decision makers output could be use for "beyond compliance" preparedness and rulemaking tools for regulatory bodies towards measures and contribution for sustainability of the system intactness, our planet and the right of future generation. In order to select between alternative technical or regulatory solutions to specific problems the first three RBA steps (HAZID, risk assessment, RCOs) can fit into the development of high-level goals (Level 1) and functional requirements (Level 2) of OBS. Equally, the last three steps (RCOs, CBA, and Recommendations) could feed into Level IV and V of OBS.

2.13 Uncertainty

Uncertainty will always be part of our activities because of limitation of knowledge of unseen in real world settings, issues associated with uncertainty are normally.

i.   Influences on recovery process
ii. Test of new advancements
iii. Influence on policy
iv. Address system changes over time
v.   services & resources

Estimating uncertainty including further validation, policy issues and rating could be obtained through the relation:

$$R(P1c) = R(E1) \times W(E1,P1) + R(E2) \times W(E2,P1) + R(E4) \times W(E4,P1)$$
$$\text{2.01}$$

Where R= rating, E= environmental factor, P= Policy factor

Uncertainty is necessary because of highly variable nature of elements and properties involved with the situation require simulate of extreme condition and model—using combination mathematical modeling and stochastic techniques while considering all factors in holistic manner that cover:

i.   Risk areas and assessment—taking all practical using historical data's and statistics that include all factors—Public health (people > other species)
ii. Mitigation of risk assessment and risk areas—This involves making permanent changes to minimize effect of a disaster—Immediacy: (Immediate threat>delayed threats)
iii. Panel of expert—Reach out to those who are capable to extend hand and do the right thing at the risk area—Uncertainty (More certain > less certain)
iv. Community participation—Educate all concern about the going and lastly place firm implementation and monitoring procedure. For adaptability (Treatable > untreatable)
v.   Emergency response—provide monitoring and information facilities and make sure necessary information is appropriately transmitted and received by all

2.14 Risk Acceptability criteria

The diagram below gives overall risk reduction areas identification and preliminary recommendation, In order to assess the risk as estimated by the risk analysis, appropriate risk acceptance criteria for crew and society for LPG tankers should be established prior to and independent of the actual risk analysis. The overall risk associated with LPG carriers should be concentrated in the reduction desired areas ALARP, where cost effective risk reduction measures should be sought in all areas. Three areas or generic accident scenarios where which together are responsible for about 90% of the total risk are: Collision, grounding and contact, and they are related in that they describe situation where by the LPG vessel can be damaged because of an impact from an external source support inland water like vessel or floating object, the sea floor or submerged objects, the quay or shore or bad weather. Figure 2.4 and 2.5 show prescription risk acceptability analysis graphs.

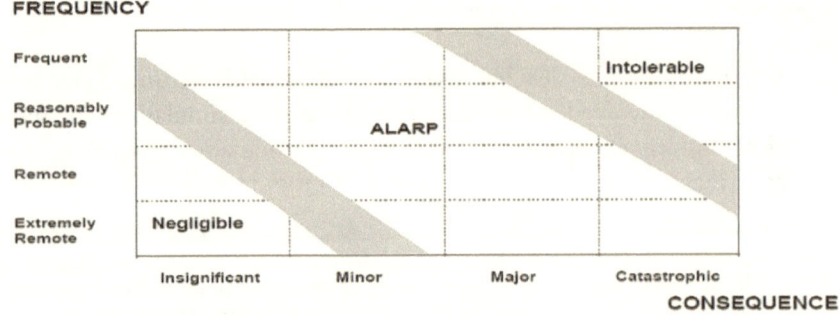

Figure 2. 4: ALARP diagram—Source [IMO]

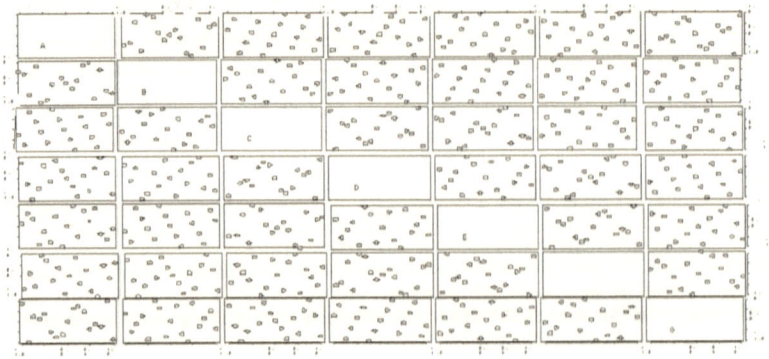

Figure 2. 5: Matrix plot analysis of system ALARP

By studying the risk models associated with these scenarios, four sub-models in particular stands out where further risk reduction could be effective. These are the accident frequency model, the cargo leakage frequency model, the survivability model and the evacuation model. Particularly, related to collision, grounding and contact, it is recommended that further efforts in step 3 of this FSA focus on measures relating to:

i. Navigational safety. improvements
ii. Maneuverability. Improved maneuverability Extended use of tugs might reduce the frequency of contact and grounding events near the terminals.
iii. Collision avoidance. i.e. warning boats in busy waters to clear the way for the LPG carrier.
iv. Cargo protection. Measures to prevent spillage through enhancing the cargo containment system's ability to maintain its integrity
v. Damage stability. Reducing the probability of sinking though enhancement of survival capabilities in damaged condition
vi. Evacuation arrangements and associated consequence through improvements relating to evacuation procedures, escape route layout or life saving appliances. Figure 2.6 shows the CBA balancing process curve for sustainable design.

Risk control options step 3 can be identified and prioritized at technical workshops, such meting could consider identification and selection of risk control options for further evaluation and cost benefit assessment. This part of the FSA also contained a high-level review of existing measures to prevent accidental release of gas.

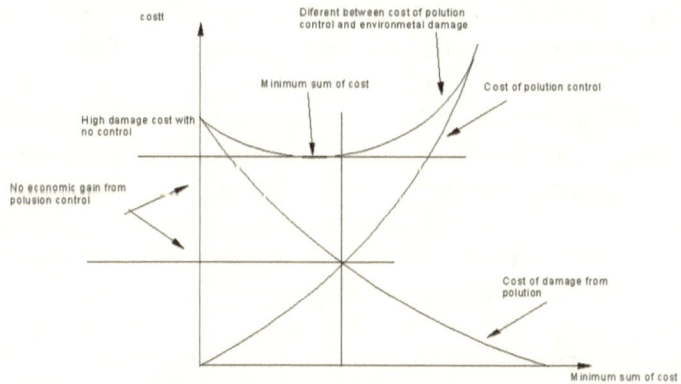

Figure 2. 6: Cost Benefit Analysis

—

$$Acceptable\ quotient = \frac{Benefit}{Risk/Cost} \qquad\qquad 2.02$$

The economic benefit and risk reduction ascribed to each risk control options should be based on the event trees developed during the risk analysis and on considerations on which accident scenarios would be affected. Figure 2.6 shows cost benefit analysis representative graph. Estimates on expected downtime and repair costs in case of accidents should be based on statistics from shipyards.

2.15 Beyond Compliance Ship Design

Existing design tools cannot, at least with any degree of reliability, be used to design a vessel to operate will ensure environmental reliability for LPG ships and operation in shallow or restricted waters. This is because of the extreme on-linearity of hull and propulsion characteristics under these conditions. In general, naval architects and marine engineers are educated and equipped with knowledge, skills, design processes that permit continuous checking balancing of constraints and design tradeoffs of vessel capabilities as the design progresses.

The intended result of the process is the best design given the basic requirements of speed, payload, and endurance. Focus is not placed on top down model of generic design based on risk where all areas of concerned are assessed at different stages of design spiral as well as risk of environmental consequence for risk involved in operability in restricted water. Operational wise, recent time has seen real attempt to fully integrate human operational practices with vessel design.

Evolving simulation technology, however give hope for assessment of extreme engineering to mitigate extreme condition as well as envisaged uncertainty. Incorporating risk assessment and goal based design for environmental protection and accident prevention as an important part of ship design spiral(shown in Figure 2.7) for LPG ship a necessary step to enabling proper tradeoffs in vessel design for reliability and other demands of time. The result is that design decisions that can compromise environment and collision are decided in favor of other factors. Only consideration of the full range of ship and terminal design and human factors relationships that affects LPG ships will produce an efficient and safe environmental friendly marine transportation system of LPG. Now that the new issue of environment is around, then we have to squeeze in more stuff in the spiral.

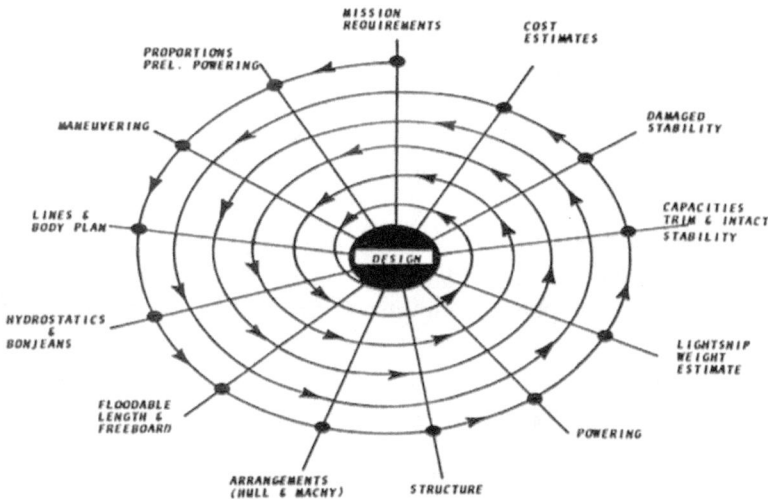

Figure 2. 7: Ship Design Spiral

In shipping and associated industries, ship protection and marine pollution are respectively interlinked in term of safety and environment, conventionally; ship safety is being deal with as its occurrence applies to environmental problem. Likewise, for many years, less attention has been given to ship life cycle, material properties, and frequency matching with the environment has resulted to corrosion. Also ship scraping, and what happen to the environment after ship scraping, yes a lot of recycling, but little or no attention is given to the residual material that find their ways to pollute the clean beautiful sea. Other areas of concern are channel design criteria ships, controllability in dredged channels, and ship maneuverability as a consideration in the Design Process. All in all, preventive and control incorporating sensible measures in ship design can only be optimized method and give us confidence on our environment. Focal areas that are will need revolutionary changes in ship design for LPG Ships are:

i. Material selection to withstand structural, weight, economical lifecycle anticorrosion and fouling
ii. Ascertain the IGC requirements for LPG carriers and special design considerations
iii. Consideration of critical load cases for each structure component as well as Corrosion

iv. Design considerations and general requirements Internal cargo pressures according to the IGC Code

v. Vertical supports, anti-rolling keys, anti-floating keys and anti-pitching keys

vi. Standard design load cases for yielding and buckling Standard design load cases for fatigue—Acceptance criteria Fatigue strength assessment

vii. Thermal stress analysis around supports

viii. Incorporating ship simulation at early stage of ship design

ix. Validation of applied loadings and the responses to structural scantly towards withstanding structural function, reliability, integrity, weight, economical lifecycle using Structural FE Analysis. Incorporation maneuvering ship simulation at early stage of design iteration

2.16 Beyond Compliance Cargo Tank Design

Pressure vessel is storage tank designed to operate at pressures above 15 p.s.i.g. Common materials held and maintained by **pressure vessels** include air, water, nitrogen, refrigerants, ammonia, propane, and reactor fuels. Due to their pressurizing capabilities, they are often used to store chemicals and elements that can change states. For this reason gas property is important in their design, the walls of pressure vessels are thicker than normal tanks providing greater protection when in use with hazardous or explosive chemicals. Important parameters to consider when specifying pressure vessels include the capacity, the maximum pressure and the temperature range.

i. The capacity is the volume of the pressure vessel—The maximum pressure is the pressure range that the vessel can withstand.

ii. The temperature ranges indicates the temperature of the material that the container can withstand—Built—in temperature control system—This helps to keep volatile chemicals in inert states. At times it may also change the state of the chemicals to make transportation easier.

Pressure vessel with temperature controls have gauges to allow for reading of internal pressures and temperatures. These gauges are available with a variety of end connections, levels of accuracy, materials of construction, and pressure ranges.

There are mainly two types of pressure vessels:

—

106

Spherical Pressure Vessel—These pressure vessels are thin walled vessels. This forms the most typical application of plane stress. Plane of stress is a class of common engineering problems involving stress in a thin plate. It can also be called as simplified 2D problems.

Cylindrical Pressure Vessel—This vessel with a fixed radius and thickness subjected to an internal gage pressure, the vessel has an axial symmetry. Analyses of LPG tanks design required of advantage of finite element modeling with fluent and other CFD software using static, dynamic, thermal and nonlinear analysis. To prove the structural integrity of the tank designs for structural and seismic loading as well as assesses leakage and burn-out scenarios.

Tank analyses should include:

  i.   Leakage and double walled piping modeling
  ii.   Prestress/post-tensioning and Burn-out modeling
  iii. Relief valve heat flux modeling Static analysis
  iv.   Wind loading and modal and seismic analysis
  v.   Temperature modeling prediction of stresses loading as well as other environmental safety
  vi.   Stress and thermal analysis of marine loading arm.

2.17 Beyond Compliance HAZOP and FMEA

Operability must follow Hazards associated with LPG ships. HAZOP and FMEA risk assessment following FSA procedure recommended. Beside this the following operational requirement are expected to exercise all the time for all operation activities.

  i.   Gas Equipment—Equipments associated with gas works that require regular look after are: Gas dryer, heat exchanger, storage and container, gas reactors, gas compressor type, gas liquefier, dust filter, air separation column, filling manifold distillation column. Expansion engines suction filter, after cooler, moisture absorber air compressor.
  ii.   Use of Personal Protective Equipment (P.P.E)—Owing to its rapid vaporisation and consequent lowering of temperature, LPG, particularly liquid, can cause severe frost burns if brought into contact with the skin. P.P.E appropriate for use with LPG must always be worn when the refueling operation is taking place.

a) Neoprene gloves, preferably gauntlets (or similar, impervious to LPG liquid)

b) Safety gear—footwear, Goggles or face shield. Long sleeved cotton overalls.

iii. Housekeeping—Housekeeping is one of the most important items influencing the safety of the Color Gas Installation. No smoking—no naked lights or other sources of ignition, including the use of mobile phones, pagers, or radio transmitters, are permitted in the vicinity of the installation.

a) Do not ignore the hazard signs or remove them. (Or put your emergency sign here).

b) The area must be kept free from long grass, weeds, rubbish, and other readily ignitable or hazardous materials.

c) All emergency exits and gangways to be kept clear at all times.

iv. Gas Storage—Gas storage facility is a vital factor in offsetting seasonal fluctuations in demand and safeguarding gas supplies at all times. Gas storage plays a vital role in maintaining the reliability of supply needed to meet the demands of consumers. LPG gases are explosive and are stores carefully and properly with extra attention and effort to avoid any kind of injury. The following are important hazard risk measured to follow for gas storage:

a) Transportable gas containers should be stored in well-defined areas and should be segregated according to the hazard presented by the contents.

b) Contents of cylinders should be easily identifiable.

c) Persons involved should receive training regarding handling of cylinder, potential risks, hazards from cylinder and contents.

d) Gases can be stored in pressure vessels, cylinders, trailer, vaporizer and tanks. These are stored away from flammable materials and electrical outlets.

e) Account should be taken of external dangers such as adjacent work operations under different managerial control or the possibility of mechanical damage due to traffic knocks.

f) The gases should not be subjected to any sort of physical damage or corrosion

g) Emergency procedures should be established

In the present times, many new next generation systems are being developed in order to cater for the growing need for operational flexibility required by various gases and gas-fired power generation customers all across the globe. The exploration, production, and transportation of gases takes time, and most of the times the gas that reaches its destination is not always needed right away, so it is injected into gas storage facilities. These gas storage facilities should have the following characteristics:

i. Low Maintenance and easy to operate
ii. Trouble Free Operation
iii. Sturdy Design and long operative life
iv. Low Working Pressure and Low Operating Cost
v. Easy availability of spare parts and Low power consumption

First Aid—Treatment must be carried out immediately by placing the casualty gently under slowly running cool water, keeping it there for at least 10 minutes or until the pain ceases or cover the affected parts with light, dampened or wet material. Encourage the affected person to exercise any fingers, toes or legs that are affected to increase circulation. In severe cases, tissue damage will take place before medical aid can be obtained. Seek professional medical treatment as required.

Inhalation—LPG vapor is mildly narcotic, inhalation of high concentrations will produce anesthesia. Prolonged inhalation of high concentrations will cause asphyxiation. The emergency treatment for inhalation is to move the casualty to fresh air, keeping them warm and at rest. In chronic cases, where there is a loss of consciousness give oxygen, or if breathing ceases give artificial respiration. Professional medical treatment should be sought as required.

i. **Eyes**—Immediately flush eyes with plenty of water for at least 15 minutes. Hold eyelids apart while flushing to rinse the entire surface of eye and lids with water. Seek medical attention immediately.
ii. **Skin**—A strong refrigerant effect is produced when liquid LPG comes into contact with the skin. This is created by the rapid evaporation of the liquid, and it can cause severe frostbite, depending on the level of exposure.

2.18 Emergency preparedness

i. In the event of fire—The fact that LPG is used as a safe and valuable heating source in millions of homes show that there are chances to controlling and preventing a fire involving LPG. To minimize the possibility of outbreak of fire, it is of key importance to provide good plant design and layout, ensure sound engineering and good operating practice, and provide proper instruction and training of personnel in routine operations and actions to be taken in an emergency. Actions required are:

a) Shut all valves on tank or cylinders and emergency control valve outside the building by turning clockwise.
b) Call the Fire Service and refer to presence of LPG tank.
c) Keep tank cool by water spray, if possible.

ii. Gas leakage—damaged vessels and cracks can result in leakage or rupture failures. Potential health and safety hazards of leaking vessels include poisonings, suffocations, fires, and explosion hazards. Rupture failures can be much more catastrophic and can cause considerable damage to life and property. The safe design, installation, operation, and maintenance of pressure vessels in accordance with the appropriate codes and standards are essential to worker safety and health. Actions required are:

a) Shut the emergency control valve outside your building
b) Extinguish all sources of ignition.
c) Shut all cylinder valves or the gas isolation valve on top of the tank
d) Do not operate electrical switches.
e) Open all doors and windows. Ventilate at low level as LPG is heavier than air.

Above all Appliances should be serviced according to the manufacturer's recommendations by a competent person

2.19 Environmental technology and beyond compliance performance prospect

Development real time simulation helps in the mitigation most of the accident and cover issues of uncertainty

i. Development in automation technology help in installation of emergency shutdown mechanism
ii. advent of advance communication technology further give hope for improvise protection prevention and control
iii. Prospect of container unitized LPG ships in inland water.

—

iv. Novel design of inland water craft to provide solution to issue of bigger ship inability to maneuver in inland water.

In line with Global warming, since air emission is linked to machineries emerging new technology for efficient and low air pollution power source for ships including LPG Ships are:

i. Alternative energy
ii. Alternative fuel and dual fuel engines
iii. Infusion of water mist with fuel and subsequent gas scrubbing units for slow speed engines
iv. Additional firing chamber
v. Potential for gas turbine complex cycle
vi. Potential for turbocharger diesel engine
vii. Compound cycle with gasified fuel, external compressor, combustion with pure oxygen
viii. Exhaust after treatment for medium speed engines

In today, environmentally conscious world there is already so much pressure on stake holder—designer operator's trainer and builders in shipping industry, especially ship carrying flammable gases like LPG / CNG to avoid accident and incident and the consequence of which could lead to catastrophic long term environmental disaster. Potential for more laws prevent and put necessary control in place is evident. However, the use of available and new technology in an innovative age of information technological and knowledge that has built through research activities related to speed, safety, reliability, miniaturization, cost, mobility and networking in most industries. Integrative utilization of which could facilitate optimization our system at design, operation and other factors of life cycle accountability process in order to come up with sustainable system. The answer to this lies on "Beyond compliance" policy using IMO FSA and GBS tool in hybrid as required to meet future law requirement and to aid effective development of rules that satisfy all concern. Functional requirements for liquid gas carrier design and operations in restricted water can be adequately developed from design, operation, human elements and construction point of view using adequate technical background as well as ergonomic design principles.

References

1. Juha Schweighofer, Petra Seiwerth, Ostereichshe Wasserstrassen, "Inland Environmental Performance" The Naval Architech, RINA,2007
2. Boitsov G. V., Partial safety factors for still water and wave loads, Ship Technology Research, Vol. 47, 2000
3. IMO, 2004, "SOLAS Consolidated edition 2004", International Maritime Organization, 2004,
4. Arthur d. Little Limited, Guideline For HAZARD Analysis as an Aid To Management of Safe Operations, 1992
5. SITTGO, Crew Safe Standards and Training For Large LNG Carrier, 203
6. IMO, 2003, "COLREG, Convention on the International Regulations for Preventing Collisions at Sea, 1972.
7. Pitblado, R. M., Baik, J., Hughes, G. J., Ferro, C., Shaw, S. J., "Consequences of LNG Marine Incidents", in Proceedings of CCPS conference, Orlando, Fl, 2004.
8. Davis, L. N., 1979, "Frozen Fire, Where Will It Happen Next?", Friends of the Earth publishers, San Francisco, Ca, USA, 1979,
9. Lakey, R., J., 1982, "LNG by Sea: How Safe is it?", Hazardous Cargo Bulletin, September, 1982,
10. IMO, 1993, "International Code for the Construction and Equipment of Ships Carrying Liquefied
11. Gases in Bulk—IGC Code 1993 edition", International Maritime Organization, 1993,
12. Lakey, R., J., Thomas, W., D., 1982, "The LNG/LPG Fleet Record", in proceedings of GASTECH 82, Paris, France, 1982.
13. IMO, 2001, "STCW with Amendments 1 & 2, 2001 Edition", International Maritime Organization, 2001,
14. Kjellstrøm, S., Borge Johansen, C., 2004, "FSA Generic Vessel Risk, Single Hull Tanker for Oil", DNV Technical Report no. 2003-1148, rev. no. 02, Det Norske Veritas, 2004.
15. IMO, 2001, "Fire Safety Systems (FSS) Code, 2001", International Maritime Organization.

CHAPTER 3

HAZOP Process for Deep Water Marine System

"It is Miracle that curiosity survives formal education"
Albert Einstein

Summary

The world of water supports the planet; therefore, its vast resources need to be fully utilized to benefit human activities in a sustainable manner. The maritime and offshore industry has made use of the ocean in a very responsible way, the challenged posed by environmental concern in the coastline is evolving new technology and new ways for technological development. In an age so dire to find alternative and sensitive ways to mitigate challenge of global warming, climate changes and its associated impact, maritime and offshore activities is loaded with requirement to build new sustainable and reliable technology for deep sea operation to fulfill alternative mitigation options for climate change and decline of coastline resources and enthophication. Expanding deep sea operation face require development of technology related to dynamic position, mobile berthing facilities, collision aversion, impact of new environment, wave, wind on marine structure, supply vessel operation coastal water transportation attracts low probability and high consequence accidents, and this makes reliability requirements for the design and operability for safety and environmental protection very necessary. Collision carries the

—

113

largest percentage of accident risk scenarios among water transportation causal risk factors. This Chapter discusses process work in risk, hazard and reliability based design and safe and efficient operability deep water operation waters. This includes a system based approach that covers proactive risk as well as holistic multi criteria assessment of required variables to deduce mitigation options and decision support for preventive, protective and control measures of risk of hazard for deep water marine offshore operation.

3.1 Introduction

Offshore operation and marine transportation service provide substantial support to various human activities; its importance has long been recognized. The clear cut advantage of inland transportation over other modes of transportation, short sea service and evolving deep sea activities are being driven by recent environmental problems and dialogues over alternative renewable ways of doing things. The criticality of offshore and marine transportation operations within the coastline and the prohibitive nature of the occurrence of accidents due to high consequence and losses have equally made it imperative and necessary to design operate and maintain sustainable, efficient and reliable deepwater offshore operation and marine transportation systems. Marine accidents fall under the scenarios of collision, fire and explosion, flooding, and grounding (Bottelberghs, P.H. 1995). This chapter discusses a model of reliability for the assessment and analysis of marine accident scenarios leading to design for the prevention and protection of the environment. The chapter will address risk process that can optimize design, existing practice, and facilitate decision support for policy accommodation for evolving offshore deepwater regime.

3.2 Risk Reliability modeling requirement

In order to build reliable deep water offshore system and supporting transportation system, it is important to understand the need analysis through examination of the components of system functionality and capability. This functionality capability of the platform, environmental loading and other support system environmental risk as well as ageing factors related to design, operation, construction, maintenance, economic, social, and disposal requirement for sustainable marine system need to be critically analysed. Risk identification work should be followed by risk analysis that include risk ranking, limit acceptability and generation of best

—

options towards development of safety and environmental risk mitigation and goal based objective for evaluation of the development of sustainable cost effective inland water transportation that fall under new generation green technology (DnV, 2004). Weighing of deductive balancing work requirement for reliable and safety through iterative components of all elements involved should include social, economic, health, ecological and technological considerations. Other concerns are related to other uses of water resources and through best practice of sediment disposal, mitigation for environmental impact, continuous management, monitoring, and compensation for uncertainty as well as preparation for future regulation beyond compliance policy or principles.

Risk assessment has been used by the business community and government, and safety cases of risk assessment have been used by United Kingdom (UK) health safety and environment (HSE). In the maritime industry, risk assessment has been used for vessel safety, marine structure, transportation of liquefied natural gas (LNG), and offshore platforms. In Europe maritime risk assessment has been used for coastal port risk analyses and pilot fatigue. International Maritime Organization (IMO) and United State Coast Guard (USCG) rulemaking have issued guidelines and procedures for risk based decision making, analysis and management under formal safety assessment [4, 12]. Risk analysis when used for rule making is called Formal Safety Assessment (FSA); while when it is used for compliance is addressed as Quantitative Risk Analysis (QRA). Contemporary time has seen risk assessment optimization using scenario based assessments, which considered the relative risks of different conditions and events. In the maritime industry, contemporary time risk assessment has been instrumental to make reliable decisions related to prediction of flood, structural reliability, intact stability, collision, grounding and fire safety. Probabilistic and stochastic risk assessment and concurrent use of virtual reality simulation that considers the broader impacts of events, conditions, scenarios on geographical, temporal impacts, risks of conditions is important to for continuous system monitoring. Additionally, sensitivity and contingency (what if) analyses can be selectively used as tools to deal with remnant reliability and uncertainty that answer hidden questions in dynamic and complex systems (Roeleven, et al., 1995).

3.3 System failure and risk based design requirement for deep water system

A basic principle for risk-based design has been formulated: the larger the losses from failure of a component, the smaller the upper bound of its hazard rate, the larger the required minimum reliability level from the component. A generalized version and analytical expression for this important principle have also been formulated for multiple failure modes. It is argued that the traditional approach based on a risk matrix is suitable only for single failure modes/scenarios. In the case of multiple failure modes (scenarios), the individual risks should be aggregated and compared with the maximum tolerable risk. In this respect, a new method for risk-based design is proposed, based on limiting the probability of system failure below a maximal acceptable level at a minimum total cost (the sum of the cost for building the system and the risk of failure).

The essence of the method can be summarized in three steps: developing a system topology with the maximum possible reliability, reducing the resultant system to a system with generic components, for each of which several alternatives exist including non-existence of the component, and a final step involving selecting a set of alternatives limiting the probability of system failure at a minimum total cost. An exact recursive algorithm for determining the set of alternatives for the components is also proposed.

i. The goal of risk based design for marine system and its goal is to enhance safety. advantage of such system in system design include:

ii. Establishment of systematic method, tools to assess operational, extreme, accidental and catastrophic scenarios and integrating human elements into the design environment

iii. Development of safety based technology for reliable operation and deign

iv. Establishment of regulatory framework to facilitate first principle approach to facilitate first principle approaches to safety

v. Development of model that can demonstrate validation and practicability

vi. Today, design shift towards knowledge intensive product, risk based design is believed to be key elements for enhancement of industrial competitiveness. The use of risk based design, operation and regulation open door to innovation and radical novel and inventive, and cost effective design solution. Risk based approach for ROV follow well established quantitative risk analysis used

—

in offshore industries. The key to successful use of risk based design require advance tool to determine the risks involved and to quantify the effects of risk preventing/reducing measures as well as to develop (evaluation criteria to judge their cost effectiveness. Components of integrated risk includes:

vii. Front End: Model potential causes, locations, sizes, and likelihoods of acid releases from System. Analysis of system capabilities: identify those releases that are mitigatable.

viii. Successful mitigation: release less than 1,500 gallons

ix. Consideration of diagnosis and response times

x. Back End: Model potential failure modes of each system design, and estimate failure likelihoods

xi. Analysis of system reliabilities reliability block diagram analysis systematic identification of failure modes: human errors, equipment failures, support system failures

xii. Analysis of consequences of unmitigatable or unmitigated releases:

xiii. Release size used as surrogate consequence measure

Risk can be calculated from:
$$R = \Sigma \ (Im \ x \ Amn \ x \ Cm) + \Sigma \ (Ji \ x \ Di) \qquad 3.01$$
Where:

R = Risk metric

Im = Annual probability of mitigatable leak at location/size m

Ji = Annual probability of unmitigatable leak at location/size i

Amn = Probability of AIES failure via moden given leak m

Cm/Di = Consequence severity of leak m

ROV operating capabilities requirement that can be investigated is under risk based design are:

i. Standardized intervention ports for all subsea BOP stacks to ensure compatibility with any available ROV.

ii. Visible mechanical indicator or redundant telemetry channel for BOP rams to give positive indication of proper functioning (e.g., a position indicator).

iii. ROV testing requirements, including subsea function testing with external hydraulic supply.

iv. An ROV interface with dual valves below the lowest ram on the BOP stack to allow well-killing operations.

v. Electrical power requirement

General requirements—refer to SOLAS requirements. Chapter II-1—outlines requirements for Ship construction sub-division and stability, machinery and electrical installations. The five part of this parts are: Part A –General, Part B—Sub-division and stability, Part C-Machinery installations, Part D—Electrical installations, Part E-Additional requirements for periodically unattended machinery spaces.

3.4 Benefits and limitation of using risk and reliability models

Rampant system failure and problems related to reliability have brought the need to adopt a new philosophy based on top down risk and life cycle model to design, operate and maintain systems based on risk and reliability. Likewise, election of alternative ways to mitigate challenges of safety and environmental risk of system deserve holistic, reliability analysis approaches that provide the following benefits:

i.   Flexibility and redundancy for innovative, alternative improvised design and concept development
ii. evaluation of risk reduction measure and transparency of decision making process
iii. systematic tool to study complex problem
iv. interaction between discipline
v.   risk and impact valuation of system
vi. facilitate proactive approach for system, safety, current design practice and management
vii. facilitate holistic touching on contributing factor in system work
viii. systematic rule making, limit acceptability and policy making development
ix. analysis of transportation system

The dynamic distributive condition, long incumbent period and complexity of marine system with limited oversight makes the process of identification and addressing human as well as organizational error including checks and balances, redundancy, and training more complicated. Other inherits draw backs associated with risk and reliability model are [5, 6]:

i.   lacks of historical data (frequency, probability, expert judgment)
ii. linking system functionality with standards requirement during analysis (total safety level vs. individual risk level, calculation of current safety level)

   iii. risk indices and evaluation criteria (individual risk acceptance criteria and sustainability balance)

   iv. quantification of human error and uncertainty

The complexity associated with human and organization requires human reliability assessment and uncertainty analysis to be modeled independently.

3.5 Marine pollution risk challenges

A group of experts on the scientific aspects of marine pollution comment on the condition of the marine environment in 1989, stated that most human product or waste ends their ways in the estuarine, seas and finally to the ocean. Chemical contamination and litter can be observed from the tropics to the poles and from beaches to abyssal depths. But the conditions in the marine environment vary widely. The open sea is still considerably clean in contrast to inland waters. However, time continue to see that the sea is being affected by man almost everywhere and encroachment on coastal areas continues worldwide, if unchecked. This trend will lead to global deterioration in the quality and productivity of the marine environment (GESAM, 1990).

This shows the extent and various ways human activities and uses water resources affect the ecological and chemical status of waterways system. Occurrence of accident within the coastline is quite prohibitive due to unimaginable consequences and effects to coastal habitats. Recent environmental performance studies on transportation mode has revealed that transportation by water provides wide advantages in term of less, low Green House Gas (GHG) release, large capacity, congestion, development initiative etc. These advantages tells about high prospect for potential modal shift of transportation and future extensive use of inland water marine transportation where risk based system will be necessary to provide efficient, sustainable and reliability safe clean waterways as well as conservation of environment. This equally shows that increase in human activities will have potential effects in coastal and marine environment, from population pressure, increasing demands for space, competition over resources, and poor economic performances that can reciprocally undermine the sustainable use of our oceans and coastal areas. Different forms of pollutants and activities that affect the quality of water, air and soil as well as coastal ecosystems are

i.   Water: pollution release directly or washed downed through ground water;
ii. Air pollution, noise population, vibration;
iii. Soil: dredge disposal and contaminated sediments;
iv. Flood risk: biochemical reaction of pollution elements with water;
v.   Collision: operational
vi. Bio diversification: endangered and threatened species, habitat;

Main sources of marine pollution:

i.   Point form pollution: toxic contaminants, marine debris and dumping.
ii. Nonpoint form: pollution: sewage, alien species, and watershed Issues.

Main sources from ships are in form of:

i.   Operational: operational activities along the shipping routes discharging waters contaminated with chemicals (whether intentionally or unintentionally)
ii. Accidental risk: Collision due to loss of propulsion or control.

Risk associated with environmental issue in the context of ship, design has impacts related to shipping trends, channel design criteria, ship and oil platform maneuverability and dynamic positioning and ship controllability.

3.6 Modeling the risk and reliability components of complex and dynamic system

The consequence of maritime accident comes with environmental problem. Marine system are dynamic system that have potential for high impact accidents which are predominately associated with equipment failure, external events, human error, economic, system complexity, environmental and reliability issues. This call for innovative methods, tools to assess operational issue, extreme accidental and catastrophic scenarios. Such method should be extensive use to integration assessment of human element, technology, policy, science and agencies to minimise damage to the environment. Risk based design entails the systematic risk analysis in the design process targeting risk prevention and reduction as a design objective. They should be integrated with design environment to facilitate and support sustainable approach to ship and waterways designs need (See Figure 3.1)

—

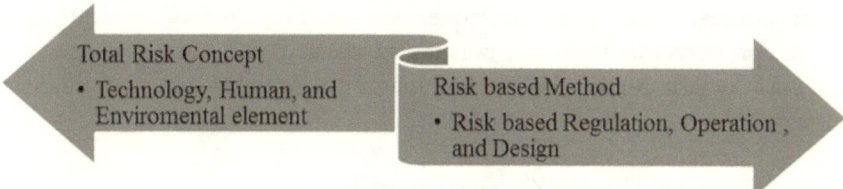

Figure 3. 1: Risk modelling process

Integrated risk based system design requires the availability of tools to predict the safety performance and system components as well as integration and hybridization of safety and environmental factors, lifecycle phases and methods. It important to develop refines, verify, validate through effective methods and tools. Such integrative and total risk tools required logical process with holistic linkage between data, individual risk, societal, organizational, system description, conventional laws, principle for system design and operation need to be incorporated in the risk process.

Verification and employment of system based approach in risk analysis should be followed with creation of database and identification of novel technologies required for implementation. Unwanted event which remain the central front of risk fight is an occurrence that has associated undesirable outcome which range from trivial to catastrophic. Depending on conditions and solution based technique in risk and reliability work, the model should be designed to protect investment, properties, citizens, natural resources the institution which has to function in sustainable manner within acceptable risk.

The risk reliability modeling process begins with definition of risk which stands for the measure of the frequency and severity of consequence of an unwanted event. Frequency at which a potential undesirable event occurs is expressed as events per unit time, often per year. Upon establishing understanding of whole system from baseline data that include elements of channel and vessel dimensioning as shown in Figure 3, the frequency can be determined from historical data. However, it is quite inherent that event that does not happen often attract severe consequence and lack data, such event is better analyzed through probabilistic and stochastic model hybrid with first principle and whatever data is available (Guedes Soares, C., A. P. Teixeira., 2001). Incidents are unwanted events that may or may not apply to accidents if necessary measure is taken according to magnitude of event and required speed of response. While accidents are unwanted events that have either immediate or delayed

–

consequences. Immediate consequences variables include injuries, loss of life, property damage and persons in peril. Point form consequences variables include further loss of life, environmental damage, and financial costs.

Risk (R) = Probability (P) X Consequence (C) 3.02

The earlier stage of the risk and reliability process involves finding cause of risk, level of impact, destination and putting a barrier by all means in the pathway of source, cause and victim. Risk and reliability process targets the following:

i. Risk analysis and reduction process: This involves analytic work through selective deterministic and probabilistic method that assures reliability in the system. Reduction process will target initial risk reduction at design stage, risk reduction after design in operation and separate analysis for residual risk for uncertainty and human reliability. Risk in complex systems can have its roots in a number of factors ranging from performance, technology, human error as well as organizational cultures, all of which may support risk taking or fail to sufficiently encourage risk aversion.

ii. Cause of risk and risk assessment: this involve system description, identifying the risk associated with the system, assessing them and organizing them according to degree of occurrence and impact in matrix form causes of risk can take many ways including the following:

a) Root cause: Inadequate operator knowledge, skills or abilities, or the lack of a safety management system in an organization.

b) Immediate cause: Failure to apply basic knowledge, skills, or abilities, or an operator impaired by drugs or alcohol.

c) Situation causal factor: Number of participants time/planning, volatility environmental factors, congestion, time of day risk associated with system can be based on.

d) Organization causal factor: Organization type, regulatory environment, organizational age management type/changes, system redundancy, system incident/accident history, individual, team training and safety management system.

To deal with difficulties of risk migration marine system (complex and dynamic by nature), reliability assessment models can be used to capture the system complex issues as well as patterns of risk migration. Historical

—

analyses of system performance is important to establish performance benchmarks in the system and to identify patterns of triggering events which may require long periods of time to develop and detect. Likewise, assessments of the role of human / organizational error and their impact on levels of risk in the system are critical in distributed, large-scale systems. This however imposes associated physical oversight linked to uncertainty during system design. Effective risk assessments required three elements: I. Framework, II. Model, III. Process:

ELEMENT I: RISK FRAMEWORK

Risk framework provides system description, risk identification, criticality, ranking, impact, possible mitigation and high level objective to provide system with what will make it reliable. The framework development involves risk identification which requires developing a structure for understanding the manner in which accidents, their initiating events and their consequences occur. This includes assessment of representative system and all linkages that are associated to the system functionality and regulatory impact.

ELEMENT II: MODEL

The challenges of risk and reliability method for complex dynamic systems like ship motion at sea require reliable risk models. Risk mitigation measures can be tested and the tradeoff between different measures or combinations of measures can be evaluated. Changes in the levels of risk in the system can be assessed under different scenarios and incorporating "what if" analyses in different risk mitigation measures. Performance trend analysis, reassessment of machinery, equipment, and personnel can be helpful in assessing the utility of different risk reduction measures. Figure 3.2 and 3.3 shows the risk components, system functionality and regulatory requirement for reliability model that can be followed for each risk scenario.

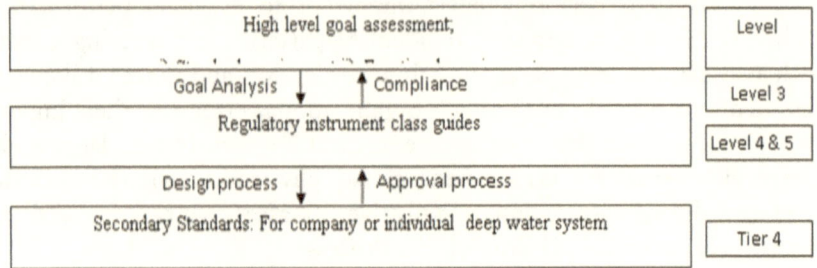

Figure 3. 2: Goal based assessment

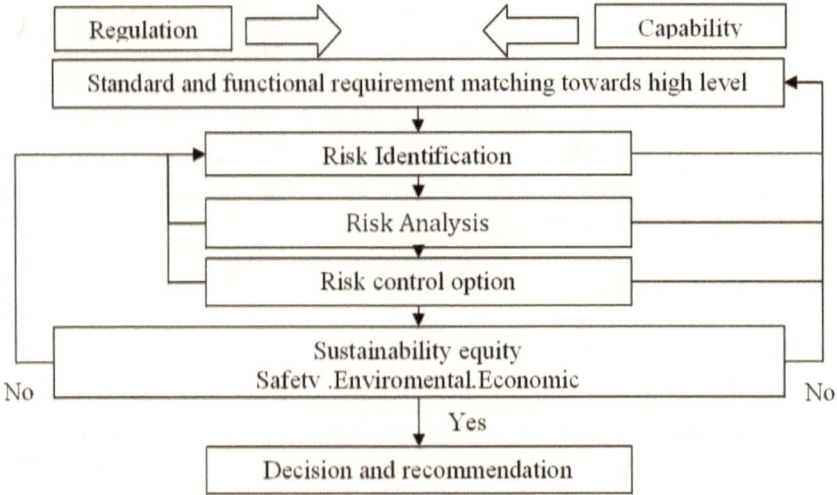

Figure 3. 3: Risk model

ELEMENT III: PROCESS

The process should be developed to provide effective and sound risk analysis where accuracy, balance information that meets high scientific standards of measurement can be used as input. This requires getting the science right and getting the right science by targeting interests of stakeholders including port, waterway community, public officials, regulators and scientists. Transparency, community participation, additional input to the risk process, checks the plausibility of assumptions could help ask the right questions of the science. Total integrated risk can be represented by:

Rt = fs (Rc, Rw, Re, Rs)   3.03

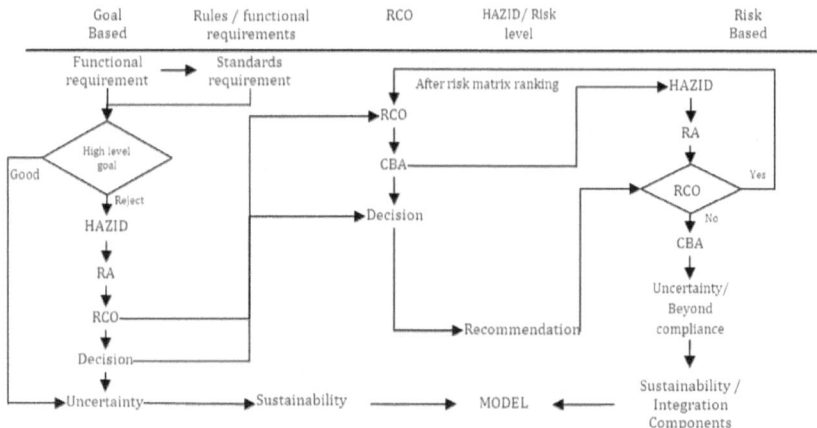

Figure 3. 4: Holistic Risk analysis Process Map

Where: Re (environment) = fe (sensitivity, advert weather . . .), Rs (ship) = fs (structural and system reliability, ship layout and cargo arrangement . . .), Rc (crew) = fc (qualification, fatigue, etc)

Holistic and integrated risk based method combined various techniques in a process as depicted in Figure 3.4, this can be applied for each level of risk for system in question. Each level is complimented by applying causal analysis (system linkage), expert analysis (expert rating) and organizational analysis (Community participation).

Table 3.1 shows models that have been used in the design system based on risks. IMO and Sirkar et al (1997) methods lack assessment of the likelihood of the event. Other models lack employment of stochastic method whose result may cover uncertainties associated with dynamic and complex components of channel, ship failure and causal factors like navigational equipment, better training and traffic control. Therefore, combination of stochastic, statistical, reliability and probabilistic together with hybrid employment of goal based, formal safety assessment methods and fuzzy multi criteria network method that use historical data of waterways, vessel environmental and traffic data could yield efficient, sustainable and reliable design product for complex and dynamic systems. The general hypothesis behind assessing physical risk model of ship in waterways is that the probability of an accident on a particular transit

—

depends on a set of risk variables which required to be analyzed for necessary conclusion of prospective reliable design (Kite Powel et al, 1996).

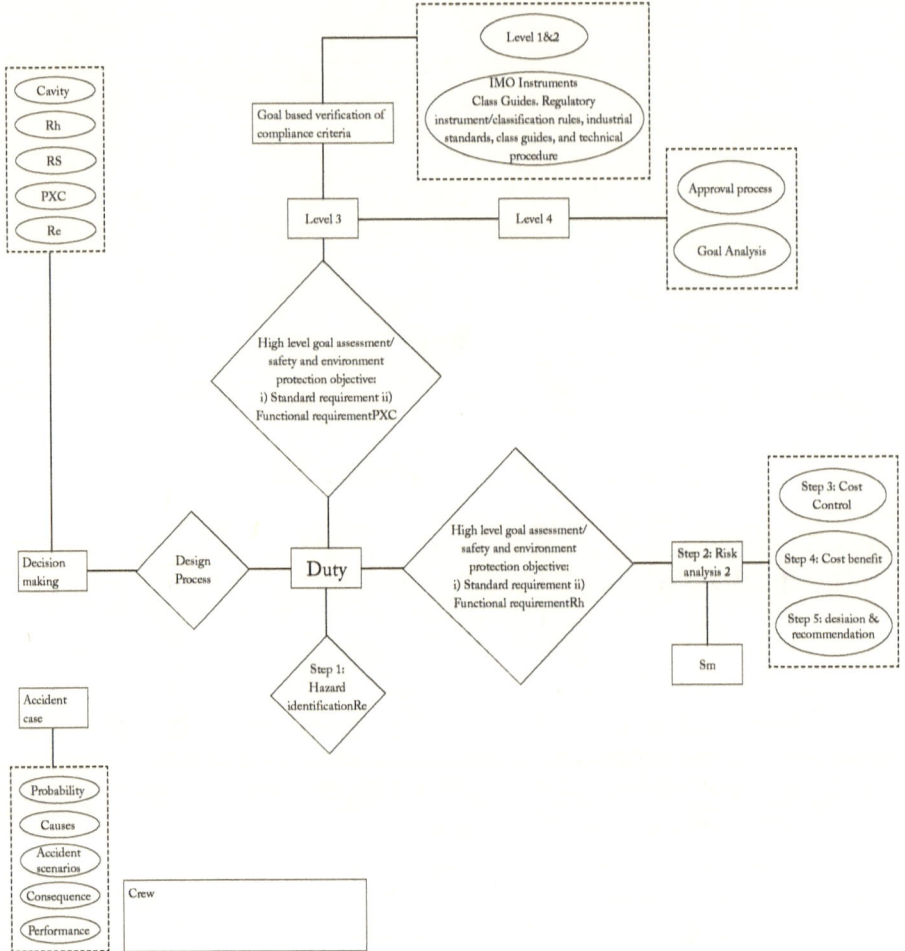

Figure 3. 5: Holistic Risk analysis Process

Table 3. 1: Risk models

Process	Suitable techniques
HAZID	HAZOP, What if analysis, FMEA, FMECA
Risk analysis	FTA, ETA
Risk evaluation	Influence diagram, decision analysis
Risk control option	Regulatory, economic, environmental and function elements matching and iteration

Cost benefit analysis	ICAF, Net Benefit
Human reliability	Simulation/ Probabilistic
Uncertainty	Simulation/probabilistic
Risk Monitoring	Simulation/ probabilistic

Risk and reliability modelling involves hazard identification, risk screening, broadly focused, narrowly focused and detailed Analysis, Table 2 shows iterative method that can be incorporated for various needs and stages of the process. Accident and incident need to be prevented as the consequence of is a result of compromise to safety leading to unforgettable losses and environmental catastrophic. Past engineering work has involved dealing with accident issues in reactive manner. System failure and unbearable environmental problems call for new proactive ways that account for equity requirement for human, technology and environment interaction in the system. The accidental categories and potential failure in waterways is shown in Figure 3.6.

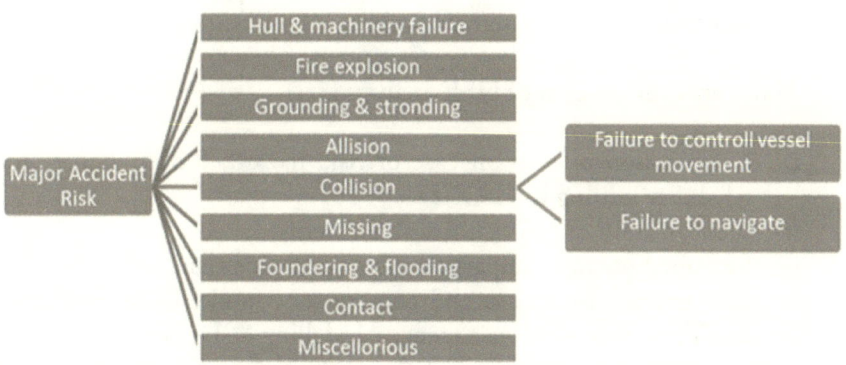

Figure 3. 6: Accident scenario

The methodologies that may be used to identify safety critical systems, subsystem and elements include:

i. Major Accident Hazard; definition, examples, compliance with regulations such as SEVESOII (COMAH) and PFEER.
ii. Qualitative method for determination of the safety risk including:—Brainstorming session methodology and example—safety criticality criteria—Required supporting documents and evidences—Action tracking

iii. Quantitative method for frequency and consequence analysis.

Quantitative risk analysis (QRA) is widely use quantitative method for offshore industry, while formal safety assessment (FSA) is use in marine industry. QRA should be simplified for to be used for determination for safety criticality criteria, safety criticality test for failure on demand and time of test/repair, HSE toolkit application, combined Event tree, Fault tree Analysis. Standards safety critical elements identification could be analyzed through development of risk matrix, regulation scope and boundary compliance, performance standard and assessment, system capability, functionality, reliability, survivability assurance and verification analysis. The dynamic risk analysis process starts with system description, functionality, regulatory determination and this is followed with analysis of (IMO, 2006):

i.   Fact gathering for understanding of contribution factor
ii. Fact analysis for check of consistency of accident history
iii. Conclusion on causation and contributing factor
iv. Countermeasures and recommendations for prevention of accident and studies of the system or project.

Major areas of concern of HSE analysis are:

i.   Examination of relevant case of risk, hazard, Process Safety and reliability leading to HAZID
ii.   Identification of Safety Critical Elements,
iii. Examination and comparison of performance standards
iv.   Examination of release and consequence model (Fire, Explosion and Toxic Release Consequence Modelling & Design)
v.   Training on fundamental of the Risk Assessment & Case Study and Implementation of HSE Management System
vi.   Conduct of HAZOP Methodology and Simultaneous Operation
vii. Risk Based Design acceptability criteria and & Integrity Assurance
viii. Applications of Dynamic Simulation in Process Safety Design
ix.   Risk management, life cycle, traceable and auditable reference different phases of the project.

Risk analysis is conducted using brainstorming worksheets, action tracking and follow-up. HAZID, HAZOP involve Process safety

Engineers, plant managers, safety supervisors, process engineers, safety Engineers and discipline engineers. Elements of QRA include:

   i.   Failure Case definition
   ii.   Consequence assessment
   iii.   Frequency analysis
   iv.   Risk calculation
   v.   ALARP demonstration
   vi.   Identification of Safety Critical Systems
   vii.   Traceability and audibility of Safety Critical Elements

3.7 Hazard Operability (HAZOP)

Hazard operability (HAZOP) is done to ensure that the systems are designed for safe operation with respect to personnel, environment and asset. In HAZOP all potential hazard and error, including operational issues related to the design is identified. A HAZOP analysis is detail HAZID, it mostly divided into section or nodes involve systemic thinking and assessment a systematic manner the hazards associated to the operation. The quality of the HAZOP depends on the participants. Good quality of HAZOP participants are (HSE, 1999):

   i.   Politeness
   ii.   To the point discussion—avoid endless discussion
   iii.   Be active and positive
   iv.   Be responsible
   v.   Allow HAZOP leader to lead

It involve How to apply the API 14C for those process hazard with potential of the Major Accident. Dynamic simulation for consequence assessment of the process deviation, failure on demand and spurious function of the safety system, alarm function and operator intervention is very important for HAZOP study. Identification of HAZOP is followed with application of combined Event tree and Fault tree analysis for determination of safety critical elements, training requirement for the operators and integrity and review of maintenance manuals. HAZOP involved use of the following:

   i.   Guide word :i.e. No pitch, No blade
   ii.   Description: I.e. No rotational energy transformed, object in water break the blade

iii. Causes: i.e. operation control mechanism
iv. Safety measurement to address implementation of propeller protection such grating, jet

The following are some of the guideword that can be used for Propulsion failure HAZOP is:

i. No pitch
ii. No blade
iii. No control bar
iv. No crank

HAZOP process is as followed:

Guide word/ brainstorming—> Deviation—> Consequence—> Safeguard—>Recommended action.

Also important HAZOP, is implementation of IEC61511 to assess the hazards associated to failure on demand and spurious trips, In HAZOP record the worksheets efficiently to cover all phases also play important role. Advance HAZOP can also e implemented through Simulation operations to identify, quantify, and evaluate the risks. SIMOP Methodology includes:

i. Consequence Assessment
ii. Frequency Analysis
iii. Risk Calculation
iv. Risk Analysis
v. Safety Criticality Elements

HAZOP is not intended to solve everything in a meeting. Identified hazard is solved in the closing process of the finding from the study. Table 3.2 shows typical HAZOP report.

Table 3. 2: Typical HAZOP report

Compression area	Fire	Hot work	3
Manfold area	Toxicity	Radioactive products	4
HP gas area	PPE		2

Separation area	Management of work permit (A)	If PTW is not followed correctly, the accident may happen	3
Compressor area	Fire & Explosion		3
Process area	Handling	Handling of proximity of process under pressure	4
Utility area	Fire fighting system	No availability of Fire Fighting system	2
Separation	Fire & Explosion	Escape routes are obstructed	3
	PPE	Contractor not using PPE	2
	PPE		3
Tank area	Fire	No Fire & Gas detection	2
Compression area	Explosion	Escape routes are obstructed	3
Compression area	Fire	Hot work	3
Manfold area	Toxicity	Radioactive products	4

Safety barrier management involves optimisation between the preventive and mitigation measures fundamental. Safety barrier management helps in determination of the safety critical elements (SCE), performance standards for the design of safety Critical Elements and in integrity assurance. Safety level integrity (SIL) involves assessment and verification according to IEC61508 and IEC61511Qualitative SIL assessment uses the risk graphs and calibration tables during the brainstorming sessions where the required SIL is assigned to the safety systems. Integrity and insurance Involve iteration of assessment of identification the credible scenarios, consequence assessment, frequency analysis, risk calculation, risk evaluation and ranking. Dynamic simulation help to identify the process hazards, measure the extent and duration of the consequences and the effect and efficiency of the safety barriers. With dynamic simulation could be optimised with greater accuracy. This saves a significant effort, time and cost for the project. It involve application of

—

i. HAZOP & SIL assessment
ii. Alarm Management
iii. Fire & Explosion Stud
iv. Case study

Subsystem analysis—Fire and explosion

Consequence modelling of Fire, Explosion and Toxic release, understanding of the fundamental and the science, governing scenarios; consequence analysis criteria. Gas dispersion & hazardous area classification, Fire zones (passive fire protection zones, the active fire protection zones, Blast Zones, blast protection zones restricted areas) Thermal & blast effect on equipment, people and environment is important to be incorporated in the risk process. Figure 3.7 shows a typical fire and explosion risk model.

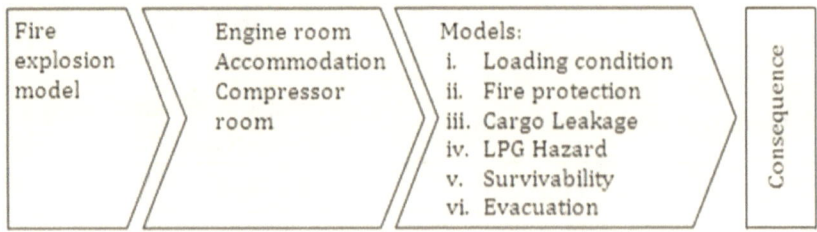

Figure 3. 7: Typical fire and explosion risk model

3.7.1 Collision scenario

Collision is the structural impact between two ships or one ship and a floating or still objects that result could to damage. Collision is considered infrequent accident occurrence whose consequence in economic, environmental and social terms can be significant. Prevention of collision damages is likely to be more cost-effective than mitigation of its consequences. Probabilistic predictions can be enhanced by analyzing operator effects, drifting and loss of power or propulsion that take into account ship and waterway systems, people and environment into consideration. Other causative factor like the probability of disabled ship as function of ship type, the probability of a disabled ship drifting towards objects also need to be accounted for. The collision model scenario also involves data that characterize of hull areas and environmental information. Figure 3.8 show a typical collision consequence situation (Cahill, R.A. 1983).

Outcome of analysis is followed by suitable Risk Control Options (RCO), where iteration of factual functionality and regulatory elements is checked with cost. The benefit realised from safety, environmental protection and effect of the probability of high level of uncertainty associated with human and organizational contributing factor to risk of collision are also important. The risk process functions to determine and deduce the idea for modest, efficient sustainable and reliable system requirement and arrangement (Emi, H. et al., 1997) Collision carried the highest statistic in respect to ship accident and associated causality. The consequences of accident are:

i. The loss of human life, impacts on the economy, safety and health, or the environment
ii. The environmental impact, especially in the case where large tankers are involved. However, even minor spills from any kind of merchant ship can form a threat to the environment
iii. Financial consequences to local communities close to the accident, the financial consequence to ship-owners, due to ship loss or penalties
iv. Damage to coastal or off shore infrastructure, for example collision with bridges

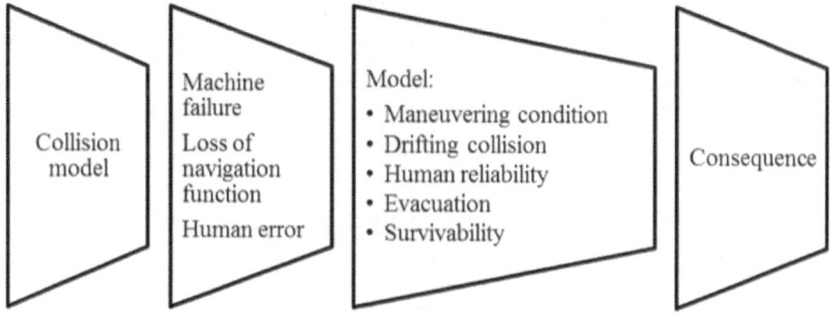

Figure 3. 8: Cause of collision

Accident events are unplanned, always possible, but effectively manageable and frequently preceded by related events that can be detected and corrected by having underlying root causes ranging from human errors, equipment failures, or external events. The result of frequency and consequence analysis is checked with risk acceptability index for industry of concerned. Table shown in Tables 3.4 and 3.4 show risk acceptability criteria for maritime industry. The analyzed influence diagram deduced from the comparison can be followed with cost control option using cost of averting fatality index or Imply Cost of Averting Fatality (ICAF) and As Low as Reasonable Possible (ALARP) principle (Parry, G. (1996)).

Table 3. 3: Frequency risk acceptability criteria for maritime industry.

Frequency Class	Quantification
Very unlikely	less than once per 10000 years (P<1/10000)
Remote	once per 100—1000 years (1/1000= P< 1/100)
Occasional	once per 10—100 years (1/100 = P< 1/10)
Probable	once per 1-10 years (1/10=P<1)
Frequent	more than once per year (P =1)

Table 3. 4: Consequence risk acceptability criteria for maritime industry

Quantification	Serenity	Occurrence	Detection	RPN
current failure that can apply to death failure, performance of mission	Catastrophic (10)	1	2	10
failure leading to degradation beyond accountable limit and causing hazard	Critical (7)	3	4	7
controllable failure leading to degradation beyond acceptable limit	Major (5)	4	6	5
Nuisance failure that do not degrade system overall performance beyond acceptable limit	Minor (1)	7	8	2

3.7.2 Failure Modes Effect Analysis (FMEA)

A Failure Modes Effect Analysis (FMEA) is a powerful bottom up tool for total risk analysis. FMEA is probably the most commonly used for qualitative analysis and is also the least complex. FMEA has been employed in the following areas:

i. The aerospace industry during the Apollo missions in the 1960s.
ii. The US Navy in 1974 developed a tool which discussed the proper use of the technique.

Today, FMEA is universally used by many different industries. There are three main types of FMEA in use today:

i. System FMEA: concept stage design system and sub-system analysis.
ii. Design FMEA: product design analysis before release to manufacturers.
iii. Process FMEA: manufacturing assembly process analysis.

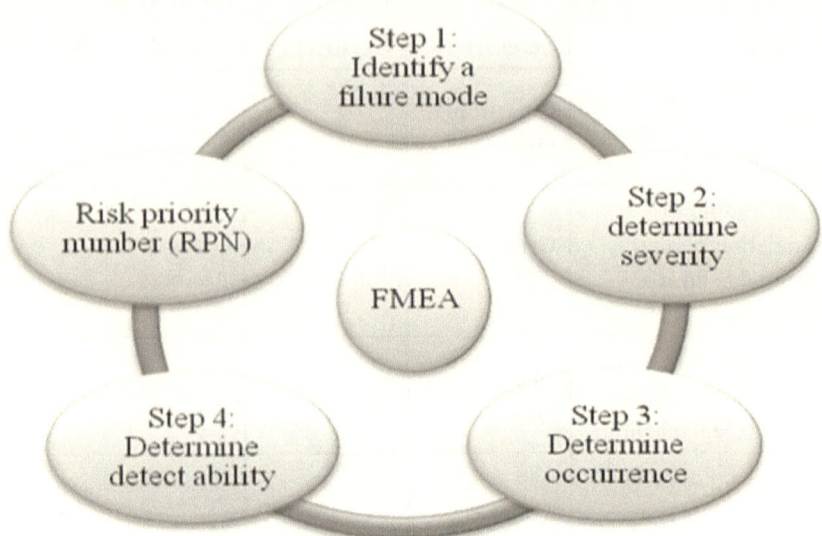

Figure 3. 9: FMEA Process

It is strongly recommended that Serenity, Occurrence and Detection (SOD) for weak control should be noted. SOD numbers is multiplied and the value is stored in RPN (risk priority number) column. This is the key number that will be used to identify where the team should focus first. If, for example, we had a severity of 10 (very severe), occurrence of 10 (happens all the time), and detection of 10 (cannot detect it) RPN is 1000. This indicates a serious situation that requires immediate attention. The consequence could further be broken down into effect for ship, human safety, oil spill, damage, ecology, emission and other environmental impacts. Number 1-10 are assigned according to level of serenity. Risk priority number (RPN) for total serenity is determining as follows Table 3.5 show typical risk matrix arrangement:

RPN = S X O X D   3.04

ALARP Principal, Risk Acceptability Criteria and Risk Control Option

Table 3. 5: Risk matrix

			Consequence Criteria				
			1 – Insignificant	2 – Minor	3 – Moderate	4 – Major	5 – Catastrophic
Likelihood	A	Consequence certain to occur	Medium (M)	High (H)	High (H)	Very High (VH)	Very High (VH)
	B	Consequence likely to occur	Medium (M)	Medium (M)	High (H)	High (H)	Very High (VH)
	C	Consequent possibly likely to occur some time	Low (L)	Medium (M)	High (H)	High (H)	High (H)
	D	consequence unlikely to occur but could happen	Low (L)	Low (L)	Medium (M)	Medium (M)	High (H)
	E	consequence may occur only in exceptional circumstances	Low (L)	Low (L)	Medium (M)	Medium (M)	High (H)

Risk acceptability criteria establishment is dynamic because of differences in environment, diversity in industries and choice of regulations requirement to limit the risk. Risk is never acceptable, but the activity implying the risk may be acceptable due to benefits of safety reduced, fatality, injury, individual risk, societal risk, environment and economy. The rationality may be debated, societal risk criteria are used by increasing number of regulators. Figure 3.10 shows ALARP diagram by IMO (Skjong et al., 2005).

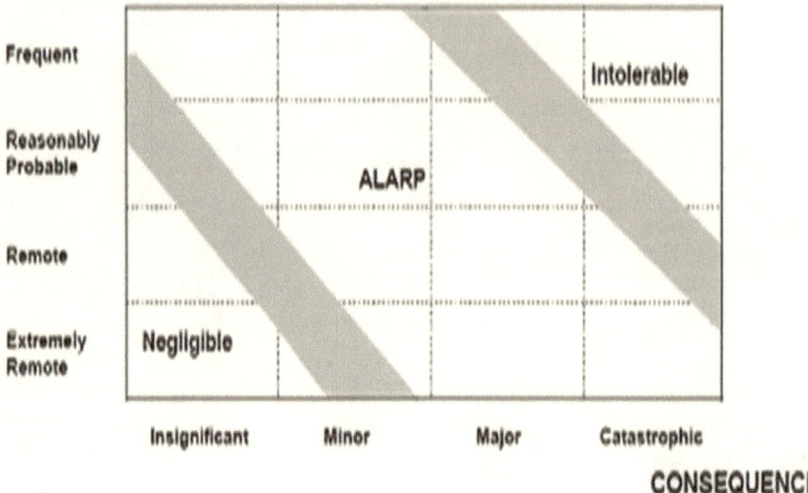

Figure 3. 10: Influence diagram; ALARP = As Low As Reasonably Practicable:
Risk level boundaries (Negligible/ALARP/Intolerable)

Figure 3.11 shows prescribed illustrative influence diagram by IMO. Based on the region where the graph falls, step for risk control option and sustainability balancing, cost benefit effectiveness towards recommendation for efficient, reliable, sustainable decision can be taken. The frequency (F) of accidents involving consequence (N) or more fatalities may be established in similar ways as individual or societal risk criteria. For risks in the unacceptable/ Intolerable risk region, the risks should be reduced at any cost. Risk matrix constructed from system and sub system level analysis can be deduced according to acceptability index and defined according to Table 3.5 and Figure 3.10 to deduced measure of As Low As Reasonably Practicable (ALARP). Within ALARP range, Cost Effectiveness Assessment (CEA) or Cost Benefit Analysis (CBA) shown in Figure 3.10 may be used to select reasonably practicable risk reduction measures.

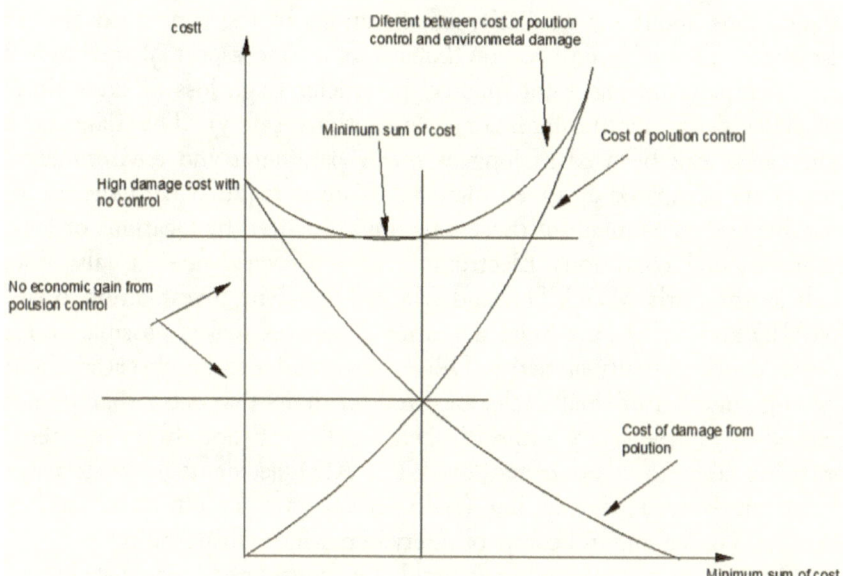

Figure 3. 11: Influence diagram

3.8 Risk Analysis Considerations

In addition to a sound process, robust risk framework and eventual deductive risk model, there are other considerations that should be factored into the design of an effective risk model. These items include the use of available data, the need to address human factors, areas of interest, stakeholder interest and approaches to treating uncertainty in risk analysis. Data required for risk work should involve information on traffic patterns, the environment (weather, sea conditions, and visibility), historical, current operational performance data, and human performance data.

The models intentions are highly dependent on appropriately selected databases that accurately represent the local situation and the effectiveness of the models However, there is always issue of missing data or data limitations especially for complex system and their allow frequency, high consequence nature. Therefore creative procedures are required to develop compensation for data relationships. The model could use probabilistic, stochastic, simulation and expert judgments couple existing deterministic and historical method for a reliable system analysis of desired design (Cooke, R.M. 1997).

When insufficient local data is available, world wide data from other areas may be referred to (e.g., Europe, south and North America), make

—

assumptions about the similarity of operations in the concerned area or elsewhere. This is to ensure how behaviour in one aspect of operational (e.g., company management quality) parameter (e.g., loss of crew time) correlates with another area (e.g., operations safety). The data from other areas can be used as long as major parameter and environmental factors are compared and well matched. Care is required with the use of worldwide data as much of those data are influenced by locations or local environmental conditions Electronic access to worldwide casualty data such as the Paris MOU, U.K.and Marine Accident Investigation Board (MAIB) and IMO Port State detention databases makes possible access to worldwide casualty statistics. Diligence should also be observed about the large number of small scale, localized incidents that occur that are not tracked by marine safety authorities, e.g small craft (not always registered or being able to be detected by VTS, AIS) accidents in waterways. American Bureau of shipping (ABS) has begun an effort to identify precursors or leading indicators of safety in marine transportation.

Human factor modeling should be considered for distributive, large scale systems with limited physical oversight. Assessing the role of human and organizational performance on levels of risk in the system is important, such error is often cited as a primary contributor to accident, which end up leaving system with many more unknown. Expert judgments and visual reality simulation can be used to fill such uncertainty gaps and others like weather data. Even when attempts are made to minimize errors from expert judgments, the data are inherently subject to distortion and bias. With an extensive list of required data, there are limits that available data can place on the accuracy, completeness and uncertainty in the risk assessment results. Expert judgments give prediction about the likelihood that failures that would occur in specific situations can be used to quantify human reliability input in risk process.

Uncertainty is always part of system behaviour. Two common uncertainties are: aleatory uncertainty (the randomness of the system itself) and epistemic uncertainty (the lack of knowledge about the system). Aleatory uncertainty is represented by probability models while epistemic uncertainty is represented by lack of knowledge concerning the parameters of the model. Aleatory uncertainty is critical, it can be addressed through probabilistic risk analysis while epistemic uncertainty is critical to allow meaningful decision-making. Simulation offers one best option to cover extreme case uncertainty beside probability.

Evaluation and comparison of baseline scenario to a set of scenarios of interest (tug escort) and operational circumstance including timelines and roles. Response Scenarios can also be analyzed for things that cannot

—

be imagined or model to be accounted for in the simulator (especially real time). A flexible critical path and slack analysis can be performed as input to the system simulation and uncertainty analysis. Human reliability is best modeled separately for a good result (Murphy, D.M. et al, 1996). Risk and reliability can be achieved by employing probability stochastic and expert rating in the risk process. A safety culture questionnaire which assesses organizational and vessel safety culture and climate can be administered to provide quantitative and qualitative input to the safety culture and environmental perception analysis for sustainable system design.

Following need for maritime activities to operate in much harsh condition, institutions are adopting system based approach that account for total risk associated with system lifecycle to protect the environment and prevent accident. Those that cannot be prevented and protected need or must be controlled under risk and reliability based design / operability platform. Employment of risk method to address each contributing factor to accident is very important. Qualitative risk in system description and hazard identification can best be tackled through HAZOP. The outcome of HAZOP can be processed in quantitative analysis which may include probabilistic and stochastic dynamic simulation process for system level analysis, while fault tree and event tree quantitative analysis can be utilized to determine risk index of the subsystem factors. Interpretation of risk index into ALARP influence diagram can provide decision support information necessary for cost control option towards sustainable, reliable, efficient technology choice for system design and operation.

The cumulative results from qualitative analysis can be made more reliable through iterative quantitative, scientific stochastic and reliability analysis. Risk methods provide valuable and effective decision support tool for application of automated system engineering analysis that facilitate inclusion of reliability, environmental protection and safety as part of the iterative design processes for new and innovative marine system designs, operability and deployment of deep sea operability system. Intelligently adoption of HAZOP and other risk processes eventually can results to safer, efficient, more reliable and sustainable system.

References

1. Bottelberghs, P.H. (1995) QRA in the Netherlands. Conference on Safety Cases. IBC/DNV. London.
2. Skjong R., Vanem E. and Endersen.(2005) Risk Evaluation Criteria. SAFEDOR. Deliverable D.4.5.2 (Submitted to MSC 81 by Denmark).
3. International Maritime Organisation (IMO). Amendments to the Guidelines for Formal Safety Assessment (FSA) for Use in the IMO Rule Making Process. 2006. MSC/ MEPC.2/Circ 5 (MSC/ Circ.1023—MEPC/Circ.392).
4. Murphy, D.M. & M.E. Paté-Cornell. (1996), The SAM Framework. A Systems Analysis Approach to Modeling the Effects of Management on Human Behavior in Risk Analysis. Risk Analysis.16:4.501-515.
5. GESAMP (1990) *The State of the Marine Environment*, IMO/ FAO/UNESCO/ WMO/WHO/IAEA/UN/UNEP (1990)Joint Group of Experts on the Scientific Aspects of Marine Pollution. Blackwell Scientific Publications London
6. Cahill, R.A. (1983). Collisions and Their Causes. London, England: Fairplay Publications
7. Cooke, R.M. (1997) Uncertainty Modeling: Examples and Issues. Safety Science. 26:1, 49-60.
8. Det Norske Veritas. (2004). Thematic Network for Safety Assessment of Waterborne Transportation. Norway.
9. Guedes Soares.C., A. P. Teixeira. (2001). Risk Assessment in Maritime Transportation. Reliability Engineering and System Safety. 74:3, 299-309.
10. Kite-Powell, H. L., D. Jin, N. M. Patrikalis, J. Jebsen, V. Papakonstantinou. (1996). Formulation of a Model for Ship Transit Risk. MIT Sea Grant Technical Report.Cambridge, MA, 96-19.
11. Emi, H. et al (1997) An Approach to Safety Assessment for Ship Structures, ETA analysis of engineroom fire, Proceedings of ESREL '97.International Conference on Safety and Reliability. Vol. 2.
12. Parry, G. (1996), The Characterization of Uncertainty in Probabilistic Risk Assessments of Complex Systems. Reliability Engineering and System Safety. 54:2-3.119-126.

—

13. Roeleven, D., M. Kok, H. L. Stipdonk, W. A. de Vries. (1995). Inland Waterway Transport: Modeling the Probabilities of Accidents. Safety Science. 19:2-3. 191-202.
14. N.Soares, C., A. P. Teixeira. (2001).Risk Assessment in Maritime Transportation. Reliability Engineering and System Safety. 74:3. 299-309.
15. UK, HSE, 1999, Offshore Technology Report" Effective Collision Risk Management for Offshore Installation, UK, London

CHAPTER 4

Environmental Risk
for Sustainable Dredging

"The important thing is not to stop questioning"
Albert Einstein

Summary

Dredging work and placement leads to changes to the environment. Environmental impact assessment has been employed to address these changes. Risk assessment which rarely covers large part of uncertainty associated with dredging work is captured in EIA. EIA focus on fixed and inflexible standards which have led to post dredging failures. This makes it necessary to do critical and dependability scientific risk analysis that quantitatively determine whether the changes are serious or irreversible. This Chapter discusses the new internationally recognized philosophy of risk analysis or formal and system risk based design that provide opportunity to focus on real concern of the dredging project. The chapter will discuss case study of failed project based on conventional EIA and best practice performance of systemic risk base design approach.

4.1 Introduction

Dredging is process of digging under water for purpose to maintain the depth in navigation channels. Dredging is required to develop and maintain navigation infrastructure, reclamation, maintenance of river flow, beach nourishment, and environmental remediation of contaminated sediments. Study on environmental impact of dredging is not new and recently there is concerned about balance between the need to dredge, economic viability, social technical approval and adequate environmental protection can be challenge. Various methods has been implemented for management of dredging activities, but choose in the best practice approach is also a bog challenge that require high level of understanding of the technical and economical aspects of the dredging process. Input from ecological experts and dredging specialists. Community participation from port authorities, regulatory agencies, the dredging industry and non-governmental organisations such as environmentalists and private sector consultancies.

4.2 The Need for and dredging requirement

Dredging is the excavation, lifting and transport of underwater sediments and soils for the construction and maintenance of ports and waterways, dikes and other infrastructures, for reclamation, maintenance of river flow, beach nourishment, to extract mineral resources, particularly sand and gravel, for use for example in the construction industry, and for the environmental remediation of contaminated sediments. Globally, many hundreds of millions of cubic meters (m3) of sediments are dredged annually, with most of this volume being handled in coastal areas. A portion of this total represents capital dredging which involves the excavation of sediments to create ports, harbors, and navigable waterways. Maintenance dredging sustains sufficient water depths for safe navigation by periodic removal of sediment accumulated due to natural and human-induced sedimentation.

Maintenance dredging may vary from an almost continuous activity throughout the year to an infrequent activity occurring only once every few years. Dredging activities offer social, economic and environmental benefits to the whole community. Hydrography chart and bathymetric map are used as guidance to vision of discrete bottom of water. Vigilant is requiring for the bottom as the, they are proned to sudden change leading to shoaling due to flood or drought. Survey of a navigation channel to locate dredging area done through drawing of isolines, or lines connecting points of equal depth, on the map so that captains and ships' pilots can get an idea of the "hills and valleys" underwater [1,3].

Remote sensing equipment is used by hydrographers' on top of the water of the water to see the bottom of the channel. Isoline are drowned

—

based on statistical data record for accuracy and reducing risk of missing important underwater features, like rocks or shoals. Dredgers: dredger is a machine that scoops or sucks sediments from under the water. There are a few different types of dredgers, the three main types of dredges are mechanical dredges, hydraulic dredges, and airlift dredges. Mechanical dredges are often used in areas protected from waves and sea swells. They work well around docks and shallow channels, but not usually in the ocean. Hydraulic dredges work by sucking a mixture of dredged material and water from the channel bottom. There are two main types of hydraulic dredges—the cuter head pipeline dredge and the hopper dredges. Airlift dredges are special-use dredges that raise material from the bottom of the waterway by air pressure. Split hull hopper dredges are self-powered, so they can move to the dredging and disposal site by themselves. Figure 4.1 shows typical hydrographic survey of a channel.

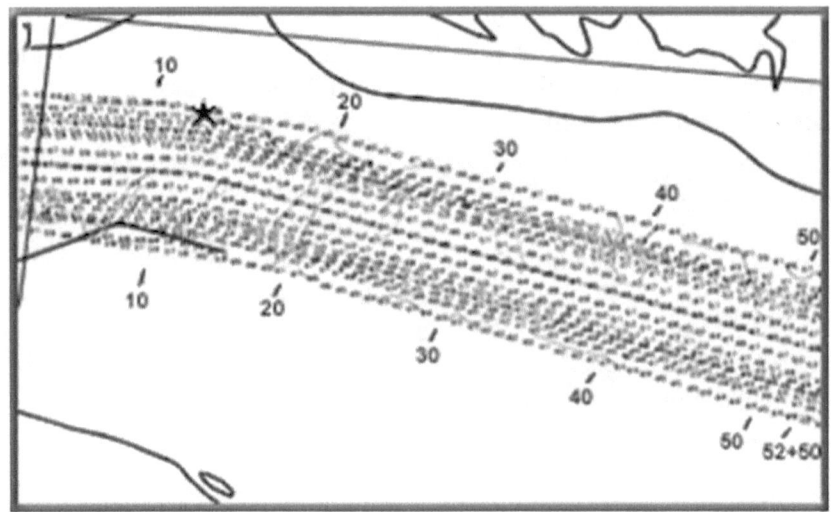

Figure 4. 1: Typical hydrography survey Channel condition survey for a channel, 53m deep, the survey lines are at 50-foot intervals. Shaded areas are shoals.(PTP, 2008)

Dredge material: Dredging is necessary to maintain waterways channel. Nearly 400 million cubic yards of material is dredged each year. Consequently, about 400 million cubic yards of material must be placed in approved disposal sites or else used for another environmentally acceptable purpose. Sustainable disposal of dredge material is very imperative as it ends up saving a lot of money and maintains reliability and efficiency use of resources advantage of sustainable beneficial disposal are [2,3]:

i.   Cost saving on money spent on finding and managing disposal sites.
ii. It avoids habitat and ecological impacts that disposal may cause.
iii. It saves capacity in existing disposal sites.
iv. It can be a low-cost alternative to purchasing expensive fill for construction projects.
v.   It can be used to enhance or restore habitat.

4.3 Environmental requirement of dredging project

The tendering of a dredging contract typically occurs after a full engineering design has been completed (i.e. after the planning and design phase). However, for other types of contracting mechanisms (e.g., design-build), the tendering of contract may occur early in the overall project process, thus requiring the Contractor to perform much of the evaluation and design work himself. Table 4.1 shows phases of dredging project and the risk control components.

The planning and design phase begins with defining overall functional requirements to meet the project objectives. This involves evaluating potential environmental impacts and any regulatory constraints, and concludes with preparing projects specifications. The planning and design phase is used to identify risk areas and risk control option in advance to help protect the environment during dredging, transport, and disposal activities and subsequent monitoring and possible remedial actions.

Table 4. 1: Dredging project phases and risk components

Project Phase	Planning and design	Construction	Post construction
Impact	Need for dredging is translated into project design Physical and biological impacts will depends on project specification.	Project construction will cause temporary or permanente physical change, Advert effect should be mitigated through best practice method	Physical change may apply to long term environmental effects that should be mitigated by appropriate project design, planning and execution

—

Scope	Functional requirements Conceptual design Potential environmental impacts Final design and specifications	Tendering and contract award Construction methods and equipments' selection Monitoring and feedback	Infrastructures in service and there may have additional mode of impact Long term monitoring and feedback may be needed to evaluate RCO
Environmental components	Planning and design decision RCO to prevent or reduce environmental impact if of the whole project	Construction decision RCO to prevent environment impact cause by physical change	Certain RCO may apply to mitigation of future impacts

Elements of project formulation include:

i.   Functional Requirements
ii.   Conceptual Design
iii. Regulatory Framework
iv.   Baseline Environment
v.   Stakeholder Input
vi.   Potential impact Review the baseline condition as a consequence of construction and post-project activities.
vii. Environmental Impact Assessment (EIA)
viii. Risk control option
ix.   Prepare Final Project Design and Specifications

A final design addresses all major elements of the project: engineering design, environmental management, construction sequencing, and construction management. The specification's level of detail will depend on the type of contract, the complexity of the project, and the experience with dredging of both the project proponent and contractor(s). Figure 4.2 shows example operational disposal control measure to limit impact of dredge disposal.

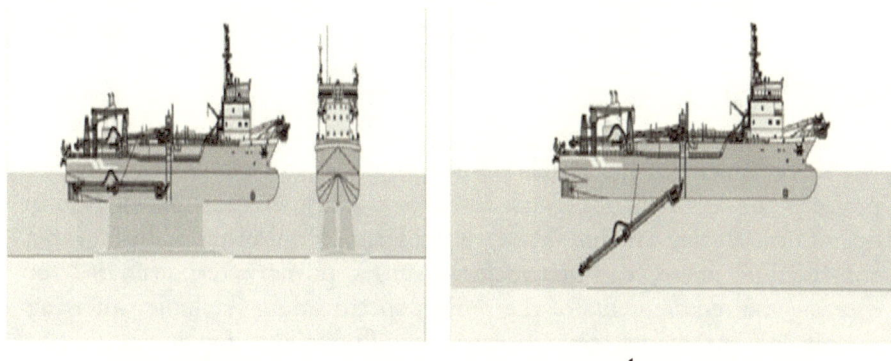

a   b

Figure 4. 2: a) uncontrolled disposal; b) control disposal (PIANC)

It is important to integrate risk control option that have been evaluated in the environmental review process to ensure that the desired balance between minimizing potential environmental impacts and constraints on construction is achieved. Additional environmental review may be required to establish that any residual risk, or actual impact, is acceptable Risk control option must be based on a clear definition of the project's technical and regulatory requirements. Studies conducted during the EIA or project planning, as well as information from regulators and stakeholders can contribute technical information for informed risk control option including [3,4]:

i. Sediment characterization (e.g., grain size distribution, level of contamination, etc.)
ii. Bathymetric/topographic surveys with design profiles, which establishes the volume of sediment to be dredged;
iii. An understanding of hydraulic/hydrodynamic/oceanographic conditions that may restrict operations;
iv. The destination or final use of the dredged material, including placement options and locations;
v. The environmental functions and value of the area to be dredged, establishing environmental boundary conditions;
vi. The environmental value of dredged material management areas (e.g., placement in confined or unconfined areas, or beneficial use options);

vii. Existing site uses (e.g., navigation, recreational use, commercial fishing, quality of life impacts [air, noise, light]) to establish reasonable operational measures;

viii. Legal conditions.

Environmental aspects related to future use and maintenance of the project's post construction condition should consider the areas of facility operations, future maintenance, long-term monitoring. During the construction phase, the contractor assumes primary responsibility for meeting the requirements of the project specifications, including meeting permit and contractual environmental conditions and implementation of risk control options. Major steps in the construction phase include:

i.   Tendering and Contract Award
ii. Contractor Defines Construction-Methods and Selects Equipment
iii. Project Execution: Risk control option should based on best practice

Figure 4. 3: Post dredging impact in Kuala Terengganu

4.4 Environmental risk requirements of dredging project

Risk analysis in a dredging project, including taking into account adherence to the Precautionary Principle. It involves methods for assessing the significance of the likely impacts and essential environmental characteristics that require consideration during both the planning and implementation phases and the mechanisms whereby impacts can occur.

4.4.1 Qualitative based environmental impact assessment

Dredging and disposal of dredged material have many potential implications for the environment, like disturbance of benthic invertebrates, disruption of their habitats and direct mortality. The scale of these impacts depends on several factors, including the magnitude, duration, frequency and methodology of the dredging activity and the sensitivity of the affected environment. The need for RCO to reduce the effects of dredging, transport, and placement depends on the results of the impact assessment process and effectiveness to meet goal of protecting sensitive environmental resources, maintenance of healthy ecosystems, and ensuring sustainable development and exploitation of resources.

Understanding the environments in which dredging and dredged material placement occur is a prerequisite of prudent decision making for environmental protection. A thorough knowledge of baseline conditions is needed so that a dredging project's environmental effects can be assessed properly and monitored against an agreed baseline. The baseline data must address natural variations, seasonal patterns and longer term trends to provide a context for determining whether a change is the result of dredging or not. As a minimum, characterization of the potentially affected environment should consist of recent surveys (performed within the last three years) and studies of the relevant environmental attributes.

For reliability the studies must be conducted by qualified scientific and engineering personnel using accepted methods. The boundary includes the physiographic, hydrologic, ecologic, social, and political boundaries of the project areas. In general, the following types of data are required for characterization of the dredging and placement sites, the transport corridor, and the areas around these sites, which could be indirectly affected, to adequately address the range of management options [5, 6]:

i. Bathymetric and adjacent topographic data
ii. Habitat and species distribution
iii. Resources such as fish populations, shellfish beds, oil and gas fields, aggregate mining and spawning grounds

iv. Physical and chemical nature of sediments
v. Water quality
vi. Hydrodynamic data
vii. Cultural resources, including archaeological and anthropological conditions
viii. Human demography and land use characteristics
ix. Users of the environmental resources, such as commercial, recreational, and subsistence fisheries; Navigation routes
x. Services in the project area, such as pipelines and cables

In addition, it is necessary to take into account any cumulative impacts. Certain ongoing activities, such as fisheries and navigation, could have impacts that in combination with the proposed dredging result in more significant effects than would result from the project activities alone. This information is generally included in the impact assessment.

4.4.2 Between environmental risk assessment and environmental impact assessment

In practice, different approach is used to evaluate and "measure" the environmental impacts of a dredging project. ERA is defined as the examination of risks resulting from the technology that threaten ecosystems, animals and people (EEA, 1998). There are three main types of ERA: human health, ecological, and applied industrial risk assessment. The origin of ERA is the assessment of risks in the industry. Then, the same approach was applied in a broader scale for assessing the risks of the release of chemicals posed to human health. The more recently developed ecological risk assessment follows the same approach as human health ERA, but extending the assessed "end-points" to species other than human beings.

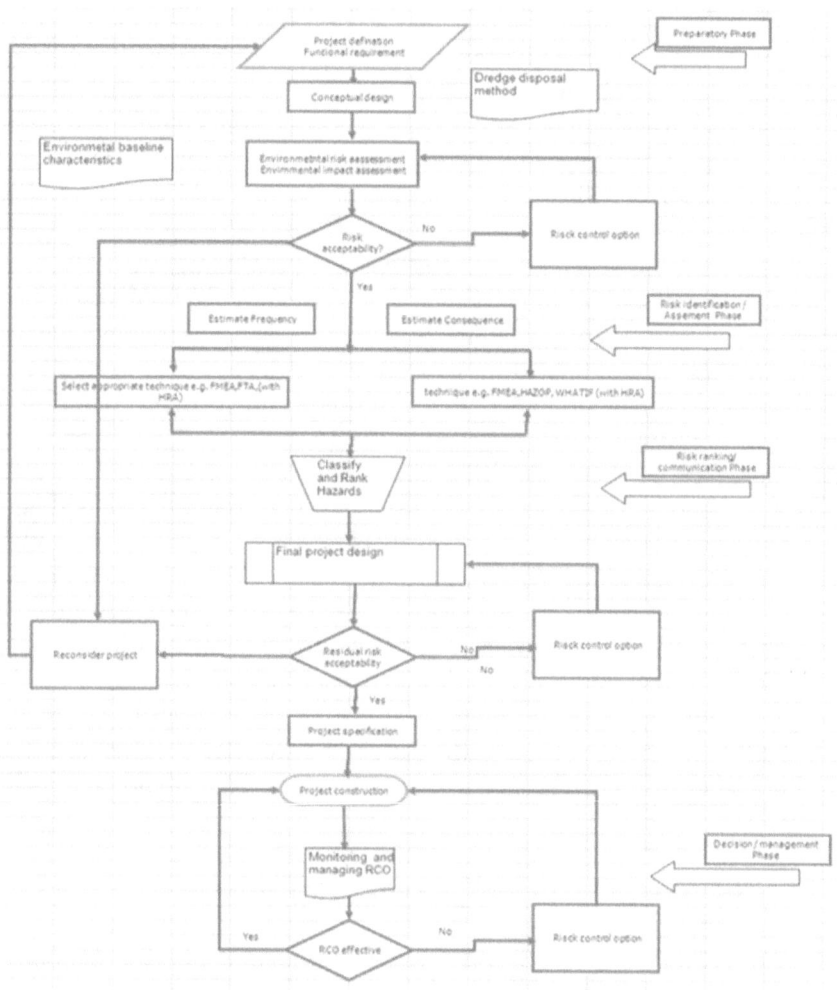

Figure 4. 4: Risk based model for dredging project

A conventional approach of an environmental risk assessment begins with the problem formulation and the identification of the hazard (or hazards). Then, the possible ways of release of the hazard are estimated, and the exposure of those chosen target species is assessed. The final steps are the consequence assessment and the estimation of the risk. Some of the steps require the use of models (e.g. the assessment of the release and the exposure), and the outcome is usually a quantitative assessment. It should be noted that many choices have to be done in the design of the risk assessment, and thus the definition and method used in each of them will be of importance to the final outcome [7].

—

It has also been common that human health and ecological ERA are normally applied for assessing the risks regarding the release of one single chemical. They would need to be adapted for assessing the impacts of dredging operations, where more than one chemical might be released together. It is cleared that the consequences of dredging might be broader than the release of chemicals .Another group of tools widely used at present is environmental impact assessment (EIA).

There are also several methodologies for performing an EIA. They depend on the assessed activity, and of course the way the final impacts are presented and aggregated. The effects considered in an EIA are very wide, from pollution effects to a wider range of ecological effects, and it is often a statutory requirement under holistic doctrine to consider all possible effects, including economic, social and political.

The difference between ERA and EIA is that the later do not treat risks as probabilities. Generally the potential impacts are predicted, and assessed quantitatively or qualitatively. However, it also uses models requires for making many decisions in the design of the assessment, which could influence the final result. Any evaluation of the impacts of a certain project has to face difficulties and uncertainties, in part due to the scientific uncertainties involved, but in part due to the decisions to be made for framing and defining the problem. The impact assessment will have to specify the range of species to include and thus get entangled in nontrivial normative (ethical, ecological and economic) issues.

4.4.3 Risk based design and precautionary principle

The "Communication from the Commission of the European Communities on the precautionary principle" (Commission of the European Communities, 2000) states that the Precautionary Principle should be applied within a structured approach to the analysis of risk. As outlined above, this comprises three elements: risk assessment, Risk analysis, risk management and risk communication. The Precautionary Principle is particularly relevant as an instrument in the management of risk.

In the context of dredging projects it can be stated that because of great natural variability there will often be a lack of full scientific certainty about the scales of potential impacts. In accordance with the Precautionary Principle decision to forego a project should be a last resort following exhaustive consideration of all reasonable RCO and reaching a conclusion that adequate environmental protection could not be achieved.

Prohibiting dredging may ensure that no impacts occur, but may also generate high risk to human safety (e.g., lack of removal of shoals that pose navigation hazards) or result in lost commerce and harm to the economy.

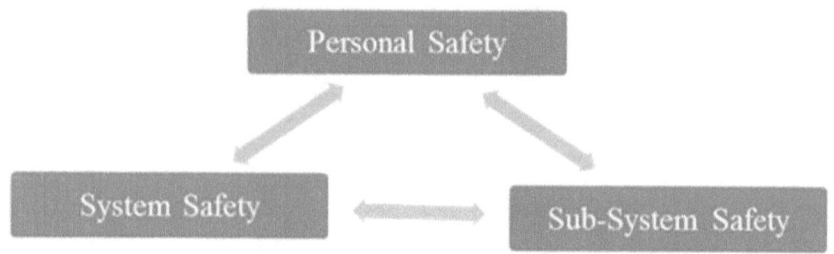

Figure 4. 5: Components of risk assessment and analysis

The RCO should be selected such that clear, defined, and ideally quantitative thresholds of protection can be achieved (e.g. to control measures of suspended sediment within a specified concentration / duration range). Figure 4.5 shows typical system risk components [8, 9].

The approach to risk assessment begins with risk analysis, a systematic process for answering the three questions posed at the beginning of this chapter: What can go wrong? How likely is it? What are the impacts? The formal definition of a risk analysis proceeds from these simple questions, where a particular answer is Si, a particular scenario; pi, the likelihood of that scenario; and Ci, the associated consequences. See Figure 4.6.

Thus cost of environmental sustainability is not cheap, but whenever we compare the benefit and longtime reliability with the cost, there is no doubt that the later will supersede the earlier. Dedication on Scientific analysis, environmental assessment work, never gets attention in the past like to today. The fact that everything stays with us, is recently calling for philosophy for minimum used of toxic material in our daily activities. Yes, scientific work or test required prudent analysis over time, but once we have information we should restrain under the doctrine that prevention is better than cure.

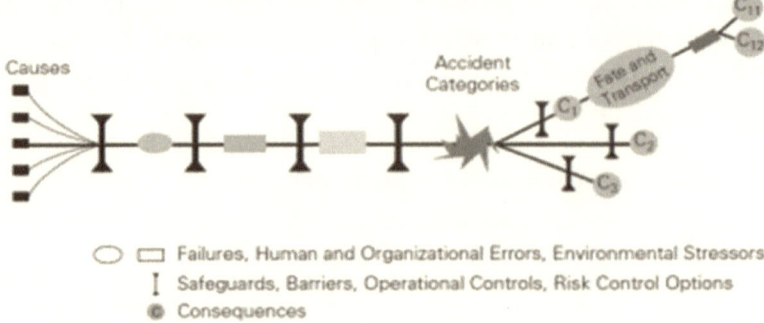

Figure 4.6: Risk based method

Work on environmental issue has always involved dispute because of impacts analysis. Global climate Change might be regarded as a primary example where this strong interlinkage between science and policy making is broadly acknowledged, Social science studies have shown how the production of scientific knowledge played a crucial part in the rise of climate change as a topic of worldwide interest and to the political arena while, on the other hand, knowledge and research on climate change issues is influenced by social factors

In most countries, the majority of dredged material is placed at sea. Land disposal options are normally much more expensive therefore, they are applied only when either transport costs to sea are inhibitory, or beneficial use is not an option, or the material is too contaminated (Burt et al., 1997a). In order to meet sustainability requirement the following describe 3 case studies where beneficial work in dredging are translated to cost [10,11].

    i.   On environmental sustainability According to US green port project, 2001, case study on Boston port navigation improvement project done in the US dredging and construction project use mitigation like Surface sediments contaminated with metals, PAHs, PCB, and other organics, Channels were over-dredged by 20 ft. Contaminated material was placed on barge and deposited into over-dredged in-channel disposal cells and covered with 3 ft. clean material, All clean material deposited in Mass Bay Disposal Site.

    ii.   Another case done in port of los Angelis use copper treatment by developing onsite system to treat copper contaminated marine sediments, Pilot study dredged, treated, and disposed of 100 tons of contaminated sediment, Full-scale project cleaned up 21,000 cubic yards of contaminated sediment, Saved $1.5 million in cleanup costs over alternative.

iii. Studies done in Europe also confirm use of processing plant for dredge material. Also regional sediment management program done by (USACE, 2003) compiled various methodologies to reduce shoaling.

4.5 Reliability and decision support framework

Various studies have been carried out to find the best hybrid supply for given areas. Results from specific studies cannot be easily applied to other situations due to area-specific resources and energy-use profiles and environmental differences. Energy supply system, with a large percentage of renewable resources varies with the size and type of area, climate, location, typical demand profiles, and available renewable resource. A decision support framework is required in order to aid the design of future renewable energy supply systems, effectively manage transitional periods, and encourages and advance state-of-the-art deployment as systems become more economically desirable. The DSS could involve the technical feasibility of possible renewable energy supply systems, economic and political issues.

Reliability based DSS can facilitate possible supply scenarios to be quickly and easily tried, to see how well the demands for electricity, heat and transport for any given area can be matched with the outputs of a wide variety of possible generation methods. This includes the generation of electricity from intermittent hybrid sources. DSS framework provide the appropriate type and sizing of spinning reserve, fuel production and energy storage to be ascertained, and support the analysis of supplies and demands for an area of any type and geographical location, to allow potential renewable energy provision on the small to medium scale to be analyzed.

DSS can provide energy provision for port and help guide the transition towards higher percentage sustainable energy provision in larger areas. The hybrid configuration of how the total energy needs of an area may be met in a sustainable manner, the problems and benefits associated with these, and the ways in which they may be used together to form reliable and efficient energy supply systems. The applicability and relevance of the decision support framework are shown through the use of a can simulate case study of the complex nature of sustainable energy supply system design.

—

4.6 Regulatory requirement and assessment

The current legal requirements have been developed based on reactive approach which leads to system failure. Reactive approach is not suitable for introduction of new technology of modern power generation systems. This call for alternative philosophy to the assessment of new power generation technologies together with associated equipment and systems from safety and reliability considerations, such system required analysis of system capability and regulatory capability. System based approaches for regulatory assessment is detailed under goal based design.

IMO has embraced the use of goal based standards for ship construction and this process can be equally well applied to machinery power plants. Figure 4.7 illustrates the goal based regulatory framework for new ship construction that could be readily adapted for marine system. Tiers of the goal base framework are shown:

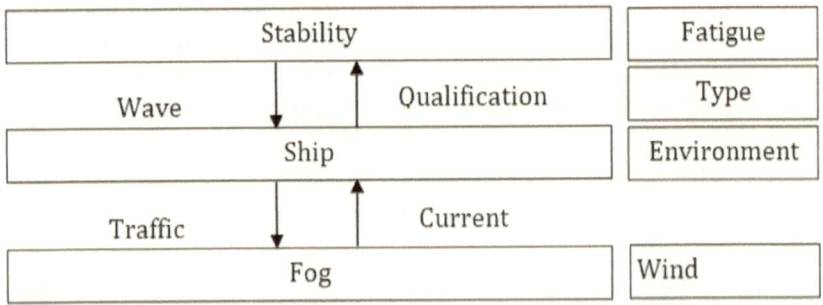

Figure 4. 7: Components of level goal standard assessment

Legal framework for dredging

The most important international agreements regarding dredging are the London Convention 10, issued in 1972 and reviewed in the 1996 Protocol 11; and the OSPAR Convention 12 from 1992. IMO also unveil Formal safety assessment for marine system. These international agreements establish frameworks within which the contracting countries are obliged to operate with respect to their handling of materials destined for placement in the sea. However, these Conventions do not include regulations of the dredging operations per se, which are mainly established at the national level, nor for the conditions of disposal of in land. Convention for the prevention of marine pollution by dumping of wastes and other matter (www.londonconvention.org).

—

A review of the Convention began in 1993 and was completed in 1996 with the acceptance of The 1996 Protocol to the London Convention. The 1996 Protocol has not yet come into force as it has not yet been ratified by a sufficient number of countries (19 out of 26). Conventions for the nProtection of the Marine Environment of the North-East Atlantic (www.ospar.org).On the other hand, dredging activities are subject to national regulations, which can vary very much among the countries. In some cases there is a specific directive regarding dredging in Malaysia the royal Malaysia navy regulate the dredging. Thus the are other agencies but there is not integration for effectiveness of the system.(personal communication .

4.7 Quantitative and formal system engineering based risk analysis

"Risk" is generally understood as an expression of the quantified link between an environmental hazard or "stressor" and the potential negative consequences it may have on targets or "receptors". When discussing risk the types of stressors as well as the targets of interest must be specified. Thus project risk can be distinguished from engineering risk, and environmental risk. But, in practice it may be very difficult to establish a quantifiable relationship between hazard and target response because of the many uncertainties in the cause-effect chain and the dynamic nature of aquatic ecosystems. Risk analysis provides a means to accommodate these uncertainties. Formal risk assessment procedures have not been adopted by many regulatory agencies or they have been applied mainly to dredging of contaminated sediments. Typically risk assessment takes the form of "professional judgment" based on the experience and expertise of parties engaged in project co-ordination [12].

Risk analysis provides an opportunity to focus on the real concerns of a project, instead of relying on fixed and inflexible standards such as threshold levels for contaminants or fixed percentages of allowable overflow of a dredger. For the purpose of this report, risk assessment is mainly captured in the EIA, whereas risk management takes the form of best management practice determination. Risk evaluation is the path from the scientific system based quantitative risk analysis is the internationally recognized best practice and modest concept of risk analysis. Table 4.2 shows components of risk analysis.

An outline about risk assessment and the application of risk-informed decision making can be found in Bridges (2007). This document does not address the scientific methods to evaluate the human health and

—

ecological risks of a project. In this case other guidance, like the PIANC Working Group Envicom 10 report on "Environmental Risk Assessment of Dredging and Disposal Operations", should be reviewed and properly qualified professionals engaged to perform the necessary work. Analysis tools that now gaining general acceptance in the marine industry is Failure Mode and Effects Analysis (FMEA). The adoption of analysis tools requires a structure and the use of agreed standards. The use of analysis tools must also recognize lessons learnt from past incidents and experience and it is vital that the background to existing requirements stemming from rules are understood. Consistent with the current assessment philosophy, there needs to be two tenets to the process—safety and dependability. A safety analysis for a hybrid power generation system and its installation on board a ship could use a hazard assessment process such as outlined in Figure 6. The hazard assessment should review all stages of a systems life cycle from design to disposal.

Table 4. 2: Components of risk analysis

Components	Purpose / Process
Risk analysis	Involve the overall process of risk assessment and risk management, including screening and scoping
Risk assessment	Involve qualitative process of identifying risk potential to quantitative risk characterization
Risk screening	Involve the specification and setup of a general framework for managing risk
Risk evaluation	Involve the scientific evaluation of risks through use of stochastic process Public/political evaluation of risks
Risk management	Involve the process of identifying and selecting measures Procedures for implementation and evaluation of measures

Figure 4.8a shows the components of risk assessment and analysis. The analysis leads to risk curve or risk profile. The risk curve is developed from the complete set of risk triplets. Table shows elements of risk analysis. The fourth column is included showing the cumulative probability, Pi (uppercase P), as shown. When the points <Ci, Pi> are plotted, the result is the staircase function. The staircase function can be considered as discrete approximation of a nearly continuous reality. If a

smooth curve is drawn through the staircase, that curve can be regarded as representing the actual risk, and it is the risk curve or risk profile that tells much about the reliability of the system. Combination of qualitative and quantitative analyses is advised to for risk estimates of complex and dynamic system.

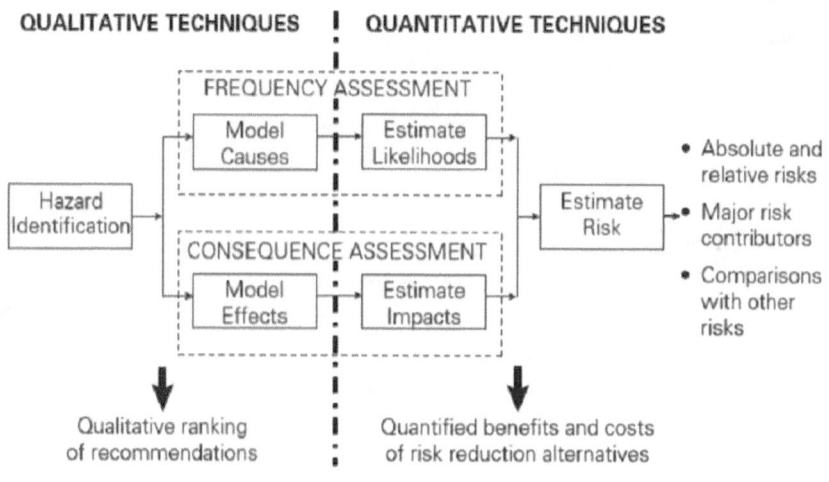

a.

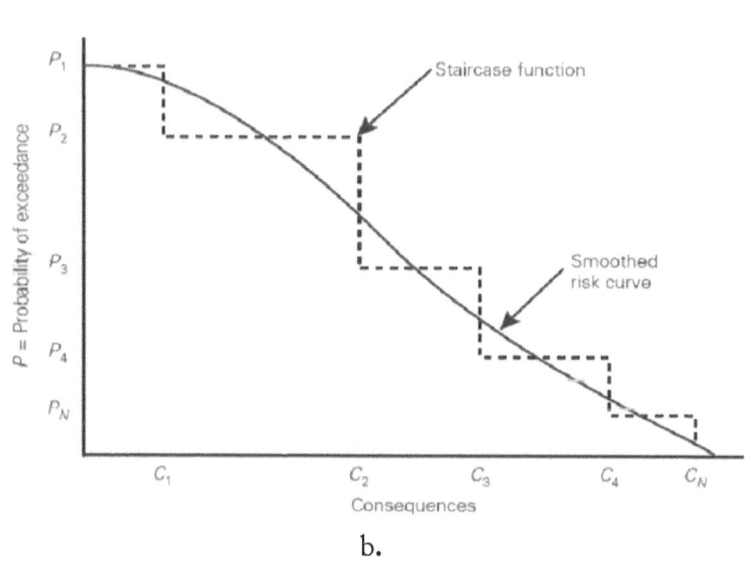

b.

Figure 4. 8: a. Components of risk and reliability analysis, **b.** Stair case risk curve

—

161

The analysis that describes and quantifies every scenario, the risk estimation of the triplets can be transformed into risk curve or risk matrix of frequency versus consequences that is shown in Figure 4.8b.

Table 4. 3: Components of risk and reliability analysis

Scenario	Probability	Consequence	Cumulative Probability
S1	P1	C1	P1=P1+P2
S2	P2	C2	P2=P3+P2
Si	Pi	Ci	Pi=Pi+3+Pi
Sn+1	Pn+1	Cn+1	Pn-1=Pn+Pn+1
Sn	Pn	Cn	Pn=Pn

The design concept needs to address the marine environment in terms of those imposed on the power plant and those that are internally controlled. It is also necessary to address the effects of fire, flooding, equipment failure and the capability of personnel required to operate the system. In carrying out a hazard assessment it is vital that there are clearly defined objectives in terms of what is to be demonstrated. The assessment should address the consequence of a hazard and possible effect on the system, its subsystems, personnel and the environment. An assessment for reliability and availability of a hybrid power generation system and its installation in a ship could use a FMEA tool. An effective FMEA needs a structured approach with clearly defined objectives

The assessment analysis processes for safety and reliability need to identify defined objectives under system functionality and capability matching. These two issues are concerned with system performance rather than compliance with a prescriptive requirement in a standard. The importance of performance and integration of systems that are related to safety and reliability is now recognized and the assessment tools now available offer such means. Formal Safety Assessment (FSA) is recognized by the IMO as being an important part of a process for developing requirements for marine regulations. IMO has approved Guidelines for Formal Safety Assessment (FSA) for use in the IMO rule-making process (MSC/Circ.1023/MEPC/ Circ.392). Further reliability and optimization can be done by using stochastic and simulation tools [8, 13].

4.8 Uncertainties and risk in dredging projects

The physical and biological characteristics of aquatic environments vary both spatially and temporally. Therefore characterizing these environments and assessing impacts and risk will always involve some uncertainty. This requires the need for basic understanding of how marine and the ecosystems function and how natural events and anthropogenic activities affect these functions. In the ideal situation, all environmental risks associated with a dredging project would be quantifiable, making the need for specific management practices clear. In reality, dredging can potentially affect diverse assemblages of organisms or their habitats on both spatial and temporal scales. Because the scales of the interactions between organisms and the dredging process are difficult to determine, often the consequences of a project are largely speculative. Some degree of uncertainty will therefore always be present indecisions regarding the need for special management practices to protect the environment.

It is important to recognize that even with extensive baseline data and input from qualified professionals, an element of uncertainty will always be associated with the results of an environmental assessment, simply due to the dynamic nature of marine and freshwater environments and the complexity of stressors and drivers apart from anthropogenic influences. Effects of dredging operations have to be seen against the background of similar natural effects. Figure 4.9 shows that in a typical dredging project the risk assessment is made after the preparation of the Conceptual Design. If at this stage it proves that the environmental risks are such that they cannot be mitigated by implementation of the appropriate best practice then the project should be reconsidered. This means that functional requirements will need to be redefined followed by a revision of the Conceptual Design.

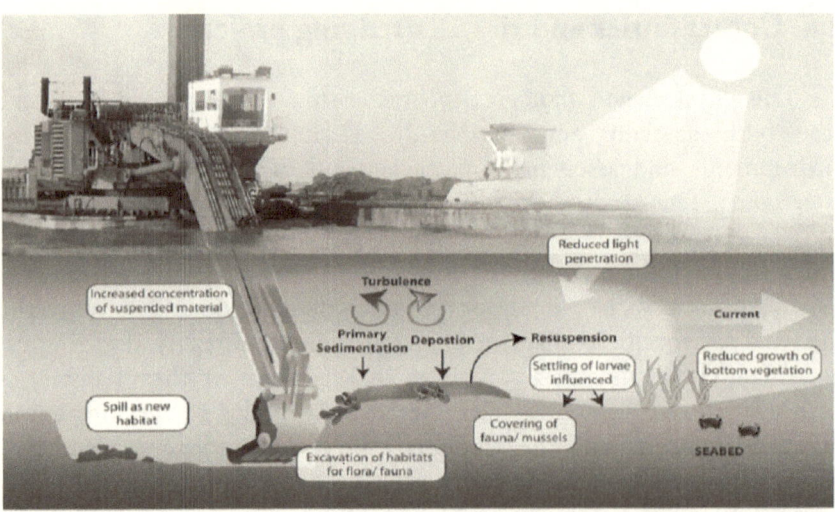

Figure 4. 9: Possible Environmental Effects of Dredging (PIANC, 2009)

In the process of risk work, newer refined RCOs may become necessary during the process of risk management. During the preparation of the final project design it is essential to establish the degree of "residual" risk. Figure 4.10 shows potential impact to marine aquatic.

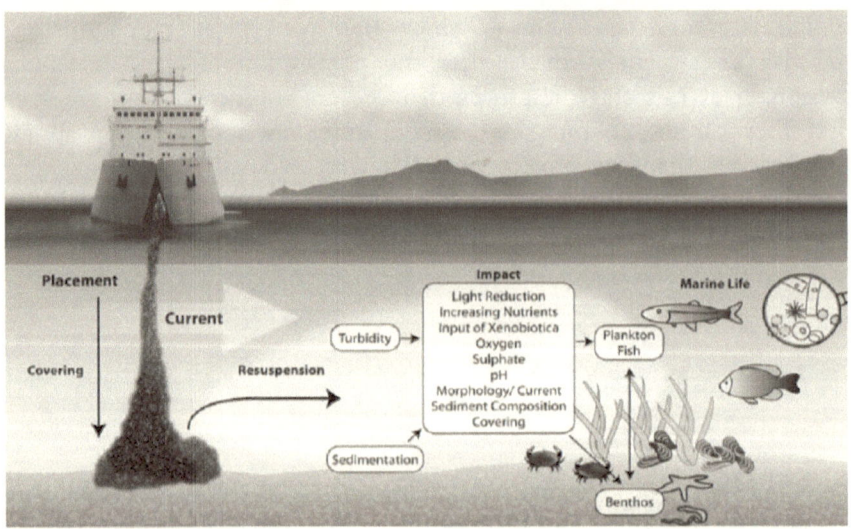

Figure 4. 10: Possible Environmental Effects of disposal (PIANC, 2009)

4.8.1 Potential Physical Changes and Environmental Impacts from Dredging and Disposal of Dredged Material

Below water, the sound from the dredge vessel could have environmental effects such as interfering with fish behavior, possibly leading to disturbed migratory routes, although fish might easily avoid temporarily disturbed areas without consequence. Other potential environmental effects not directly related to dredging but associated with the presence of the dredger include spills of oil and fuel, exhaust emissions, and the possible introduction of invasive species via the release of ballast water. One of the less understood areas of concern is the impact of sediment released into the aquatic environment that may occur at any of the stages from excavation to placement. A high concentration of sand in suspension will have very low turbidity while a relatively low concentration of fine silt or clay in suspension will have a high turbidity. Also sediment effect on the flora and fauna, concentration, the turbidity, the total amount of loss of sediments or the spatial distribution of a sediment plume are other impacts.

Sediment re-suspended in the water column in high concentrations can directly lead to physical abrasion of, for example, filter-feeding organs or gill membranes of fish and shellfish. Indirectly, if present for sufficient duration, high turbidity (i.e. reduced light penetration), can result in decreased growth potential or total loss of submerged aquatic vegetation. The resuspension of sediments can also release toxic chemicals or nutrients such as phosphates and nitrates, which may increase the atrophic status of the system (this reinforces the need for appropriate sediment characterization).Release of anaerobic sediment and organic matter in high concentrations may in some cases deplete the dissolved oxygen. Subsequent sedimentation around the dredging site can smother benthic flora and fauna or compromise habitat quality.

4.8.2 Spatial and Temporal Scales of Effects

The environmental effects vary spatially and temporally from project to project. When the effects are considered to have a significant adverse impact it is necessary to investigate means to reduce or mitigate them. The significance of the environmental effects depends on site-specific factors that govern the vulnerability and sensitivity of environmental resources in the project area. When the sediment being moved is chemically contaminated, the need for environmental protection is generally recognized by all stakeholders. Complexity with respect to uncertainty has made necessity for several efforts to find tools for the assessment and management of different types of uncertainty. As mentioned before, the word uncertainty is used in many different situations for expressing a lack of certain, clear knowledge for taking a decision. Uncertainty is any departure from the unachievable ideal of complete determinism. In the case presented here, uncertainty signifies that is not possible to provide a unique, undisputable, objective assessment of a certain action (for example an environmental risk assessment of the dredging). However, depending on the actor (e.g. the modeller, the policy-maker, or stakeholders), the perception of the nature, kind, object and meaning of uncertainty can be very different. This will be clear when presenting the perception of uncertainty of the stakeholders involved in the case. Nevertheless, the simple definition presented above gets more complicated when trying to describe the sources, or the sorts or dimensions of uncertainty.

Typology approaches adopted for characterization and assessment of uncertainty by this group focus on uncertainties encountered from the point of view of the modeler that assesses policy-makers (which they call model based decision support). Therefore, their proposal aims to be useful for expressing the uncertainty involved in the use of models, perhaps rather than expressing uncertainty from the point of view of the policy-makers or stakeholders. The typology is based in the distinction of three dimensions of uncertainty:

i. The location of uncertainty (where within the model)
ii. The level of uncertainty (from deterministic knowledge to total ignorance)
iii. The nature of uncertainty (whether the uncertainty is due to the imperfection of our knowledge or is due to the inherent variability of the phenomena being described).

4.9 Risk communication and management

Parties involved in a dredging project view the process differently depending on their individual perceptions of these risks and rewards, as well as their individual tolerance of the perceived risk. In this sense there may be several types of risk in a project. For the proponent the consequences of failure of the whole project may be very severe and will usually be measured in economical terms. For an environmentalist the potential effects on the environment may be recognized as the highest priority risk. Communication is an essential component of sharing concerns and identifying means to mitigate them to the fullest extent reasonably possible. During the risk analysis, it is important to balance the identified environmental effects and risks against the economic and social consequences of the project. Figure 4.11 shows stake holder involvement in dredging decision

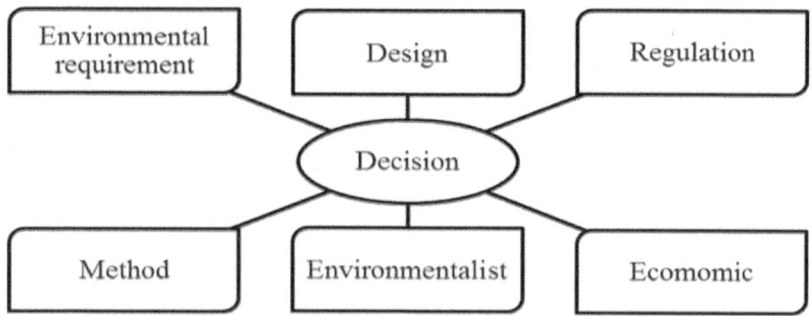

Figure 4. 11: Stake holder involvement in dredging decision

Complete and transparent communications are therefore essential throughout the process from beginning to end. This refers to all parties involved. Communication should address uncertainties and natural variability in the environment. Seldom does an actual project present a clear choice between unbiased, neutral, and generally accepted options. Rather, the choice among options is frequently driven by values and perceptions. This tension can best be reduced through open lines of communication that include:

i.   A transparent process;
ii. Outreach that begins during the earliest possible stage of the project and continues throughout all phases;
iii. An open and honest process; and
iv. Proactive engagement of local and/or regional media, because their influence on public opinion can be large.

Risk perception is very much influenced by the social, political and historical contexts. Environmental Protection Agency's (USEPA) can be found giving some generic recommendations. Figure 4.12 shows expected impact in dredging project. The figure also shows the interrelation ships between physical changes above and below the water and their potential to cause environmental effects. The figure also illustrate physical changes can create multiple environmental effects.

—

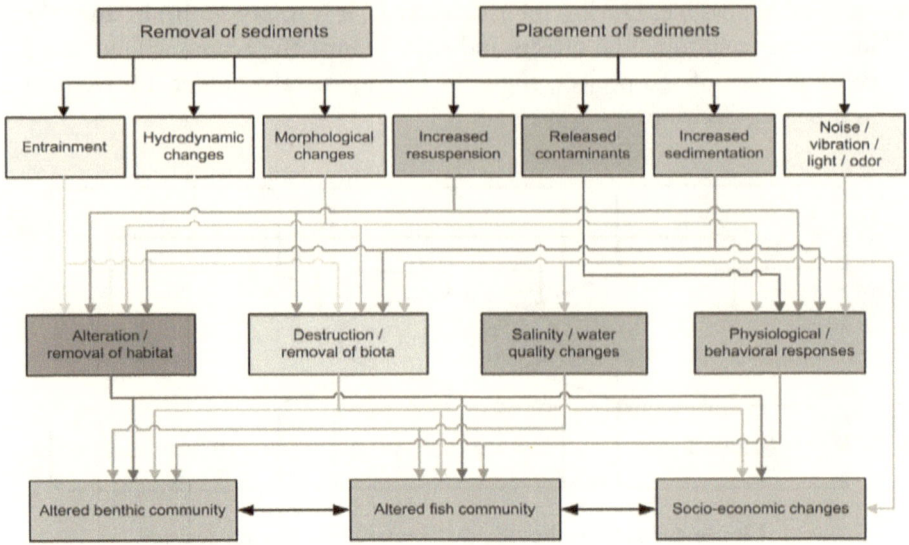

Figure 4. 12: Conceptual Model of Physical Changes and Ecological Effects from Dredging—Related Activities

4.10 Selecting evaluation and risk control option for dredging project

For a monitoring program to be fully effective, it must include a timely communication of results and related actions. Stakeholders should be involved to help build overall program credibility. Action might be taken to adjust the monitoring program itself or as a direct response to the monitoring results. Based on the monitoring data, adjustments to the monitoring program could include:

i. Reducing the level of monitoring because no effect was observed;
ii. Continuing with the existing monitoring program to gain further clarification of the response; or
iii. Expanding the monitoring program to include additional parameters or sites.

So that responding can be quick and effective, it is necessary to establish hierarchy of options to adverse monitoring results. The level of response can be targeted to the receptor and its sensitivity. Options could include:

i. Continuing with dredging under the existing regime

—

ii. Modifying the dredging regime to reduce the actual effect on a sensitive parameter

iii. Ceasing dredging within an area until further information is gathered

iv. Ceasing dredging within an area altogether or Ceasing dredging and implementing recovery measures.

Risk control options are meant to improve the environmental performance of a dredging project. Some form of environmental evaluation or Environmental Impact Assessment (EIA) is normally required by international conventions. One example is the London Convention, which establishes a framework for the evaluation of placement of dredged material at sea. The "Specific Guidelines for Assessment of Dredged Material" (International Maritime Organization, 2000) comprises the following steps:

i. Dredged material characterization

ii. Waste prevention audit and evaluation of disposal options

iii. Is the material acceptable for marine placement?

iv. Identify and characterize the placement site

v. Determine potential impacts and prepare impact hypothesis(ies)

vi. Issue permit

vii. Implement project and monitor compliance

viii. Field monitoring and assessment.

Within the LC-DMAF guidance it is stated that assessment of potential effects should lead to a concise statement of the expected consequences of the sea or land disposal options, i.e., the "Impact Hypothesis". When applying these Guidelines uncertainties in relation to assessments of impacts on the marine environment will need to be considered and a precautionary approach applied in addressing these uncertainties. Figure 4.13 shows risk matrix for risk measure of risk based design.

All dredging and placement projects will cause some changes to the environment. It is therefore necessary to determine whether these can be considered serious and or irreversible. Because adequate information is rarely available to answer these questions with absolute certainty, an evaluation of the relative risk of permanent detriment to the environment is required. Many factors affect this assessment of the general environmental risk including the scale of the project, the natural variability of all of the elements of the system likely to be affected, possible

—

contamination levels, and the timing of the project. Preparation of an EIA involves collaboration among environmental scientists and engineers in consultation with port authorities, dredging companies, and often a diverse assemblage of stakeholders. The amount of technical information available will be important, but should be used in tandem with the perceptions and knowledge held by the engaged stakeholders. Risk evaluation is a value judgment reached by consideration of the total body of evidence offered by all interested parties.

Frequency of Occurrence (or Likelihood)	Consequences (Severity of Accident)				
	Incidental (1)	Minor (2)	Serious (3)	Major (4)	Catastrophic (5)
Frequent (5)	M	H	VH	VH	VH
Occasional (4)	M	M	H	Risk without measure	VH
Seldom (3)	L	M	H	H	VH
Remote (2)	L	L	Risk after measure	H	H
Unlikely (1)	L	L	M	M	H

L = low risk; M _ moderate risk; H = high risk; VH _ very high risk.

Figure 4. 13: Risk priority matrix

An overview of the selection process is shown in Figure 3-1, which also shows how the process can be repeated to achieve a project that optimally conforms to acceptable environmental risks. The flow chart shows that there are multiple stages within the process that allow feedback and repetition in order to achieve a project that is in full compliance with acceptable environmental risk. For example, if the application of certain RCO does not reduce the risk to an acceptable level then the project would need to be reevaluated to determine if other alternatives could be used or the project design modified to reduce or remove such unacceptable risks. Feedback loops also occur following the analysis of monitoring results against the required objectives. The effectiveness of the RCO is assessed against the degree of derived protection of the environmental

—

resource and if found to be ineffective then further RCO may be necessary. Monitoring to measure the effectiveness of the selected RCO provides adaptive feedback that can be applied to future projects, and is always a prudent strategy.

These constrains are very important to bear in mind when we think of environmental management at the local or regional level with projects with are used limited time and budget of money. Therefore the lack of knowledge that can be experienced by both managers and citizens in assessing a concrete project may have more to do with limited resources than general scientific ignorance. Benefit-Cost Analysis (BCA) is a tool for organizing information on the relative value of alternative public investments like environmental restoration projects. When the value of all significant benefits and costs can be expressed in monetary terms, the net value (benefits minus costs) of the alternatives under consideration can be computed and used to identify the alternative that yields the greatest increase in public welfare. However, since environmental goods and services are not commonly bought or sold in the marketplace, it can be difficult to express the outputs of an environmental restoration project in monetary terms.

4.11 Risk monitoring

It is acknowledged that monitoring can take many forms and fulfill various objectives before, during, and after any dredging and placement project. This document does not provide an exhaustive description of monitoring technology but rather focuses on the role of monitoring as a necessary element in the context of BMP application. In particular, monitoring can be proposed as a management practice in itself or used to assess the effect of other management practices. Monitoring is the first step-in determining whether corrective actions will be necessary to ensure the required outcomes [13, 14].

One of the key issues related to any environmental monitoring program is the scope for combining broad monitoring objectives for separate parameters into a single survey. Monitoring programmes can be categorized into three types:

i. Surveillance monitoring
ii. Feedback monitoring
iii. Compliance monitoring

—

Formulating a suitable monitoring strategy requires the following elements:

i. Targeted objective
ii. Beeline condition
iii. Monitoring criteria
iv. Methodology for measuring change
v. Threshold values
vi. Timely review procedure

Requirements for monitoring are site-specific and based on the findings of the baseline surveys. For example, surveys could be necessary to record:

i. The abundance and distribution of species, which is needed to determine the rate of species and community recovery within the study area
ii. The effect of dredging on seabed morphology
iii. The effect of dredging on the concentration of suspended sediments in the water column
iv. The type of substrate remaining following dredging
v. Use of the area by fish
vi. Actual effects on any sensitive species or communities within the study area

Sometimes, model studies can be used to determine the appropriate locations for monitoring. Monitoring involves many uncertainties and difficulties that need to be considered. Models are generally not well validated or calibrated and so it is not easy to quantify the results with certainty though they are continually improving. After the monitoring criteria have been selected, the methodology for measuring change against those criteria needs to be determined.

A number of biological, physical, and chemical variables need to be considered when defining a monitoring scheme. The variety of possible effects depends on the Characteristics of the dredging and placement areas and the dredged material itself, therefore, the monitoring programmed sign must be site—and case-specific and proportional to the extent of the environmental concern. It is also important to understand the possible causes of the environmental problem to identify the source of the problem. There should be a specific hypothesis that can be tested using easily acquired data. The monitoring could be in the water column, on the seabed, on land or in the air. It could be physical, chemical, or biological or a combination. Key considerations in establishing the monitoring methodology are summarized below:

i.   The methodology used to monitor environmental effects should be the same as that used to determine the characteristics of the relevant parameter during the baseline survey, to ensure comparability.

ii. The sampling stations should be the same, although there are likely to be fewer stations (e.g., the feature of interest may require a more targeted approach than was adopted for the baseline survey).

iii. For parameters where timing is critical (e.g., benthic and fish sampling), repeat surveys should be undertaken at the same time of year as the baseline survey to ensure that seasonal changes in abundance and distribution do not affect the results.

iv. The frequency of sampling is determined based on the monitoring objectives and criteria. The expected impact is also a factor to consider when determining frequency of sampling. For some parameters (e.g., impacts on geology), changes occur over a long time scale and therefore require less frequent monitoring, possibly post project.

It is important to identify a level above or below which an effect is considered unacceptable, referred to as an environmental threshold. If the monitoring shows that the threshold level is close to being reached then remedial action is required to reduce the level of effect. In the absence of a threshold value, monitoring of many parameters is justified to improve the knowledge base of the particular effect. Timely review of monitoring results is essential to ensure the success of the program. It is recommended that the results of monitoring should be reviewed at times that will allow for meaningful adjustments to the dredging and placement activities.

Dredging provides economic and social benefits for the whole community. However dredging can and often will have an impact on the environment outside of the desired change, of say deepening a channel. To assess the significance of these effects an environmental impact study often needs to be undertaken. During such a study, cumulative and in-combination effects should be considered as it is important to place the dredging activity into context with other activities, e.g., fisheries, navigation, etc. Previous regulatory work for system design has been prescriptive by nature. Performance based standards that make use of alternative methods of assessment for safety and reliability of component design, manufacture and testing is recommended for hybrid alternative energy system installation.

System failure and carefree of environment in past project poised all field of human endeavor to adopt precautionary principle by providing tools to conduct dredging projects in an environmentally sound manner and design based on comprehensive system based scientific method discussed in This chapter. Properly applied the precautionary principle provides incentives to develop better solutions. The chapter present structured approach and strives for an objective means of selecting the most appropriate Risk control option for that lead to the best protection of the environment and meet sustainable development requirement. Absolute Reliability of the dredge work can be realise by using predictive statistical tools and the data collected.

Reference

Erftemeijer PLA and Lewis III RRL (2006): *Environmental impacts of dredging on seagrasses*: A review. Marine Pollution Bulletin 52, pp. 1553-1512

Herbich JB (2000): Handbook of Dredging Engineering, 2nd Edition. McGraw-Hill Professional, 992 pages

International Maritime Organization (IMO) 2000): *Specific guidelines for assessment of dredged material*

John SA, Challinor SL, Simpson M, BurtTN and Spearman J (2000). *Scoping the assessment of sediment plumes from dredging*. CIRIA Report C547, London,190 pages

Keevin TM, (1998): *A review of Natural Resource Agency Recommendations for Mitigating the Impacts of Underwater Blasting*. Reviews in Fishery Science, pp. 281-313

New Delta Project (2007): Final report of Theme 6 'Sustainable Dredging Strategies'. Framework for a sustainable dredging strategy. 2007, 48 pages www.newDelta.org

OSPAR (2007): *Draft literature review on the impacts of dredged sediment disposal at sea. Document* Nr. EIHA 07/2/2-E

International Maritime Organization (IMO). *Amendments to the Guidelines for Formal Safety Assessment (FSA) for Use in the IMO Rule Making Process*. MSC—MEPC.2/Circ 5 (MSC/Circ.1023—MEPC/Cir. 2006

PIANC (2006): Working Group Envicom 10: *Environmental risk assessment of dredging and disposal operations.*

PIANC (2008): Working Group Envicom14: *Dredged material as a resource options and constraints*

PIANC Working Group Envicom 16:*Dredging and port construction around Coral Reefs* t.b.p.

Rees HL, Murray LA, Waldock R, BolamSG, Limpenny DS and Mason CE (2002): *Dredged material from port developments*: A case study of options for effective environmental management.

EPA (2001): Guidelines for dredging. *Best practice environmental management*. Environment Protection Authority, Victoria, Australia. Publication 691. 116pages.

PIANC (1998) Working Group PTC I-17: *Handling and treatment of contaminated dredged material from ports and inland waterways* "CDM"

—

CHAPTER 5

Risk Based Multi—Hybrid Alternative Energy for Marine System

"The most incomprehensive thing about the world is that it is
all comprehensible"
Albert Einstein

Summary

Sources of alternative energy are natural. There has been a lot of research about the use of free fall energy from the sun to the use of reverse electrolysis to produce fuel cell. For one reason or the other these sources of energy are not economical to produce. Most of the problems lie on efficiency and storage capability. Early human civilization use nature facilities of soil, inland waterways, waterpower which are renewable for various human needs. Modern technology eventually replaces renewable nature with non-renewable sources which requires more energy and produces more waste. Energy, Economic and Efficiency (EEE) have been the main driving force to technological advancement in shipping. Environmental problem linkage to source of energy poses need and challenge for new energy source. The chapter discuss risk based iterative and integrative sustainability balancing work required between the 4 Es in order to enhance and incorporate use of right hybrid combination of alternative energy source (solar and hydrogen) with existing energy source

—

(steam diesel or steam) to meet marine system energy demands (port powering).

The chapter communicates environmental challenges facing the maritime industry. Effort in the use of available world of human technocrat to integrate sources of alternative energy with existing system through holistic proactive risk based analysis and assessment requirement of associated environmental degradation, mitigation of greenhouse pollution. The chapter will also discuss alternative selection acceptable for hybrid of conventional power with compactable renewable source solar / hydrogen for reliable port powering. And hope that the Decision Support System (DSS) for hybrid alternative energy communicated in This chapter to improve on on-going quest of the time to balance environmental treat that is currently facing the planet and contribution to recent effort to preserve the earth for the privilege of the children of tomorrow.

5.1 Introduction

Scale, transportation, language, art, matter and energy remain keys to human civilization. The reality of integration of science and system lies in holistically investigation of efficiency of hybridizing alternative energy source with conventional energy source. This can be achieved with scalable control switching system that can assure reliability, safety and environmental protection. Option for such sustainable system is required to be based on risk, cost, efficiency benefit assessment and probabilistic application.

Green House Gas (GHG) pollution is linked to energy source. Large amount of pollution affecting air quality is prone by reckless industrial development. GHG release has exhausted oxygen, quality of minerals that support human life on earth, reduction in the ozone layer that is protecting the planetary system form excess sunlight. This is due to lack of cogent risk assessment and reliability analysis of systems before building. Moreso, because conventional assessment focus more on economics while environment and its associated cycle is not much considered [1, 8]. Human activities are altering the atmosphere, and the planet is warming. It is now clear that the costs of inaction are far greater than the costs of action. Aversion of catastrophic impacts can be achieved by moving rapidly to transform the global energy system. Sustainability requirement that can be solved through energy conservation (cf. IPCC 2007: 13) are energy and associated efficiency, development, environment, poverty. Stakeholder from government's consumers, industry transportation, buildings, product designs (equipment networks and infrastructures) must participate in the decision work for sustainable system.

—

Recently the marine industry is getting the following compliance pressure regarding environmental issues related to emission to air under IMO MARPOL Annex 6. A world without port means a lot to economy transfer of goods, availability of ships and many things. Large volume of hinterland transportation activities import tells a lot about intolerant to air quality in port area. Adopting new energy system will make a lot of difference large number of people residing and working in the port. Most port facilities are powered by diesel plant. Integrating hybrid of hydrogen and solar into the existing system will be a good way for the port community to adapt to new emerging clean energy concept. Hybrid use of alternative source of energy remains the next in line for the port and ship power. Public acceptability of hybrid energy will continue to grow especially if awareness is drawn to risk cost benefit analysis result from energy source comparison and visual reality simulation of the system for effectiveness to curb climate change contributing factor, price of oil, reducing treat of depletion of global oil reserve.

Combined extraction of heat from entire system seems very promising to deliver the requirement for future energy for ports. This chapter discusses available marine environmental issues, source of energy today, evolution of alternative energy due to the needs of the time and the barrier of storage requirement, system matching of hybrid design feasibility, regulations consideration and environmental stewardship. The chapter also discusses holistic assessment requirement, stochastic evaluation, using system based doctrine, recycling and integrated approach to produce energy. With hope to contribute to the ongoing strives towards reducing green house gases, ozone gas depletion agents and depletion of oxygen for safety of the planet in order to sustain it for the right of future generation.

5.2 Energy, Environment and Sustainable Development

Since the discovery of fire, and the harnessing of animal power, mankind has captured and used energy in various forms for different purposes. This include the use of animal for transportation, use of fire, fuelled by wood, biomass, waste for cooking, heating, the melting of metals, windmills, waterwheels and animals to produce mechanical work. Extensive reliance on energy started during industrial revolution. For years there has been increased understanding of the environmental effects of burning fossil fuels has led to stringent international agreements, policies and legislation regarding the control of the harmful emissions related to their use. Despite this knowledge, global energy consumption continues to increase due to rapid population growth and increased global industrialization. In order to

—

meet the emission target, various measures must be taken, greater awareness of energy efficiency among domestic and industrial users throughout the world will be required, and domestic, commercial and industrial buildings, industrial processes, and vehicles will need to be designed to keep energy use at a minimum. Figure 5.1 shows that the use of fossil fuels (coal, oil and gas) accounted continue to increase [1, 2].

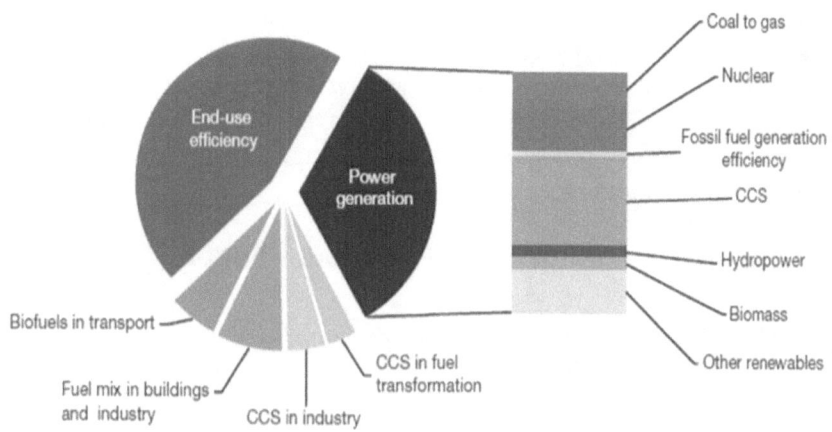

Figure 5. 1: GHG Emissions Reductions through 2050,
by Consuming Sector [EIA, 2007]

Various measures must be taken to reduce emission targets. The current reliance on fossil fuels for electricity generation, heating and transport must be greatly reduced, and alternative generation methods and fuels for heating and transport must be developed and used. Sustainable design can be described as system work that which enhances ecological, social and economic wellbeing, both now and in the future. The global requirement for sustainable energy provision is become increasingly important over the next fifty years as the environmental effects of fossil fuel use become apparent. As new and renewable energy supply technologies become more cost effective and attractive, a greater level of both small scale and large scale deployment of these technologies will become evident. Currently there is increasing global energy use of potential alternative energy supply system options, complex integration and switching for design requirement for sustainable, reliable and efficient system.

The issues surrounding integration of renewable energy supplies need to be considered carefully. Proactive risk based Decision support system is important to help the technical design of sustainable energy systems, in order to encourage planning for future development for the supply of electricity, heat, hot water and fuel for transportation. Renewable energy systems have intermittence source, this make assurance reliability of the supply and subsequent storage and back-up generation a necessity. Generic algorithms of the behavior of plant types and methods for producing derived fuels to be modeled, available process and manufacturers' data must be taken into consideration. Today, simulation tool for analysis that allow informed decisions to be made about the technical feasibility of integrated renewable energy systems are available. Tool that permit use of supply mix and control strategies, plant type and sizing, suitable fuel production, and fuel and energy storage sizing, for any given area and range of supply should be adopted.

5.3 Energy Consumption, Demand and Supply

Energy is considered essential for economic development, Malaysia has taken aggressive step in recent year to face challenges of the world of tomorrow, and this includes research activities strategic partnership. One example is partnership with the Japanese Government for construction on sustainable energy power station in the Port Klang power station, Pasir Gudang power station, Terengganu Hydro-electric power station and Batang Ai Hydro-electric power station which are main supply to major Malysian port. The above enumerated power stations are constructed with energy-efficient and resource-efficient technologies. Where power station are upgraded the power station by demolishing the existing aging, inefficient and high emission conventional natural gas/oil-fired plant (360MW) and installing new 750MW high efficiency and environment friendly combined cycle gas fired power plant built at amount of JPY 102.9 billion.

The combined-cycle generation plant is estimated to reduce the power station's environmental impact, raise generation efficiency and make the system more stable. The total capacity of power generation of 1,500MW is equal to 14% of total capacity of TNB in peninsula of 10,835MW and indeed this power station is one of the best thermal power stations with highest generation efficiency in Malaysia of more than 55%. The rehabilitation, the emissions of Nitride oxide (NOx) is reduced by 60%, Sulfur dioxide (SO_2) per unit is reduced by almost 100% and Carbon dioxide (CO_2) emission is reduced by 30%. Port operation energy demands are for transportation, hot water and heat. This third generation plan can easily be integrated with alternative energy [3, 4]. Table 5.1 shows Malaysia energy environment outlook.

—

The energy use in all sectors has increased in recent years, most especially the energy use for transport has almost doubled it continues to grow and becoming problem. This trend is being experienced in industrialized and developing world. Energy demand for port work is supply from grids which are well established in most developed world. The method and sitting of generating conventional energy and renewable energy determine system configuration. Hierarchy systems that can be deduced from these two variables are:

i.   Limited capacity energy
ii. Limited energy plant
iii. Intermittent energy plant

Table 5. 1: Malaysia environmental review

Energy-Related Carbon Dioxide Emissions (2006E)	163.5 million Metric tons, of which Oil (44%), Natural Gas (41%), Coal (15%)
Per-Capita, Energy-Related Carbon Dioxide Emissions ((Metric Tons of Carbon Dioxide) (2006E)	6.7 Metric tons
Carbon Dioxide Intensity (2006E)	0.6 Metric tons per thousand $2000-PPP** 96.0 billion kilowatt hours

5.4 Emerging renewable energy system

The design of integrated sustainable energy supply technology systems that are reliable and efficient for transport, heat, hot water and electricity demands can be facilitated by harnessing weather related sources of energy (e.g. wind, sunlight, waves, and rainfall). In order to provide a reliable electricity supply, reduce energy wastage, and enable the energy requirements for heat and transport to be met, the outputs of these intermittent sources may be supplemented by various means [5,7]. The intermittent nature of most easily exploited sources of alternative energy remains the major problem for the supply the electricity network. This has implications for the management of this transitional period as the balance between supply and demand must be maintained as efficiently and reliably as possible while the system moves towards the ultimate goal of a 100% renewable energy supply over the next fifty to one hundred

—

years. It important to take the of amount intermittent electricity sources that can be integrated into a larger-scale electricity supply network into consideration. Excess supply could be supplied by plant run on fuels derived from biomass and waste.

The renewable hybrid age require utilities, local authorities and other decision makers to be able to optimization that beat constraints, potentials, and other energy requirements from port powering. The sizing and type of storage system required depends on the relationship between the supply and demand profiles. For excess amount electricity produced this could be used to make hydrogen via the electrolysis of water. This hydrogen could then be stored, used in heaters or converted back into electricity via a fuel cell later as required. Using excess electricity, this hydrogen could be produced centrally and piped to for port or produced at vehicle filling stations for haulage, or at individual facilities in the port. Alternatively, excess electricity could be used directly to fuel electric haulage trucks, recharging at times of low electricity demand, or use for HVAC system or water heating for immediate use, or to be stored as hot water or in storage heaters [6,8].

5.5 Energy Supply and Demand Matching

Fossil fuel use for transportation and port activities has increased dramatically over the past decade, and shows little signs of abating. This has caused concern about related environmental and health effects. There is need for to develop alternatively fuel system that produces little or no pollution. The main fuels that can be used in a variety of land, sea and air vehicles are biogas in natural gas and fuel cell vehicles, biodiesel in diesel vehicles, ethanol and methanol in adapted petrol and fuel cell. Biogas can be converted to run on natural gas and in some fuel cell. It must be cleaned first to create a high heating value gas (around 95% methane, a minimum of heavy gases, and no water or other particles).

Fuel cell powered engine can run on pure hydrogen, producing clean water as the only emission. Biodiesel can be used directly in a diesel engine with little or no modifications, and burns much more cleanly and thoroughly than diesel, giving a substantial reduction in unburned hydrocarbons, carbon monoxide and particulate matter. The main barriers to the implementation of alternative fuels is the requirement for a choice of fuel at a national level, the necessity to create a suitable refueling infrastructure, the length of time it will take to replace or convert existing vehicles, and the need for a strong public incentive to change [9, 10, 11]. Choice of conventional energy source could be: Internal Combustion and Diesel Engines: Steam Turbines, Stirling Engines, Gas Turbines

—

5.5.1 Choice of alternative energy

Fuel Cells: The principle of the fuel cell was discovered over 150 years ago. NASA has improved the system in their emission free operation for spacecraft. Recent years has also seen improvement in vehicles, stationary and portable applications. As a result of this increased interest, stationary power plants from 200W to 2 MW are now commercially available, with efficiencies ranging from 30 to 50% and heat to electricity ratios from 0.5:1 to 2:1. Fuel cell re load follower energy, the efficiency of a fuel cell typically increases at lower loadings. Fuel cell system also has fast response. This make them well suited to load following and transport applications. Fuel cell is advanced alternative energy technology with electrochemical conversion of fuel directly into electricity without intermediate stage, the combustion of fuel; hence by-pass the restriction of second law of thermodynamic .the basic fuel supply in the fuel cell systems is hydrogen and carbon dioxide. The simplified fuel cell is exact opposite of electrolysis.

The four basic element of the system are hydrogen fuel, the oxidant, the electrodes and the electrolyte chemicals. The fuel is supplied in the form of hydrogen and carbon dioxide which represent electrode and oxidant cathode, the electrolyte material that conduct the electric current can be acid or alkaline solid or liquid. Cycle of operation begin with hydrogen carbon dioxide to the anode, where hydrogen ion are formed, releasing a flow of electron to the cathode through the electrolyte medium. The cathodes take oxygen from the air and transform it into ion state in combination with anode electron. The oxygen carrying ion migrates back to the anode, completing the process of energy conversion by producing a flow of direct current electricity and water as a by-product.

$$2H_2 -> 4e^- + 4H^+ \qquad\qquad 5.01$$

$$4h^+ + 4e^- + O_2 -> 2H_2O \qquad\qquad 5.02$$

$$2H_2 + O_2 —> 2H_2O + Heat \qquad\qquad 5.03$$

The fact that it is made from water has promise for its unlimited supply the fact that water is the by-product also guarantee vast reduction of pollution on earth, solving problem of green house gas release and global warming. Fuel cell system involve combination of groups of small chemical reactions and physical actions that are combined in a number of ways and a in a number of different sections of the generator. This energy source uses the principles of thermodynamics, physical chemistry, and

—

physics. The net result is a non-polluting, environmentally sound energy source using air or even water cooling with a minimum temperature rise of 20 C above ambient and no emissions. The chemicals, metals, and metal alloys involved are non-regulated. The chemical reactions are encased within the process unit where they are recovered, regenerated, and recycled. This process produces no discharge or emission [4, 7].

Fuel cells are classified by the type of electrolyte they use, and this dictates the type of fuel and operating temperature that are required. The most commonly used fuel cell for small scale due to its low operating temperature, and compact and lightweight form, is the Proton Exchange Membrane fuel cell (PEMFC). Phosphoric Acid and Molten Carbonate fuel cells (PAFC and MCFC) are also available for larger scale applications, and require higher operating temperatures (roughly 200°C and 650°C), which means they must be kept at this temperature if fast start-up is required. All of these fuel cells may be run on pure hydrogen, natural gas or biogas. Certain PAFCs may also use methanol or ethanol as a fuel. If pure hydrogen is used, the only emission from a fuel cell is pure, clean water. If other fuels are used, some emissions are given off, though the amounts are lower due to the better efficiencies achievable with fuel cells. Table 5.2 shows the types of fuel cell and their characteristics [7, 11].

Table 5. 2: Type of electrolyte fuel cell

Types	Electrolyte	Operating temperature
Alkaline	Potassium hydroxide	50-200
Polymer	Polymer membrane	50-100
Direct methanol	Polymer membrane	50-200
Phosphoric acid	Phosphoric acid	160-210
Molten carbonate	Lithium and potassium carbonate	600-800
Solid oxide	Ceramic compose of calcium	500-1000

Comparing the efficiency of fuel cell to other source of alternative energy source, fuel cell is the most promising and economical source that guarantee future replacement of fossil fuel. However efficiency maximization of fuel cell power plant remains important issue that needs consideration for its commercialization. As a result the following are important consideration for efficient fuel cell power plant—Efficiency calculation can be done through the following formula:

—

$$E_C = g\,\frac{G}{nF}$$

5.04

$$G = H*T*Si$$

5.05

Where: Ec=EMF, G =Gibbs function nF=Number of Faraday transfer in the reaction, H= Enthalpy, T=Absolute temperature, S=Entropy change i=Ideal efficiency

Advantages of fuel cell include size, weigh, flexibility, efficiency, safety, topography, cleanliness. Mostly use as catalyst in PAFC, and however recovery of platinum from worn—out cell can reduce the cost and market of the use of P ACF economical. It has cost advantage over conventional fossil fuel energy and alternative energy. Disadvantages of fuel cell are adaptation, training, and cost of disposal. Fuel cell has found application in transportation, commercial facility, residential faculty, space craft and battery

5.5.2 Solar Energy System

Photovoltaic (PV) solar system use silicon photovoltaic cell to convert sunlight to electricity using evolving unique characteristic of silicon semiconductor material and accommodating market price of silicon is god advantage for PV fuel cell. Silicon is grown in large single crystal, wafer like silicon strip are cut with diamond coated with material like boron to create electrical layer, through doping the elementary energy particle of sunlight—photon strike the silicon cell. They are converted to electron in the P-N junction, where the p accepts the electron and the n reject the electron thus setting into motion direct current and subsequent inversion to AC current as needed. Electrical conductor embedded in the surface layer in turn diverts the current into electrical wire [3, 12]:

i.   Collector module need to face south for case of photovoltaic, this depends on modular or central unit's modular
ii. Module storage unit need maintenance
iii. The system need power inverter if the load requires AC current
iv. Highlight of relevant procedural differences from regular projects of this type will be needed
v.   Discuss requirements benefits and issues of using new procedures, and incorporating that into the total cost

vi. Procedure to build on will be described, hybrid system and integration system will be described and analyzed from the results and

vii. System successful complied with all regulations

viii. Efficiency penalty caused by extra power control equipment

Sola collector can be plate or dish type. Stefan' law relates the radiated power to temperature and types of surface:

$$\frac{P}{T} = \varepsilon \sigma T^4$$

5.06

Where P/A is the power in watts radiated per square meter, ε is surface emissivity, σ is Stefan-Boltzmann constant= $5.67 \times 10e8 \ W/m^2.K^4$

The maximum intensity point of the spectrum of emitted radiation is given by:

$$\lambda_{max} = \frac{2898}{T(K)}$$

5.07

5.6 Hybrid System

With a focus on developing applications for clean, renewable, non-fossil fuel, energy systems. Our final emphasis is on maritime related activities; however, as marine engineers we are devoted to promoting all types of alternative & sustainable energy technologies. Various types of engine, turbine and fuel cell may be run on a variety of fuels for combined heat and power production. Hybrid system can provide control over power needs, green and sustainable energy that delivers a price that is acceptable and competitive.

The power plants can be located where it is needed less high power lines are required, not only reducing costs but assisting health by reducing magnetic fields that people are so worried about, Global warming is addressed d by direct action by providing power that does not release any emissions or discharges of any kind. The technology associated with the design, manufacture and operation of marine equipment is changing rapidly.

The traditional manner in which regulatory requirements for marine electrical power supply systems have developed, based largely on incidents and failures, is no longer acceptable. Current international requirements for marine electrical power supply equipment and machinery such as engines, turbines and batteries have evolved over decades and their applicability to new technologies and operating regimes is now being questioned by organizations responsible for the regulation of safety and reliability of ships. Figure 5.2 and 5.3 shows hybrid configuration for conventional power, solar and hydrogen, and figure 5.4 shows physical model of hybrid of solar, wind and hydrogen being experimenting in UMT campus.

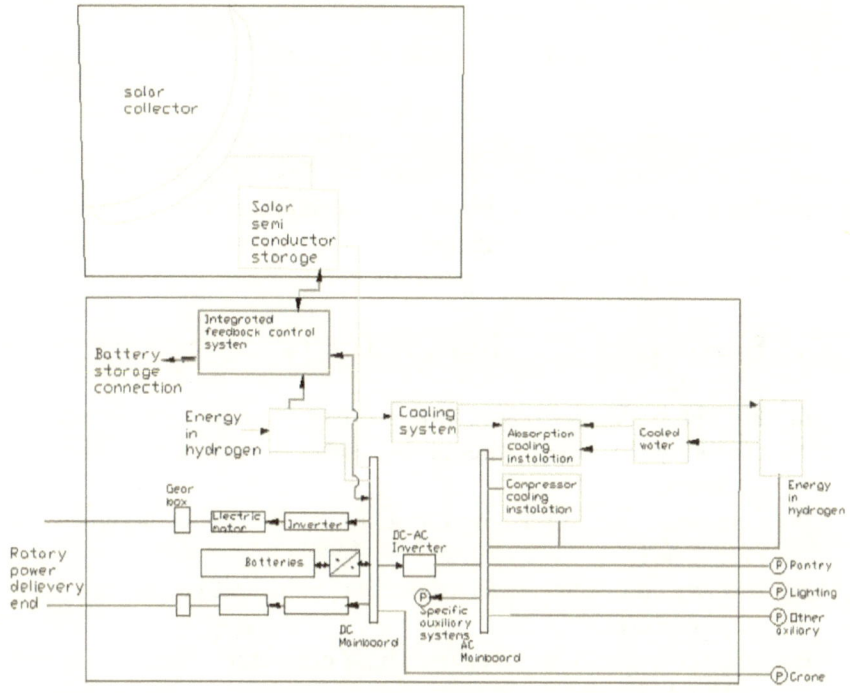

Figure 5. 2: Hybrid configuration

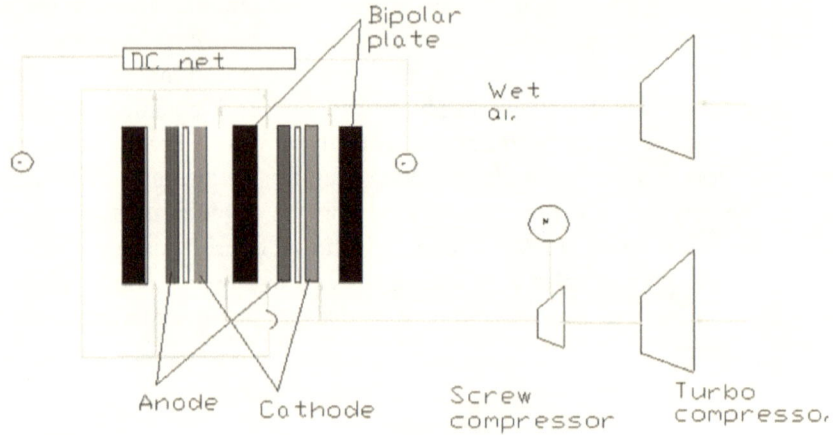

Figure 5. 3: Hybrid configuration

Various technologies have been employed towards the use of alternative free energy of the sun since the first discovery in the 18th century. Improvement and development has been made towards making it available for use like existing reigning source of energy. Major equipment and hardware for the hybrid configuration are:

i.   Semiconductor solar with high efficient storage capability will be designed
ii. Hybrid back—up power will be design based with integrative capability to other alternative power source like wind and hydrogen
iii. Controller design for power synchronization will be designed and prototyped
iv. Inverter and other power conversion units will be selected based on power needs
v.   Solar collector or receiver with high efficiency collection capacity will be designed
vi. Software development and simulation
vii. Steam will be used as energy transfer medium

Figure 5. 4: Physical model of hybrid system under experimentation in UMT

The power plants can be built in small units combined, which allow greater control over the output and maintains full operational output 100% of the time. The plant produces fewer emissions, the plant can be located close to the areas where the power is required cutting down on the need for expensive high power lines. Excess energy produced can be connected to the grid under power purchase arrangement. The system can be built in independent power configuration and user will be free from supply cut out. One of the unique features of hybrid system is the sustainable, clean energy system that uses a hydrogen storage system as opposed to traditional battery. Its design construction and functionality are inspired by the theme of regeneration and the philosophy of reuse. High efficiency solar panels works with an electrolyser to generate the hydrogen for fuel cell.

The hybrid system can provide means to by—pass and overcome limitation posed by past work in generating replaceable natural energy of the sun and other renewable energy source that can be designed in hybrid system. Reliable deployment of hybrid system developments of mathematical model follow by prototyping, experimentation and simulation of the system are key to the design and its implementation. The main advantages of hybrid configurations are: Redundancy and modularity, high reliability of hybrid circuitry embedded control system, improve emergency energy switching and transfer, low operating cost through integrated design, low environmental impacts due to nature of the energy source [3, 12, and 13].

5.7 Reliability and Decision Support Framework

Various studies have been carried out to find the best hybrid supply for given areas. Results from specific studies cannot be easily applied to other situations due to area-specific resources and energy-use profiles and environmental differences. Energy supply system, with a large percentage of renewable resources varies with the size and type of area, climate, location, typical demand profiles, and available renewable resource. A decision support framework is required in order to aid the design of future renewable energy supply systems, effectively manage transitional periods, and encourages and advance state-of-the-art deployment as systems become more economically desirable.

The DSS could involve the technical feasibility of possible renewable energy supply systems, economic and political issues. Reliability based DSS can facilitate possible supply scenarios to be quickly and easily tried, to see how well the demands for electricity, heat and transport for any given area can be matched with the outputs of a wide variety of

—

possible generation methods. DSS can provide energy provision for port and help guide the transition towards higher percentage sustainable energy provision in larger areas. The hybrid configuration of how the total energy needs of an area may be met in a sustainable manner, the problems and benefits associated with these, and the ways in which they may be used together to form reliable and efficient energy supply systems. The applicability and relevance of the decision support framework can be shown through the use of a can simulate case study of the complex nature of sustainable energy supply system design.

5.7.1 Regulatory Requirement and Assessment

The Unifies International association of classification society (IACS) unified requirements are applicable to marine power plant and electrical installations. A listing of the applicable requirements to marine power plants is shown in appendix of This chapter. They IACS requirement provide prescriptive statements that provide a definition or identify what has to be done and in some cases how to do it. They relate to safety and reliability of marine power plant and support systems and arrangements. The current requirements have been developed based on reactive approach which leads to system failure. Reactive approach is not suitable for introduction of new technology of modern power generation systems. This call for alternative philosophy to the assessment of new power generation technologies together with associated equipment and systems from safety and reliability considerations, such system required analysis of system capability and regulatory capability [5,14]. System based approaches for regulatory assessment is detailed under goal based design as shown in figure.

International Maritime Organisation (IMO) has embraced the use of goal based standards for ship construction and this process can be equally well applied to machinery power plants. Figure 5.5 illustrates the goal based regulatory framework for new ship construction that could be readily adapted for marine power plant application. Tiers of the goal base framework are shown in figure 5.5.

—

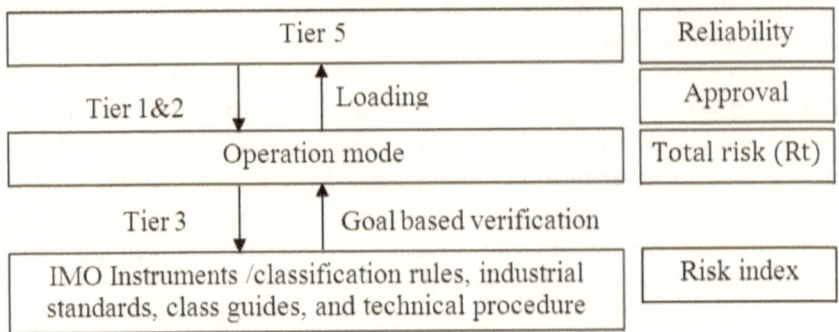

Figure 5. 5: Components of level goal standard assessment

5.7.2 Risk based design

The approach to risk assessment begins with risk analysis, a systematic process for answering the three questions posed at the beginning of this chapter: What can go wrong? How likely is it? What are the impacts. The analysis that describes and quantifies every scenario, the risk estimation of the triplets can be transformed into risk curve or risk matrix of frequency versus consequences that is shown in Figure 5.7 and 5.8.

5.7.3 Quantitative risk assessments

Analysis tools that now gaining general acceptance in the marine industry is Failure Mode and Effects Analysis (FMEA). The adoption of analysis tools requires a structure and the use of agreed standards. The use of analysis tools must also recognise lessons learnt from past incidents and experience and it is vital that the background to existing requirements stemming from SOLAS or IACS are understood. Consistent with the current assessment philosophy, there needs to be two tenets to the process—safety and dependability. A safety analysis for a hybrid power generation system and its installation on board a ship could use a hazard assessment process such as outlined in Figure 5.6. The hazard assessment should review all stages of a systems life cycle from design to disposal.

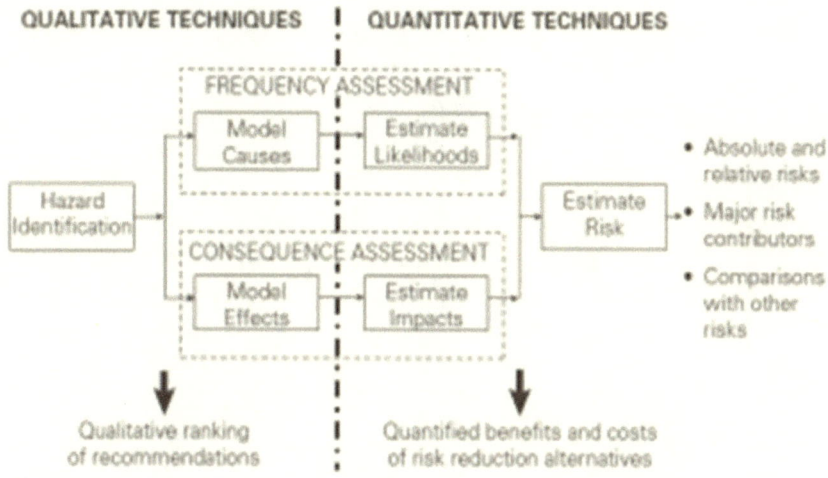

Figure 5. 6: Components of risk and reliability analysis

Figure 5.6 shows the components of risk assessment and analysis. The analysis leads to risk curve or risk profile. The risk curve is developed from the complete set of risk triplets. The triplets are presented in a list of scenarios rearranged in order of increasing consequences, that is, C1, C2, C3, CN, with the corresponding probabilities as shown in table 5.3. A fourth column is included showing the cumulative probability, Pi (uppercase P), as shown.

When the points <Ci, Pi> are plotted, the result is the staircase function. The staircase function can be considered as discrete approximation of a nearly continuous reality. If a smooth curve is drawn through the staircase, that curve can be regarded as representing the actual risk, and it is the risk curve or risk profile that tells much about the reliability of the system. Combination of qualitative and quantitative analyses is advised to for risk estimates of complex and dynamic system.

Table 5. 3: Components of risk and reliability analysis

Scenario	Probability	Consequence	Cumulative Probability
S1	P1	C1	P1=P1+P2
S2	P2	C2	P2=P3+P2
Si	Pi	Ci	Pi=Pi+3+Pi
Sn+1	Pn+1	Cn+1	Pn-1=Pn+Pn+1
Sn	Pn	Cn	Pn=Pn

$$n=N$$

$$-$$

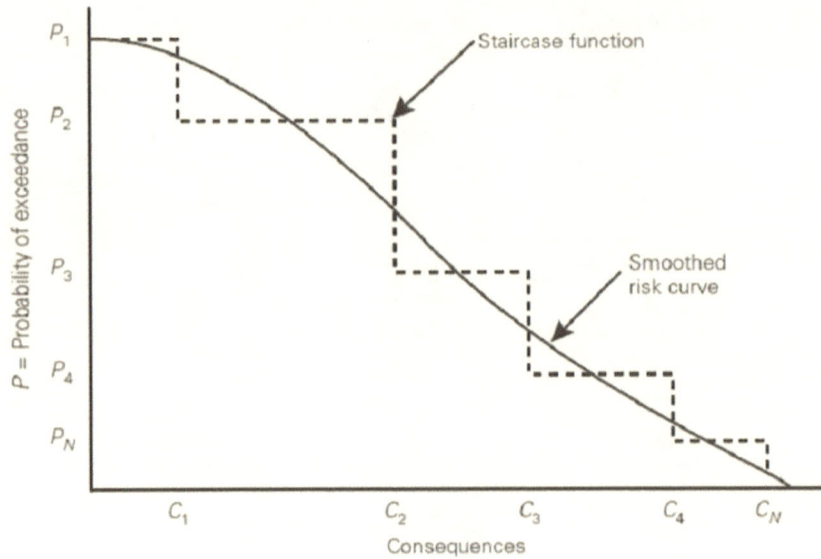

Figure 5. 7: Stair case risk curve

Frequency of Occurrence (or Likelihood)	Consequences (Severity of Accident)				
	Incidental (1)	Minor (2)	Serious (3)	Major (4)	Catastrophic (5)
Frequent (5)	M	H	VH	VH	VH
Occasional (4)	M	M	H	Risk without measure	VH
Seldom (3)	L	M	H	H	VH
Remote (2)	L	L	Risk after measure	H	H
Unlikely (1)	L	L	M	M	H

Figure 5. 8: Risk priority matrix
L = low risk; M _ moderate risk; H = high risk; VH _ very high risk.

The design concept needs to address the marine environment in terms of those imposed on the power plant and those that are internally controlled. It is also necessary to address the effects of fire, flooding, equipment failure and the capability of personnel required to operate the system. In carrying out a hazard assessment it is vital that there are clearly defined objectives in terms of what is to be demonstrated. The assessment should address the consequence of a hazard and possible effect on the system, its subsystems, personnel and the environment.

An assessment for reliability and availability of a hybrid power generation system and its installation in a ship could use a FMEA tool. An effective FMEA needs a structured approach with clearly defined objectives and IACS is currently developing standards that can uniformly be applied to marine systems and equipment where an analysis is required. The work currently being undertaken by IACS will identify those systems and machinery that require analysis. For a hazard and failure mode analysis it is necessary to use recognised standards and there are a number of generic standards that can be applied and adapted for analysis of a hybrid system:

i.   IEC 61882, Hazard and operability studies (HAZOP) studies,
ii. IEC 60812, Analysis techniques for system reliability, application guide, Procedure for failure mode and effects analysis (FMEA).
iii. IEC 61508, Functional safety of electrical/electronic/ programmable electronic safety-related systems.

The assessment analysis processes for safety and reliability need to identify defined objectives under system functionality and capability matching. These two issues are concerned with system performance rather than compliance with a prescriptive requirement in a standard. The importance of performance and integration of systems that are related to safety and reliability is now recognised and the assessment tools now available offer such means. Formal Safety Assessment (FSA) is recognised by the IMO as being an important part of a process for developing requirements for marine regulations. IMO has approved Guidelines for Formal Safety Assessment (FSA) for use in the IMO rule-making process (MSC/Circ.1023/MEPC/ Circ.392).

Further reliability and optimization can be done by using stochastic and simulation tools [13, 14]. The development of requirements for fuel cells in the marine environment power plant application could usefully recognize the benefits of adopting a goal-based approach. In order to determine the power supply capacity and system architectural arrangements required and to give specific requirements for services

—

195

that affect the propulsion and safety of the vessel the various services are grouped under a number of headings.

It is clear that stand alone alternative energy system may not survive until more tests has been performed on available technology. Hybrid system represents the next in line for energy system optimization for low pollution and high efficiency. However it is recommended that the system configuration, control switching, intermittent requirement should be designed and developed based on risk before their deployment. The use of solar, hydrogen, wind, wave and Waste base hybrid also promise more environmental and cost saving for future hybrid. Each source has their unique strength where science needs to tap more to hybrid them with each other and with existing system.

Energy, environment, economic and efficiency and safety are the main technology driver today. Issue of energy and environment has been address. Problem associated with choice of energy system in the face of current environmental challenges has been discussed. The chapter also discussed Standards and issues that are applicable to marine power generation systems. Alternative methods of assessment that can be applied to technology for which the current standards do not fit a recognized design and operating scenario and matter of lessons learnt from experience and from failures need to be understood before using alternative methods.

Thus, solar energy has been existing for a long time, different parties have done various research programs on to solar energy and hydrogen energy in different ways, a lot have been achieved in alternative energy technology. The state of the planet, surrounded with issue of energy pollutant shows current need for development of reliable production of alternative energy, since, previous work has shown lack of reliability on standalone system. Incorporating risk based DSS scheme for hybrid system that integrate conventional system with new system could bring a break through to counter problem associated with production of alternative energy. Previous regulatory work for system design has been prescriptive by nature. Performance based standards that make use of alternative methods of assessment for safety and reliability of component design, manufacture and testing is recommended for hybrid alternative energy system installation.

References

1. M. J. Grubb, 'The integration of renewable electricity sources', Energy Policy, 1991, Vol. 19, No. 7, pp 670-688
2. F. R. McLarnon, E. J. Cairns, 'Energy Storage', Review of Energy, Vol. 14, pp 241-271
3. Yun J, Back N and Yu C, 2001, An Overview of R&D in the Field of Solar Building and System in Korea, *Journal of Korea ArchitectureaInstitute,* Chungnam, Deajeon.
4. The European Commission, "Energy technology—the next steps. Summary findings from the ATLAS project", December 1997.
5. IMO marine environmental protection committee 44th session available at: http: *www.imo.org/meeting/44.html,* 2000
6. H. Lund, P. A. Ostergaard, "Electric grid and heat planning scenarios with centralised and distributed sources of conventional, CHP and wind generation", Energy 25, 2000, pp 299-312
7. D. McEvoy, D. C. Gibbs, J. W. S. Longhurst, "City-regions and the development of sustainable energy supply systems", Int. J. Energy Res. 2000, 24, pp 215-237.
8. Henningsen, R.F. *Study of Greenhouse Gas Emissions from Ships.* Final report to the International Maritime Organization. MARINTEK, Trondheim, Norway, 2000.
9. Ronald O' Rouske, 2006 NAVY ship propulsion Technologies, Congressional Research Service Reportfor Congress
10. S. Rozakis, P. G. Soldatos, G. Papadakis, S. Kyritsis, D. Papantonis, 'Evaluation of an integrated renewable energy system for electricity generation in rural areas', Energy Policy, 1997, Vol. 25, No. 3, pp 337-347
11. G. C. Seeling-Hochmuth, 'A combined optimisation concept for the design and operation strategy of hybrid-PV energy systems', Solar Energy, 1997, Vol. 61, No. 2, pp 77-87.
12. J.O. Flower, An Experimental integrated switched reluctance propulsion unit: design, construction and preliminary results, Trans IMarE, 1996, Vol 108, part 2, pp 127—140.
13. R. Chedid, S. Rahman, 'Unit sizing and control of hybrid wind-solar power systems', IEEE Transactions on Energy Conversion, 1997, Vol. 12, No. 1, pp 79-85

14. International Maritime Organization (IMO). Amendments to the Guidelines for Formal Safety Assessment (FSA) for Use in the IMO Rule Making Process. MSC—MEPC.2/Circ 5 (MSC/ Circ.1023—MEPC/Cir. 2006. Yun J, Back N and Yu C, 2003, Design and Analysis of KIER Zero Energy Solar House, *Proc. ISES Solar World Congress*, Goteborg, Sweden.

CHAPTER 6

Cost Benefit of risk analysis for marine system

"Truth is what stands the test of experience"
Albert Einstein

Summary

Inland water transportation project is considered today as one of the mitigation option available for humanity to curb carbon footage. Collision in inland water transportation represents the biggest treat to inland water transportation, its occurrence is very infrequent but his occurrence has grave consequence that makes its avoidance a very imperative factor. The nature of the threat of collision can be worrisome, as they can lead to loss of life, damage to environment, disruption of operation, injuries, instantaneous and point form release of harmful substance to water, air and soil and longtime ecological impact. However, the development of complex system like inland water transportation and collision avoidance system also needs to meet economic sustainability for decision requirement related to collision. This makes analysing and quantifying occurrence scenarios, consequence of accident very imperative for reliable and sustainable design for exercise of technocrat stewardship of safety and safeguard of environmental. This chapter presents the cost benefit analysis for risk control option for required for operational, societal and technological change decision for sustainable inland water transportation

—

199

system. The chapter presents the result of predictive cost for collision aversion for in River Langat waterways development.

6.1 Introduction

Collision risk is a product of the probability of the physical event occurrence as well as losses that include damage, loss of life and economic losses. Accident These accidents represent a risk because they expose vessel owners and operators as well as the public to the possibility of losses such as vessel, cargo damage, injuries, loss of life, environmental damage, and obstruction of waterways. Collision accident scenarios carry heavy consequence, thus its occurrence is infrequence. Complete risk and reliability modelling require frequency estimation, consequence quantification, uncertainties and cost benefit analysis of the holistic system [1, 2, 3].

Like the frequency and consequence analysis, collision cost data are hard to come by, however, whatever little data's that is available should be made meaningful as much as possible through available tools especially system based predictive tools required for necessary mitigation decision for sustainable and reliable waterways. Inherently, accident data for waterway are few that make probabilistic and stochastic methods the best preliminary method to analyze the risk in waterways. Other information relating channel vessel and environment employed in the risk process, lacking information about the distribution of transits during the year, or about the joint distribution of ship size, flag particular and environmental conditions become derivative from probabilistic and stochastic estimation in the model. Result from such model could further be enhanced through simulation methods as required. This chapter discusses cost benefit analysis to support risk control option for waterways collision risk aversion model. [4, 11].

6.2 Background

The case study considered for this study is Langat River, 220m long navigable inland water that has been considered underutilized due to lack of use of the water up to its capacity. Personal communication, survey and river cruise on Langat River revealed that collision remain the main threat of the waterways despite less traffic in the waterways.

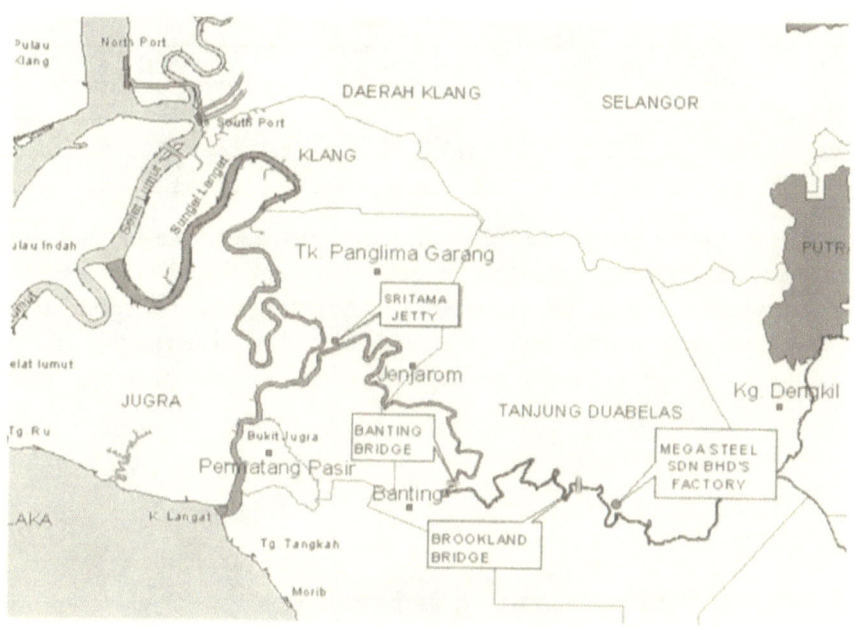

Figure 6. 1: Langat map

Risk and reliability collision aversion model leading to cost control option for sustainable development of the Langat waterways is important, this include cost benefit analysis for decision support based on safety and conservation [6]. Data related to historical accidents, transits, and environmental conditions were collected. Barge and tug of capacity 5000T and 2000T are currently plying this waterway at draft of 9 and 15 respectively. Safety associated with small craft is not taken into account. Collisions (including contact between two vessels and between a vessel and a fixed structure), causes of collision are linked to navigation system failure, mechanical failure and vessel motion failure are considered in this risk worked towards sustainable, reliable and efficient use of the river for transportation [5, 6]. Figure 6.1 shows figurative data of channel width parameter required for damage analysis.

Table 6.1 shows waterways parameters, estuarine data: Sungai Langat: 135.7 km => North Estuary 44.2 km, South Estuary: 9.9 km,

Table 6. 1: Waterway parameter

Basic Maneuvering Lane	1.5B
Addition for cross wind (less 15 knots)	0.0B
Addition for cross current (negligible <0.2 knots)	0.0B

—

Addition for bank suction clearance	1.0B
Addition for aids to navigation (Excellent)	0.0B
Addition for cargo hazard (medium)	0.0B
Channel width for Inland Waterways, (B= Beam of the ship)	2B=53 m
At Bend, Channel width	3.0B = 64 m

Vessel width parameter plays a very important role in collision scenario and potential damage. Vessel movement for the case under consideration currently has no vessel separation system. However, there is traffic movement from both inbound and outbound navigation in the channel. The same type of barge size is considered for the estimation work.

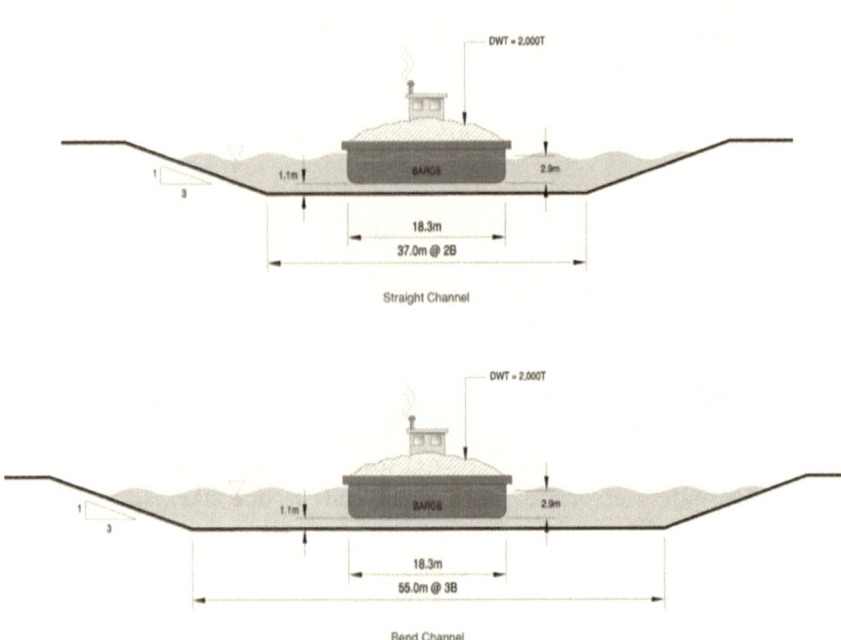

Figure 6. 2: Langat vessel particular

Figure 6. 3: Accidents at Langat

6.3 Safety and Environmental Risk for IWT

Risk and reliability based model aim to develop innovative methods and tools to assess operational, accidental and catastrophic scenarios. It requires accounting for the human element, and integrates them as required into the design environment. Risk based design entails the systematic integration of risk analysis in the design process. It target safety and environment risk prevention and reduction as a design objective. To pursue this activity effectively, an integrated design environment to facilitate and support a holistic risk approach to ship and channel design is needed. A total risk approaches which enable appropriate trade off for advanced sustainable decision making. Waterways accident falls under scenario of collision, fire and explosion, flooding, grounding. Collision carries the highest percentage, more frequent it is cause by (Cahill, 1983):

i. Loss of propulsion
ii. Loss of navigation system
iii. Loss of mooring function and
iv. Loss of Other accident from the ship or waterways

Risk based design entails the systematic risk analysis in the design process targeting risk preventive reduction. It facilitates support for total risk approach to ship and waterways design. Integrated risk based system design requires the availability of tools to predict the safety, performance and system components as well as integration and hybridisation of safety

—
203

element and system lifecycle phases. Figure 6.4 shows the conceptual components of risk and reliability based design.

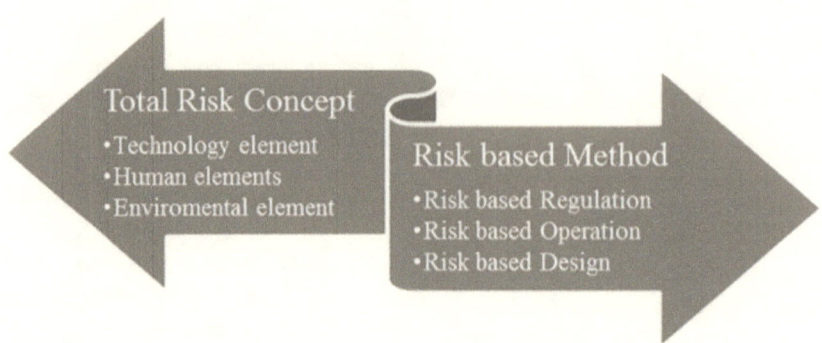

Figure 6. 4: Risk based modelling concept

Therefore, it becomes imperative to develop, refine, verify, validate reliable model through effective methods and tools. The risk process begins with definition of risk which stands for the measure of the frequency and severity of consequence of an unwanted event (damage, energy, oil spill). Frequency at which potential undesirable event occurs is expressed as events per unit time, often per year. The frequency can be determined from historical data. However, it is quite inherent that event that don't happen often attract severe consequence and such event are better analyzed through risk based and reliability model.

Risk (R) = Probability (P) X Consequence (C)   6.01

Incidents are unwanted events that may or may not apply to accidents. Necessary measures should be taken according to magnitude of event and required speed of response should be given. Accidents are unwanted events that have either immediate or delayed consequences. Immediate consequences variables include injuries, loss of life, property damage, and persons in peril. Point form consequences variables could apply to further loss of life, environmental damage and financial costs. The earlier stage of the process involves finding the cause of risk, level of impact, destination and putting a barrier by all mean in the pathway. Risk work process targets the following:

i.   Cause of risk and risk assessment, this involve system description, identifying the risk associated with the system, assessing them and organising them in degree or matrix. IWT risk can be as a result of the following:

a) Root cause: Inadequate operator knowledge, skills or abilities, or the lack of a safety management system in an organization.

b) Immediate cause: failure to apply basic knowledge, skills, or abilities, or an operator impaired by drugs or alcohol.

c) Situation causal factor: Number of participants, time, planning, volatility environmental factors, congestion and time of day risk associated with the system.

d) Organization causal factor: Organization type, regulatory environment, organizational experience management type, changes, system redundancy, system incident, accident, history, individual, team training and safety management system.

ii. Risk analysis and reduction process, this involve analytic work through deterministic and probabilistic method that strengthen can reliability in system. Reduction process that targets initial risk reduction at design stage, risk reduction after design in operation and separate analysis for residual risk for uncertainty as well as human reliability factor.

Uncertainty risk in complex systems can have its roots in a number of factors ranging from performance, new technology usage, human error as well as organizational cultures. They may support risk taking, or fail to sufficiently encourage risk aversion. To deal with difficulties of uncertainty risk migration in marine system dynamic, risk analysis models can be used to capture the system complex issues, as well as the patterns of risk migration.

Historical analyses of system performance are important to establish system performance benchmarks that can identify patterns of triggering events, this may require long periods of time to develop and detect. Assessments of the role of human and organizational error, and its impact on levels of risk in the system, are critical in distributed, large scale dynamic systems like IWT couple with associated limited physical oversight. Effective risk assessments and analysis required three elements highlighted in the relation below.

Risk modeling = Framework + Models + Process   6.02

–

Reliability based verification and validation of system in risk analysis should be followed with creation of database and identification of novel technologies required for implementation of sustainable system.

6.4 Risk Framework

Risk framework provides system description, risk identification, criticality, ranking, impact, possible mitigation and high level objective to provide system with what will make it reliable. The framework development involves risk identification which requires developing understanding the manner in which accidents, their initiating events and their consequences occur. This includes assessment of representation of system and all linkage associated risk related to system functionality and regulatory impact (Akita et al, 1972). Risk framework should be developed to provide effective and sound risk assessment and analysis. The process requires accuracy, balance, and information that meet high scientific standards of measurement. The information should meet requirement to get the science right and getting the right science. The process requires targeting interest of stakeholder including members of the port and waterway community, public officials, regulators and scientists. Transparency and community participation helps ask the right questions of the science and remain important input to the risk process, it help checks the plausibility of assumptions and ensures that synthesis is both balanced and informative. Employment of quantitative analysis with required insertion of scientific and natural requirements provide analytical process to estimate risk levels, and evaluating whether various measures for risk are reduction are effective.

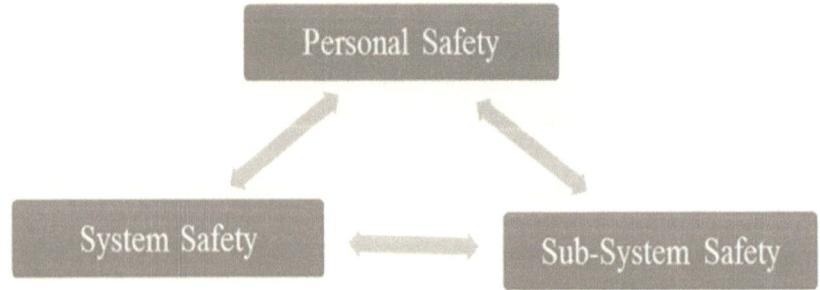

Figure 6. 5: System risk areas

6.5 Safety and Environmental Risk and Reliability Model (SERM)

There is various risk and reliability tools available for risk based methods that fall under quantitative and qualitative analysis. Choice of best methods for reliability objective depends on data availability, system type and purpose. However employment of hybrid of methods of selected tool can always give the best of what is expect of system reliability and reduced risk. Figure 6.6 shows generic risk model flowchart.

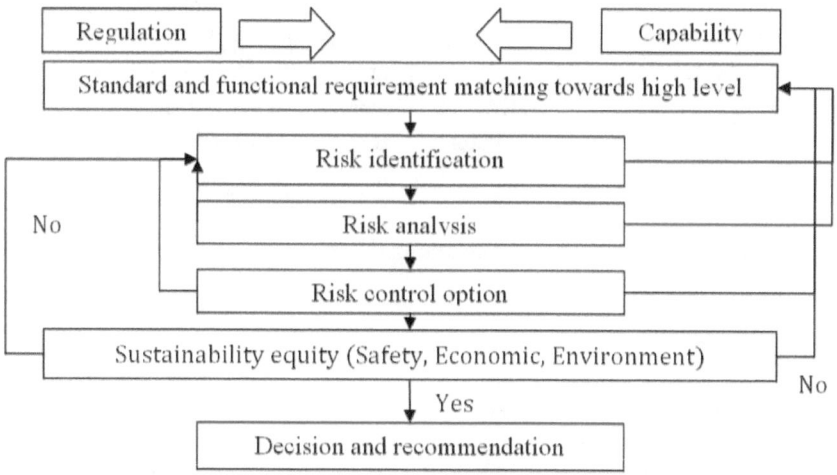

Figure 6. 6: Risk and Reliability model flowcharts

6.6 SERM Process

SERM intend to address risks over the entire life of the complex system like IWT system where the risks are high or the potential for risk reduction is greatest. SERM address quantitatively, accident frequency and consequence of IWT. Other risk and reliability components including human reliability assessment which is recommended to be carried out separately as part of integrated risk process. Other waterways and vessel requirement factors that are considered in SERM model are:

i.   Construction
ii. Towing operations and abandonment of ship
iii. Installation, hook-up and commissioning
iv. Development and major modifications

Integrated risk based method combined various technique as required in a process. Table 6.2 shows available risk based design for techniques. This can be applied for each level of risk. Each level can be complimented by applying causal analysis (system linkage), expert analysis (expert rating), and organizational analysis (Community participation) in the risk process. Figure 6.5 shows stakes holder that should be considered in risk process.

Table 6. 2: Risk based design techniques

Process	Suitable techniques
HAZID	HAZOP, What if analysis, FMEA, FMECA
Risk analysis	Frequency, consequence, FTA, ETA
Risk evaluation	Influence diagram, decision analysis
Risk control option	Regulatory, economic, environmental, function elements matching and iteration
Cost benefit analysis	ICAF, Net Benefit
Human reliability	Simulation/ Probabilistic
Uncertainty	
Risk monitoring	

Technically, the process of risk and reliability study involves the following four areas:

i. System definition of high goal objective: This requires defining the waterways by capturing gap between system functionality and standards. The scope of work for safely and environment risk and reliability analysis should define the boundaries for the study. Identifying which activities are included and which are excluded, and which phases of the system's life are to deal with.

ii. Qualitative hazard identification and assessment: It involve hazard identification through qualitative review and assessment of possible accidents that may occur, based on previous accident as well as experience or judgment of system users where necessary. Though, using selective and appropriate technique depends on the range, magnitude of hazards and indicates appropriate mitigation measures.

iii. Quantitative hazard frequency and consequence analysis: once the hazards have been identified and assessed qualitatively. Frequency analysis involves estimation of how likely it is for the accidents to occur. The frequencies are usually obtained from analysis of previous accident experience, or by some form of analytic modeling employed in this thesis. In parallel with the frequency

—

analysis and consequence modeling evaluates the resulting effects of the accidents, their impact on personnel, equipment and structures, the environment or business.

iv. Risk acceptability, sustainability and evaluation: Is the yardsticks to indicate whether the risks are acceptable, in order to make some other judgment about their significance. This begins by introducing non technical issues of risk acceptability and decision making. In order to make the risks acceptable. The benefits from these measures can be evaluated by iterative process of the risk analysis. The economic costs of the measures can be compared with their risk benefits using cost benefit analysis leading to results of risk based analysis. This input necessities to the design or ongoing safety management of the installation, to meet goal and objectives of the study.

The process of risk work can further be broken down into the following elements:

i. Definition and problem identification: functionality and standards requirement, gap identification and setting high level objective for proactive method.

ii. Hazard and consequence identification: risk ranking and setting acceptance standards for the risks.

iii. Analysing the likelihood's of occurrence: and risks of possible events, normally express in rate of accident per year.

iv. Analyzing consequences: Impact, lose of life, damage to infrastructure, economic loss, penalty, injury, oil spill, environmental degradation.

v. Evaluation of uncertainty: other channel complexity and human reliability analysis.

vi. Risk control option (RCO) and risk control measure (RCM): devising or confirming arrangements to prevent or mitigate the events, and respond to them if they do occur.

vii. Sustainability and risk acceptability criteria: Checking that the residual risks for acceptable established RCO and RCM sustainability analysis.

viii. Reliability based model verification and validation: statistical software, triangulation, iteration.

ix. Recommendation for implementation: Implement, establishing performance standards to verify that the arrangements are working satisfactorily and continuous monitoring, reviewing and auditing the arrangements

—

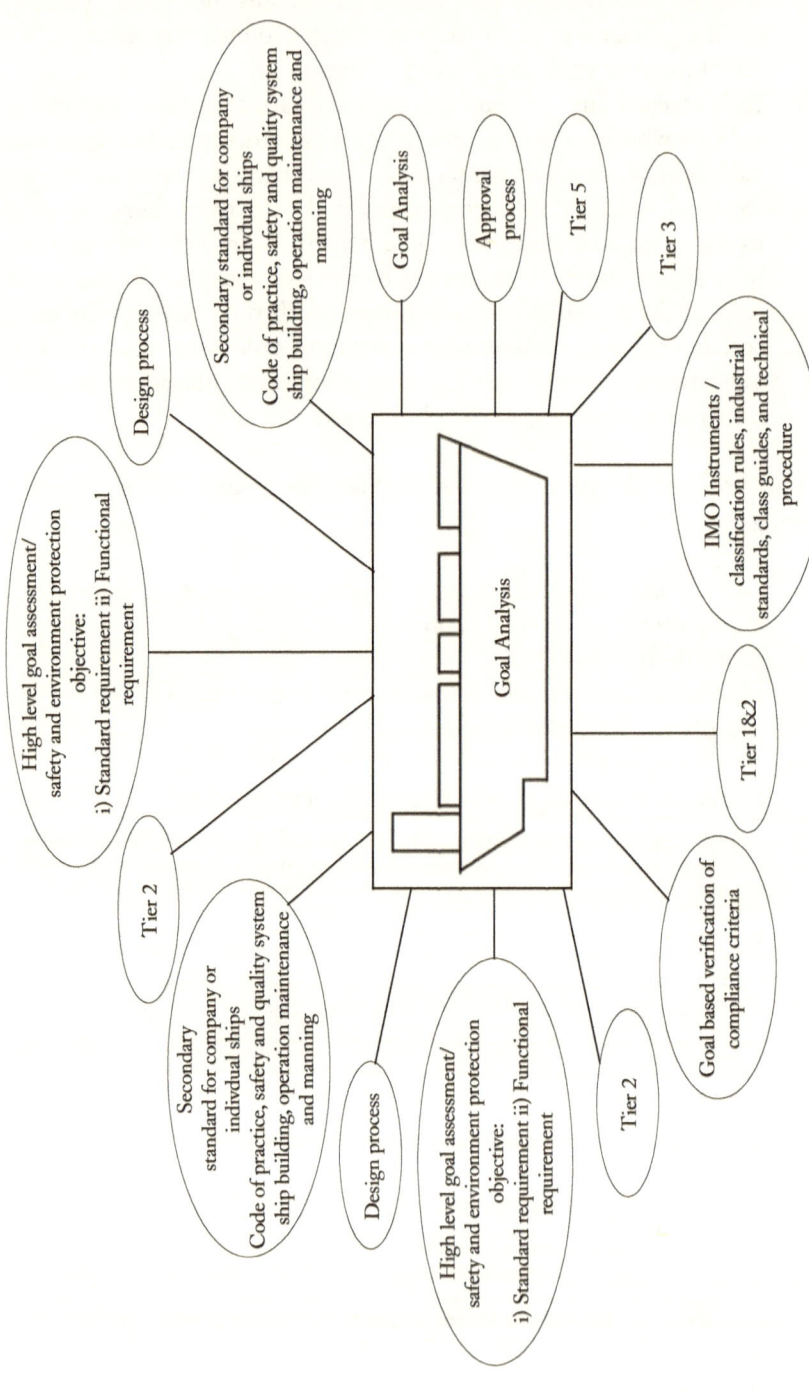

Figure 6. 7: Ships and waterways interest group that are concern about safety

Employment of these benefit provide a rational. Formal environmental protection structure and process for decision support guidance and monitoring about safety issues. The scope of sustainable risk based design under consideration involves stochastic, analytic and predictive process work leading to avoidance the harms in waterways. Figure 6.8 shows block diagram of SERM components for IWT.Safety and Environmental Risk and Reliability Model (SERM) for IWT required having clear definition of the following issues:

i.   Personnel, attendance
ii. Identify activities
iii. Vessel accidents including passing vessel accident, crossing, random
iv. Vessel location and waterway geography on station and in transit to shore.
v.   Impairment of safety functions through determination of likelihood of loss of key safety functions lifeboats, propulsion temporary refuge being made ineffectiveness by an accident.
vi. Risk of fatalities, hazard or loss of life through measure of harm to people and sickness.
vii. Property damage through estimation of the cost of clean-up and property replacement.
viii. Business interruption through estimation of cost of delays in production.
ix. Environmental pollution may be measured as quantities of oil spilled onto the shore, or as likelihood's of defined categories of environmental impact or damage to infrastructures.

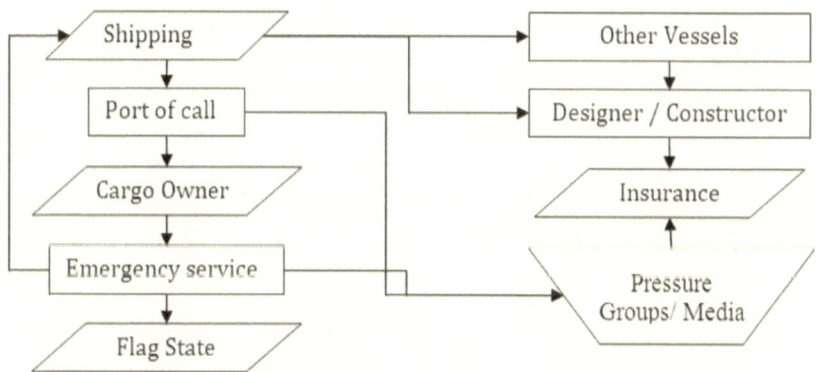

Figure 6. 8: IWT safety and environmental risk model components block diagram

Allowance should be made to introduce new issue defining the boundary in the port from time to time. The choice of appropriate types of risk tool required for the model depend on the objectives, criteria and parameter that are to be used. Many offshore risk based design model consider loss of life or impairment of safety functions. There is also much focus on comprehensive evaluation of acceptability and cost benefit that address all the risk components. Figure 6.9 shows the risk and reliability model combined process diagram. The analysis is a purely technical risk analysis. When the frequencies and consequences of each modelled event have been estimated, they can be combined to form measures of overall risk including damage, loss of life or propulsion, oil spill. Various forms of risk presentation may be used. Risk to life is often expressed in two complementary forms. The risk experienced by an individual person and societal risk. The risk experienced by the whole group of people exposed to the hazard (damage or oil spill). Accident and incident are required to be prevented not to happen at all. The consequence of no safety is a result of compromise to safety leading to unforgettable loses and environmental catastrophic. Past engineering work has involved dealing with accident issues in reactive manner. System failure and unbearable environmental problem call for new proactive ways that account for equity requirement for human, technology and environment interaction. The whole risk assessment and analysis process starts with system description, functionality and regulatory determination and this is followed by analysis of:

i. Fact gathering for understanding of contribution factor
ii. Fact analysis of check consistency of accident history
iii. Conclusion drawing about causation and contributing factor
iv. Countermeasure and recommendation for prevention of accident
v. Most risk based methods define risk as:

Risk = Probability (Pa) x Consequence (Ca) 6.03 or in a more elaborate expression risk can be defined as:

Risk = Threat x Vulnerability x
{direct (short-term) consequences + (broad) Consequences}   6.04

In risk analysis, serenity and probability of adverse consequence hazard are deal with through systematic process that quantitatively measure, perceive risk and value of system using input from all concerned waterway users and experts.

—

Risk can also be expressed as:

Risk = Hazard x Exposure   6.05

Where hazard is anything that can cause harm (e.g. chemicals, electricity, Natural disasters), while exposure is an estimate on probability that certain toxicity will be realized. Severity may be measured by No. of people affected, monetary loss, equipment downtime and area affected by nature of credible accident. Risk management is the evaluation of alternative risk reduction measures and the implementation of those that appear cost effective where:

Zero discharge or negative damage = Zero risk   6.06

The risk and reliability model subsystem in this thesis focus on the following identified four risks assessment and analysis application areas that cover hybrid use of technique ranging from qualitative to qualitative analysis (John, 2000): i. Failure Modes Identification Qualitative Approaches ii. Index Prioritisation Approaches iii. Portfolio Risk Assessment Approaches, and iv. Detailed Quantitative Risk Assessment Approaches.

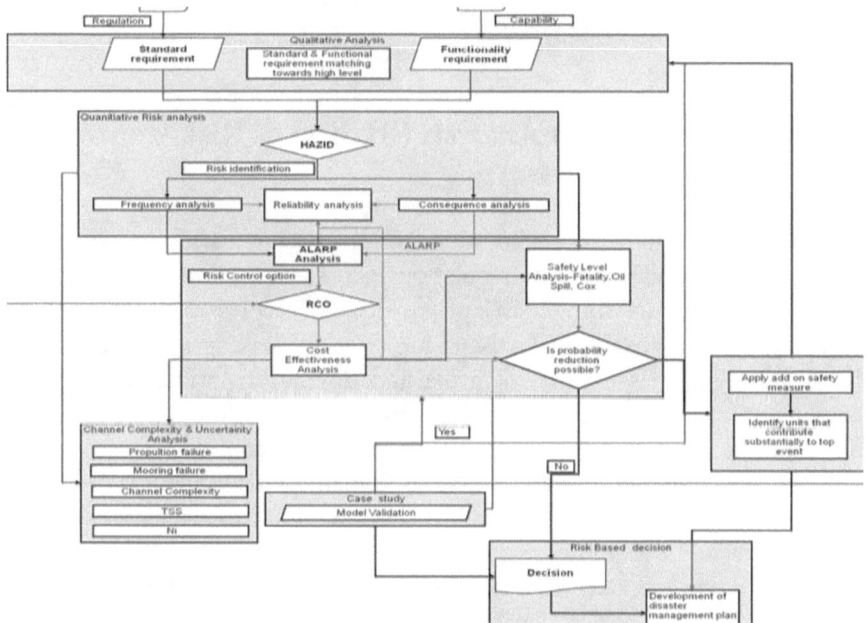

Figure 6. 9: Risk based model process map

6.7 Reliability and Validation Analysis:

System reliability could be determined through the following analysis:

i.   Standard Deviation: Accident means, variance and standard deviation from normal distribution
ii.   Stochastic Analysis: Accident average and projection rates per year calculation can be reliability projection for the model. Poison distribution, standards distribution for and binomial distribution could be analyzed for required prediction and system capability. Poison distribution involves the likelihood of observing k event in time interval T is poison distribution.
iii. Comparing the model behaviour apply to other rivers of relative profile and vessel particular.
iv.   Triangulating analysis of sum of probability of failure from subsystem level failure analysis
v.   System improvement, for example Traffic Separation Scheme (TSS) Implementation effectiveness, could achieve reduction in head collision. This can be done through integration of normal distribution along width of the waterways and subsequent implementation frequency model.
vi.   Comparing the model behaviour applied to other rivers of relative profile and vessel particular

6.8 Risk Cost Benefit Analysis (RCBA) & Risk Control option Model Process

RCBA is use to deduce mitigation, options selection and proposed need for technology, reliability, new regulations and sustainability required to be modeled for effective mitigation options. RCBA involves quantification of cost effectiveness that provides basis for decision making about identified RCO. This includes the net or gross and discounting values for cost of equipment, redesign and construction, documentation, training, inspection maintenance drills, auditing, regulation, reduced commercial used and operational limitation (speed, loads). Benefit could include reduced probability of fatality, injuries, serenity, negative effects on health, severity of pollution and economic losses. Identified types of cost and benefits for each risk control option according to RCBA for the entities which are influenced by each option can be deduced. And also identification of the cost effectiveness expressed in terms of cost per unit risk reduction.

6.9 Risk Cost Option (RCO) and Cost Effective Analysis (CEA)

Risk control measures are used to group risk into a limited number of well practical regulatory and capability options. Risk Control Option (RCO) aimed to achieve (David, 1996):

i.   Preventive: reduce probability of occurrence
ii. Mitigation: reduce severity of consequence

RCO could follow the following generic approach:

i.   General approach: controlling the likelihood of initiation of accidents. be effective in preventing several different accident sequences; and
ii. Distributed approach: control of escalation of accidents and the possibility of influencing the later stages of escalation of other unrelated, accidents.

The economic benefit and risk reduction ascribed to each risk control options is be based on the event trees developed during the risk analysis and on considerations on which accident scenarios would be affected. Estimates on expected downtime and repair costs in case of accidents should be based on statistics from shipyards or responsible government institution for repair or construction. This CBA is then followed by assessment of the control options as a function of their effectiveness against risk reduction. In estimating RCO, the following are taken into consideration:

i.   DALY (Disability Adjusted Life Years) or QALY (Quality Adjusted Life Years)
ii.   LQI (Life Quality Index)
iii. GCAF (Gross Cost of Averting a Fatality)
iv.   NCAF (Net Cost of Averting a Fatality)

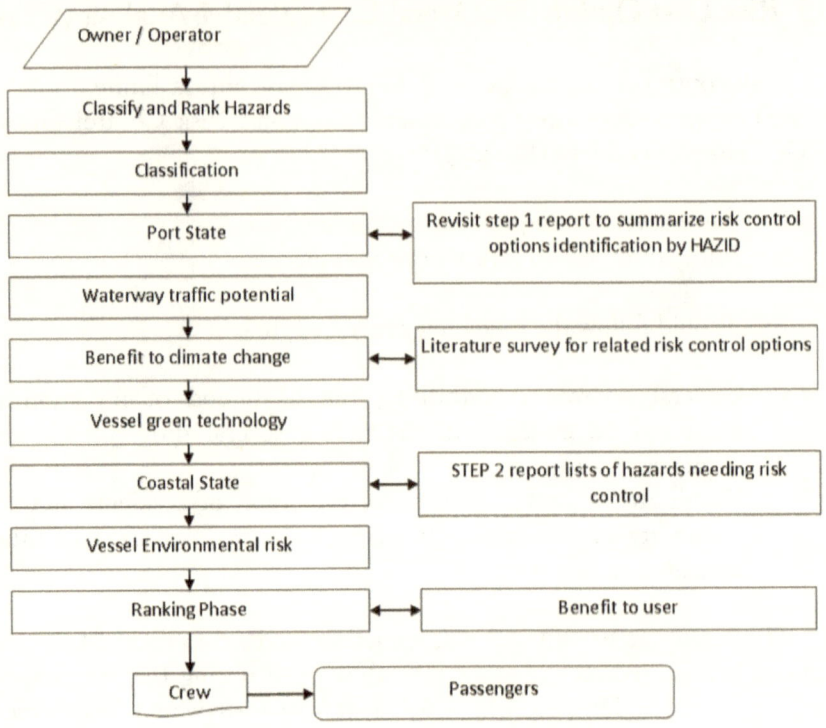

Figure 6. 10: RCO and CBA model

The common criteria used for estimating the cost effectiveness of risk reduction measures are NCAF and GCAF which can be calculated with the following equation:

$$Gross\ CAF = \frac{\Delta C}{\Delta R}$$

6.07

$$GCAF = \frac{Cost-Benefit}{Reduction}$$

16.08

$$NET\ CAF = \frac{\Delta C - \Delta B}{\Delta R}$$

6.09

$$NCAF = GCAF - Change\ in\ Benefit$$

6.10

$$ICAF = \frac{NCAF}{R}$$

6.11

Where ΔR is Reduction in annual fatality rate, ΔR is Risk reduction in term of averted number of fatality implied by the risk control option. ΔR is also economic benefit resulting from implementing the risk control option

NCAF and GCAF depend on the following criteria:
i. Observation of the willingness to pay to avert a fatality;
ii. Observation of past decisions and the costs involved with them;
iii. Consideration of societal indicators such as the Life Quality Index (LQI).

In RCO, It is important to address the following:
i. Primary cause or accident scenario, number of accident
ii. Number of losses, number of life loss per accident
iii. Cost of fatality per accident, average total cost per accident

Cost per unit risk reduction, $\text{CURR} = \dfrac{\text{Cost} - \text{Benefit}}{\text{Reduction}} = \dfrac{\text{NPV}}{\text{Benefit}}$ 6.12

Where 50 minor injuries = 10 serious injuries = 1 life = property or damage = loss or degradation of environment.

6.10 Net Present Value (NPV)

The NPV can be calculated from:

$$NPV = \sum_{t=1}^{n} [(C_t + B_t)(1 + (1+r)^{-t}]$$

6.13

or

$$R[\frac{(1+i)^n - 1}{i.(1+i)^n}]$$

6.14

Where: t = Time horizon for assessment, starting in year 1, Number of year in vessel life time, B = the sum of benefit in period, r = the discount rate per period, Ct—sum of cost in period.

The estimated risk is represented by:

R_0 = Accident frequency Na or P (Number of ships per year) x
    Consequence C x (Cost of damage per accident)
    Risk after implementation of safety measure.   6.15

R_1 = Accident frequency P (Number of ships per year)
    x Consequence C x (Cost of damage per accident).   6.16

Benefit of reduced risk (R) = R_0—R_1 6.17

NPV of the benefit for estimated risk and implemented safety measure is calculated and ratio of cost of C to benefit B is compared and expected to be < 1.

6.11 Implied Cost of Averting Fatality (ICAF)

ICAF represent estimation of benefit of avoiding damage or fatality. It plays important role in cost benefit analysis of risk. ICAF can be estimated using the following means (DnV, 2005):

Ronold Life quality index (L) = $\gamma^w . \varepsilon^{1-w}$

                                                                6.18

Where: L = life quality index, γ = Gross domestic product per person per year ε = Life expectancy (year), w = Proportion of life spent in economic activities in developing countries is approximately 1/8.

$$\frac{\Delta \varepsilon}{\varepsilon} > \frac{\Delta \gamma}{\gamma} . \frac{w}{1-w}$$

                                                                6.19

Where $\Delta \varepsilon = \frac{\varepsilon}{2} = \frac{1}{2}$ of life expectancy, largest change in GDP,

$$|\Delta \gamma|_{max} = \frac{-y(1-w)}{2w}$$

Optimal acceptable ICAF $|\Delta \gamma|_{max}$ $\Delta \varepsilon = \frac{\gamma - \varepsilon}{4} . \frac{1-w}{w}$ 6.20

—
218

Where, Social cost: $NC^{(1+i)^t}$, t<6000, Social cost = $NC^{(1+i)^{6000}}$, t>6000, N = number of injuries or fatalities, C = Cost of damage per day depends on types and countries, I = daily rate of interest, T = Duration of damage or sick leave in day, 6000 days is equivalent with a fatality, DNV = US$ 3 million = cost effective ICAF rate = 2GBP million = developed country.

6.12 Damage or Loss of Life Quantification

Ship collision is rare and independent random event in time. The event can be considered as poison events where time to first occurrence is exponentially distributed (Emietal, 1997).

$$f(t) = \gamma \, e^{-\gamma t}$$

6.21

Where: γ = Annual rate of exceeding of consequence energy capacity, t = the time to the future loss

$$C_0 = C_f \int_0^T e^{-it} . \gamma^{-dt} \, dt$$

6.22

$$C_0 = \frac{C_f \cdot \gamma}{i + \gamma}$$

6.23

Total cost C_t = present value of future cost (C_0) + Cost of protective measure (C_c)

$$C_t = C_0 + C_c$$

6.24

In prioritizing alternative under (RCBA), it is important to address the following: Available concept, consequence energy capacity (MJ), Return for accidence T (Year) as well as:

$$Annual \ rate \ of \ excedance = (\gamma = \tfrac{1}{\tau}.(-))$$

6.25

And C_o, C_c, and C_t (US$)

6.26

The cost effective risk reduction measures should be sought in all areas. It is represented by followed:

$$Acceptable\ quotient = \frac{Benefit}{Risk/Cost}$$

6.27

6.13 Cost Benefit of RCO for Optimal Energy Capacity

Benefit of River alignment, implementation of (TSS), dual propulsion shaft, VTS, AIS or bridge clearance can be quantified into cost of safety and protection of environment (IMO, MSC 2001).

Energy capacity that can be absorbed catastrophic energy without collapse is 10^6 MJ @ Catastrophic collapse. Where, economic loss of bridge structure + loss for use of infrastructure = US\$ 2 X 10^8, I= rate of interest = 3 % per year.

$$Present\ value\ of\ loss => C_0 = C_f . e^{-it}$$

6.28

Where: C_o-Present value of future loss, C_f—loss at deficient time in future or future loss in present monetary unit (notificated), i =real interest rate of future loss, t= the time to the future lost. At random variable and occurrence time, loss is stochastic:

$$C_o = C_f . \int_o^T e^{-it} f(t) dt$$

6.29

E (t) = the expected value function, f (t) = probability density function for time t), the time to the occurrence of catastrophic collapse. $C_f = 2x10^8$ Where: and I at 3% = 0.03

A relationship between safety, environmental risk components and ageing/failure factor of IWT should be expected achievement of the analysis. Based on this relationship a generic model based on system engineering, scientific and stochastic analysis useful for decision support on damage, maintenance and management of reliable, safety and environmental friendly IWT can be developed. The model may also have future allowance for conversion into software for universal and smart IWT monitoring and management.

6.14 Sustainability Analysis

Sustainability is defined as development work that meets needs of the present generation without compromising the ability of the future generations to meet their own needs. It requires balancing work between technical, developments, economic, community participation, information sharing, environment and safety. Suitability principle calls on all fields of human activities to review and adjust the way things are done. At its 21st session in February 2001, the UNEP governing council adopted a decision to investigate the feasibility of a "Global Assessment of the State of the Marine Environment" UNEP GC Decision 21/13. Figure 6.11 shows typical sustainability analysis graph that can be for decision support for the system.

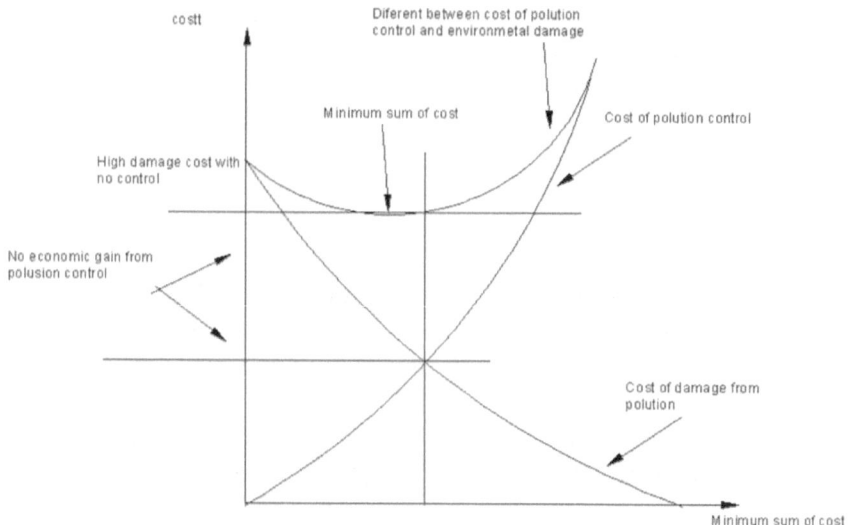

Figure 6. 11: Sustainability balance

It was recognised that a global marine assessment (GMA) was needed and feasible. The scope of an assessment process was outlined. It was agreed that GMA activities should include cost, technical, environmental, socio economic considerations, together with the relevant work, approaches and experience of national, regional and global so that the project in question can touch more of living thing and environmental consideration and impacts (UNEP, 2008). The case presented in this thesis is examples of the daily practice of environmental management. Being anything but a straightforward application towards reliable management

—

work derived from sound decision making which involve judgment and criteria. In order to interpret, assess pros and cons, measure the different interests possibly confronted, and finally choose the best decision according to social interests and the common good.

6.15 Decision Making

Decision making involves discussion of hazard and associated risks, review of RCO that keep ALARP curve in acceptable region, compare and rank RCO based on associated cost and benefit. It also involves specification of recommendation for decision makers towards beyond compliance preparedness. And rulemaking tools for regulatory bodies towards measures and contribution for sustainable system design. RCO provide measures, outcome of objective comparison of alternative option, and subsequent contribution recommendation for sustainable implementation need of the system intactness, the planet and the right of future generation.

6.16 Case application

6.16.1 ALARP risk curve for changing

Figure 6.12a shows accident consequence accident energy and accident occurrence frequency against. The maximum speed that should not be exceed is at Fa = 2.06E-06, and Ea = 72E—MJ. For the current speed of 3 knot, Ea= 14MJ, and Ea = 3.8e-5.Figure 6.12b show the limit definition for risk when the graph of accident frequency, drifting and powered collision are combined. At the point E1 is 64MJ, E2 is 82MJ and 2.8 E-6.

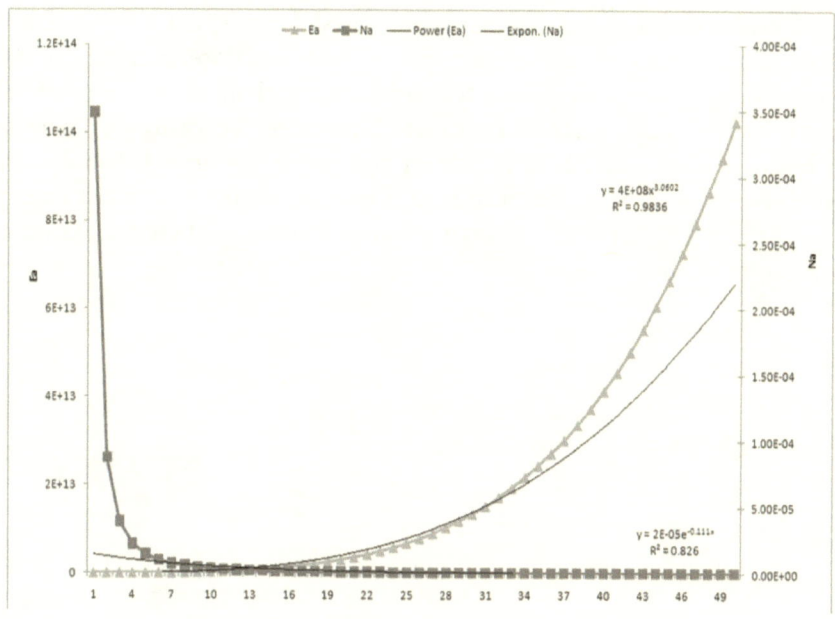

a.

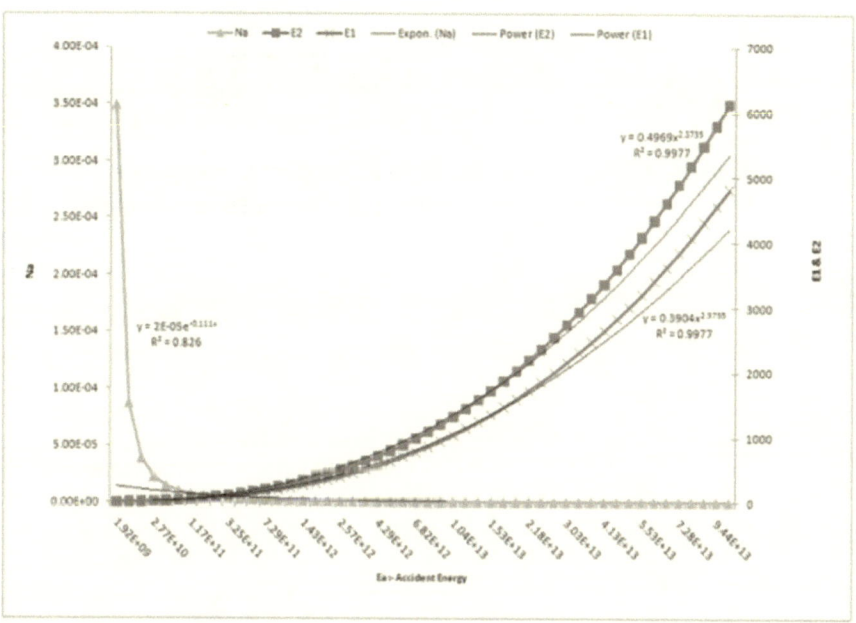

b.

Figure 6. 12: Accident energy Vs consequence energy

Current width of the channel is only good for the next 100 years; it is unacceptable for maritime industry risk acceptability criteria of per 10, 000 years. With this, risk definition for width of the channel can be provided (refer Figure 6.13). Current beam of ship should be maintain, increase in beam of ship could lead to rise in risk (refer Figure 6.13). Ship risk becomes higher at large beam of ship. Figure 6.14 shows accident energy Vs Accident Frequency @ Changing beam of ship, risk become higher at large beam of ship.

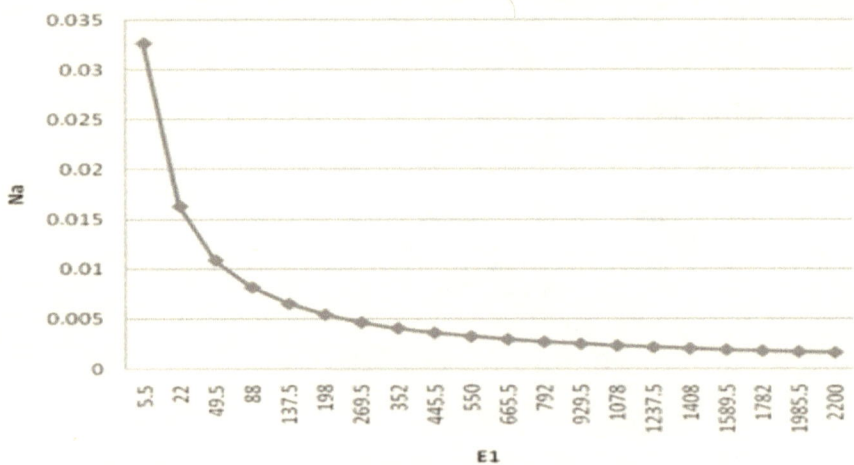

Figure 6. 13: Accident Frequency and Accident Consequence@ different energy @ different width of channel and beam of ship.

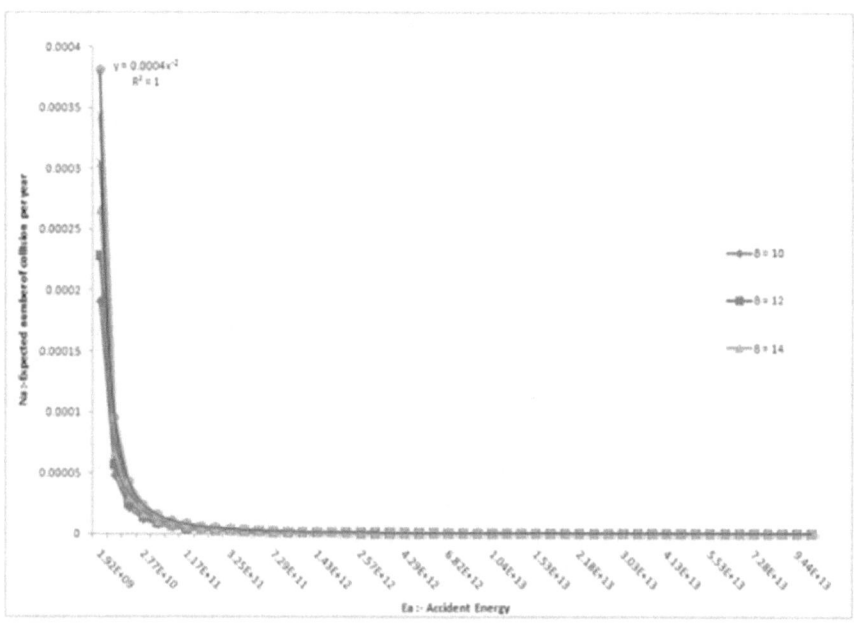

Figure 6. 14: Accident frequency and consequence energy at changing beam of ship

Figure 6.12 present curve fit quite well, and the graph show region where V, Nm, W, and B is unacceptable.

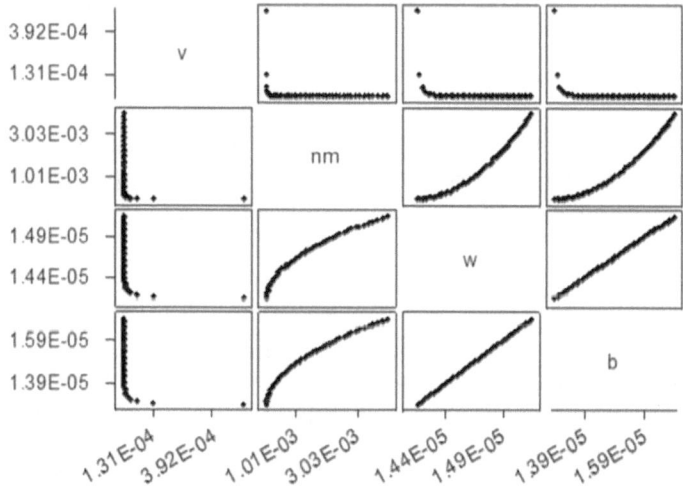

Figure 6. 15: Matrix plot for accident frequency Vs consequence, curve fit quite well with lower region of ALARP

6.16.2 Identified Risk Control Options to Reduce the Collision, Grounding and Contact

RCO for each collision situation has to be more clearly defined. In order to identify new RCO, generated result from the analysis of frequency and consequence, cost and benefit is weighted. It is important to support this with expert rating to contribute to possible risk prioritisation control options for IWT of on Sungai Langat. The descriptions of the major hazards and corresponding risk control options from the hazard identification and the results from the risk analysis which are summarized could be presented to the group of experts for further validation. From the risk study, prioritized RCO that were selected for further evaluation in terms of cost effectiveness assessment are discussed in the next paragraph. Even thus this research is about collision, the impact to collision is not far from contact and grounding collision scenario. Therefore some of the measure that will be taken could benefit curbing accident from contact and grounding. The main RCO'S are:

i.   Improved navigational safety.
ii. Redundant propulsion system: two shaft lines.
iii. Required maintenance plan for critical items as well design requirement for increase double hull width, increase double bottom depth or increase hull strength.
iv. Human factor and human reliability is quite critical in risk work, it need to be done separately.

6.16.3 RCO 1: Improved Navigational Safety

Improved navigational safety can be achieved in a number of different ways. From various identified risk control options, five cost effective risk control options for navigation improvement that could potentially reduce the frequency of collision and grounding which are:

i.   TSS
ii. ECDIS (Electronic Chart Display and Information System)
iii. track control system
iv. AIS (Automatic Identification System) integration with radar
v.   Improved bridge design

The risk control options related to navigational safety in the list above might be promising alternatives for Langat River. The cost effectiveness

—

of implementing this measure for Langat River is evaluated in this study. Hence, the risk control option for improved navigational safety is defined as implementation of one or more of the above alternatives. Electronics Chart Display Information System (ECDIS) is a navigational aid that can be an alternative to nautical paper charts and publications. It is use to plan and display the ship's route as well as to plot and monitor ship positions throughout the voyage. It is capable to determining a vessel's position in real time. The use of this aid would reduce the navigation officer's workload compared to using paper charts.

Although not mandatory, the use of ECDIS onboard current vessels is quite common and the proposed risk control option is to make ECDIS mandatory for vessel plying Langat River. Automatic Identification System (AIS) is designed to send and receive information in relation to a vessel's identity, course and cargo. Current regulations require such information to be presented in an AIS display and the minimum requirements are three lines of data consisting basic information of a selected target. The AIS can also be integrated with the radar's Automatic Radar Plotting Aid (ARPA) function so that the additional data is available in the radar display. Benefits from the AIS—ARPA integration will include enhanced situation and traffic condition awareness and improved ability to make early decisions based on real-time data for avoidance of potential collisions. Other improvement of radar performance and navigation can be achieved in the following ways:

i. Identification of ship names for radar targets and clarify the target's intentions
ii. Detection of targets which are in radar shadow areas and extends radar range
iii. Easier access of AIS information
iv. Accounting for the ship's rate of turn to predict accurately the target's path

Installation of valve control radar can reduce risk of oil spill due to overfilling, malfunction of a valve or human failure among other causes. The levels of storage tanks on board must be continuously monitored since overfilling or product discharge on deck could have consequences for human life and for property.

6.16.4 RCO 2: Redundant Propulsion System

Machinery failure is a significant causal factor in collision accident. Collision can be avoided if the ships had redundant propulsion or steering systems. The redundant propulsion and steering system must ensure that, irrespective of the ship's loading condition, when a failure in a propulsion or steering system occurs:

i.   The maneuverability of the ship can be maintained.
ii.   A minimum speed can be maintained to keep the ship under control.
iii. The ship can maintain operation with a redundant propulsion or steering system so that a vessel can ride out the storm or slow navigation in port.
iv.   The propulsion and steering functions are quickly re established.

Redundant propulsion and steering systems can greatly reduce the risk of vessel disability and subsequent loss of life or cargo and damage to the environment. Operational advantage of redundant propulsion is:

i.   Greater reliability.
ii.   Improved safety levels.
iii. Higher vessel availability.
iv.   Protection of environment

Cost effectiveness assessment for redundant propulsion systems will be achieved by installation of independent engines and two shaft lines. The use of all electric propulsion could be a good advantage for optional navigation mode. This would also have effect on different hull forms compared to ships with single propellers.

6.16.5 RCO 3: Human Capital Development

Discussion with waterways authority revealed that only the captain's qualification and competency is being screened and regulated. It is recommended to institutional screen on certification and competency of all officers on the vessels and to undergo simulation for normalization of behavior. Accident analysis and assessment result and reliability analysis done for watch keeping subsystem analysis using fault thee analysis. Human reliability analysis can best be tackled using DELPHY method recommended by IMO.

This risk control option aims at increasing the bridge team's ability to handle difficult maneuvering tasks and crisis situations by increased

—

use of simulator training. The effect of such training could provide better navigational safety and a reduced risk of collision, grounding and contact events. The simulator training could be specially designed for particular port environments, underwater topography, and particular bridge layouts on specific vessels and would give the participants exercises in handling challenging situations from different positions of the bridge team. Important parts in such exercises might be passage planning, situation awareness and operation during malfunction of critical technical equipment. The risk control option suggested herein goes beyond the basic training requirements defined by IMO's International Convention on Standards of Training, Certification and Watch keeping for Seafarers (STCW), (IMO, 1996).

6.16.6 Sustainability Analysis and Cost Effectiveness of Selected RCOs

Risk Cost Benefit Analysis to deduce and proposed need for new regulations based on mitigation and options selection. RCBA involve quantification of cost effectiveness that provides basis for decision making about RCO identified, this include the net or gross and discounting values. Table 6.3 depicts losses associated with Langat River accidents. Consideration is also given to cost of equipment, redesign and construction, documentation, training, inspection maintenance, auditing, regulation, reduced commercial used, operational limitation like speed and loads. Benefit could include reduced probability of fatality, injuries, serenity, negative effects on health, severity of pollution, economic losses.

Table 6. 3: Cost of Failure for Langat River
(Personal communication (LUAS, 2008))

Primary cause	Number of accident	Number of life loss	Number of life loss per accident	Cost of fatality per accident	Average total cost per accident
Loss of navigation control	2	2	1	N/A	700 K RM
Loss of propulsion	2	2	1	N/A	900 K RM
External factor	2	2	1	N/A	100 K RM

—

Cause from other vessels	2	2	1	N/A	100 K RM
Human error	2	2	1	N/A	200 K RM

The result of risk analysis carried out on WTS demonstrate that the risk is below the ALARP region, there may be no need to conduct cost effectiveness analysis, but the fact that the waterways of that magnitude is underutilized and lack of basic infrastructure and system, regulation and monitoring has contributed to historical increased in yearly accident recorded. This necessitate need to conduct RCO, Cost effectiveness measure and sustainability for future installation of safety and environmental infrastructure.

6.17 Gross CAF

Cost work is model in different way, and translation of quantity is allowed between benefit, damage, oils spill, fatality. For this case, based on estimate on the level of damage, 1 fatality is considered due to frequency of accident. Figure 6.16 shows that at minimum energy of 20MJ, less 700, 000 RM gross costs will be required to avert fatality. Whereas, at 400 MJ energy of impact 2.07 million RM gross cost will be required to avert fatality

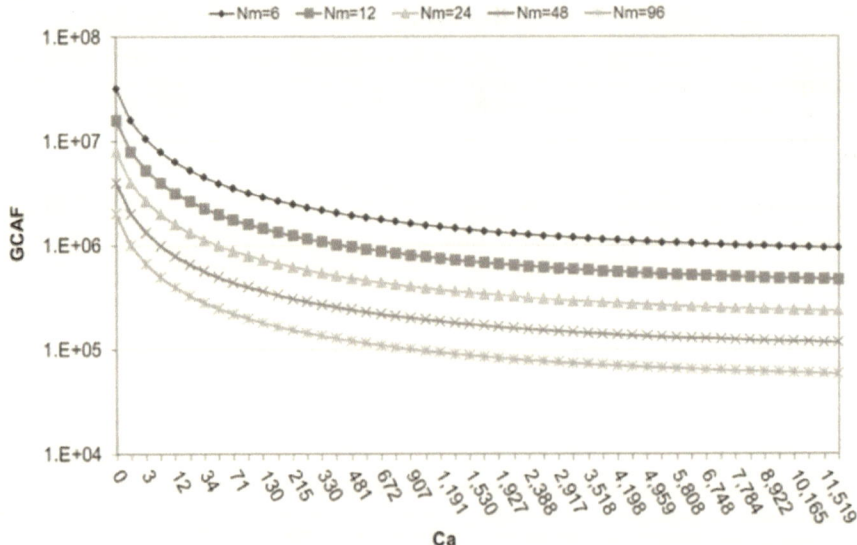

Figure 6. 16: Gross Cost of Averting Fatality vs Accident Consequence

6.18 Net CAF

Figure 6.17 shows that 1.47 million net cost will be required to avert fatality at minimum energy of 20MJ, 2,8 million RM will be required to avert fatality at catastrophic accident energy of 453MJ.

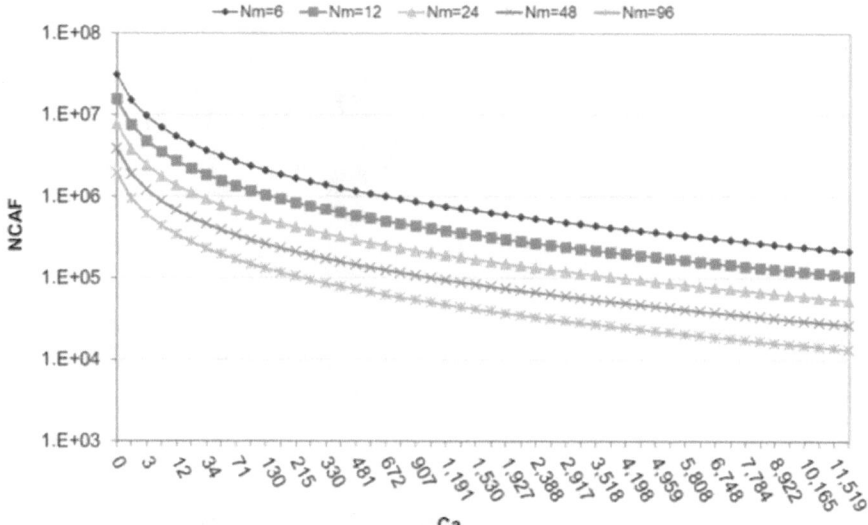

Figure 6. 17: Net Cost of Averting Fatality Accident Consequence

Figure 6.18 depicts the cost of losses per accident causal factors. Propulsion failure carry the highest (RM 2,000,000) follow by loss of navigation function which require about RM 700, 000 and about 400, 000 will be require to fix human error problem. These costs are still acceptable as long as they are less than 3 million.

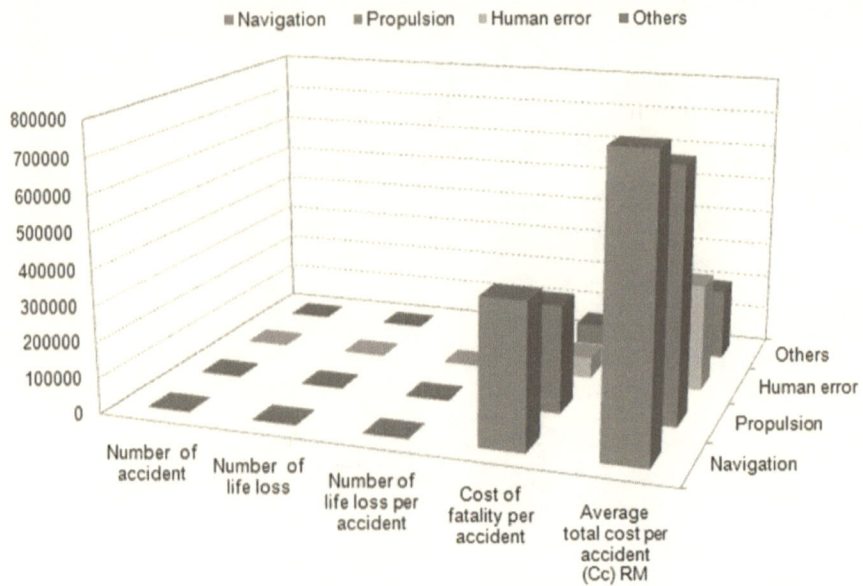

Figure 6. 18: Cost of losses per accident causal factors

All numbers are based on introduction of one RCO. Introduction of more than one RCO will lead to higher NCAF and GCAFs for other RCOs addressing the same risks. High GCAF and NCAF values indicate that the considered RCO is not a cost effective measure. A negative NCAF indicates that the RCO is economically beneficial in itself, For example the costs of implementing the RCO are less than the economical benefit of implementing it. From the Figure, number of accident and loss of life are considered low. According to current practice within IMO and selected criteria for this study, a risk control option will be regarded as cost effective if it is associated with GCAF ≤ USD 3 million or NCAF ≤ USD 3 million. Cost effective measures that can be demonstrated to have a high potential for risk reduction will consequently be recommended for implementation.

6.19 Implied Cost of Averting Fatality ICAF

ICAF represent estimation of benefit of avoiding damage or fatality and it ply important role in cost benefit analysis of risk. This can be estimated using the following means. See Figure 6.19 for ICAF for Langat River. Figure 6.19 shows that 731000 RM will be ICAF required at minimum accident energy released of 20MJ while, 2.53 million will be the ICAF of released energy of 453MJ.

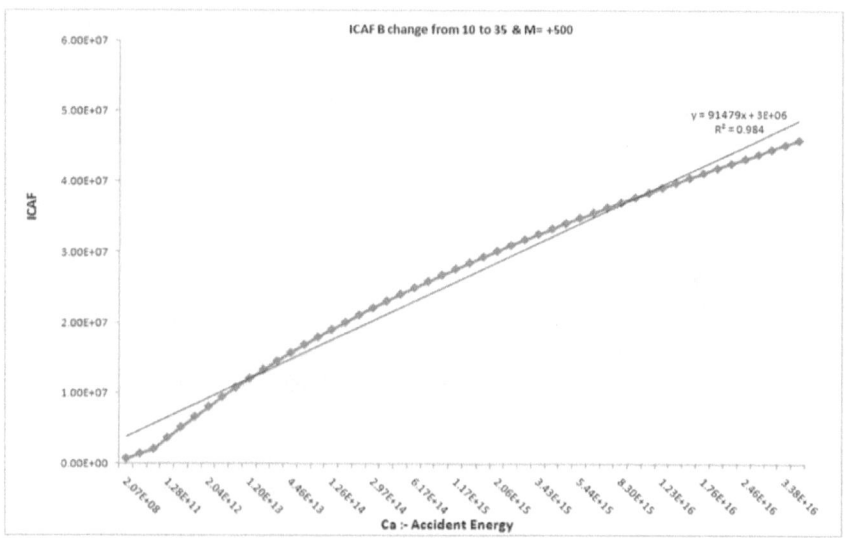

Figure 6. 19: Implied Cost of Averting Collision

6.19.1 Cost Benefit of Damage for Optimal Energy Capacity

E(t) is the expected value function, F(T) : Probability density function for time t (time to the occurrence of catastrophic collapse). Figure 6.20 shows RCO's analysis for total cost of damage. Figure 6.21 shows cost of unit cost risk reduction (CURR), 1.47 million is required to reduce the risk for minimum release of accident energy of 2Omj and 2.85 million for catastrophic accident of 453MJ. Figure 6.22 shows RC0 analysis for alternative option.

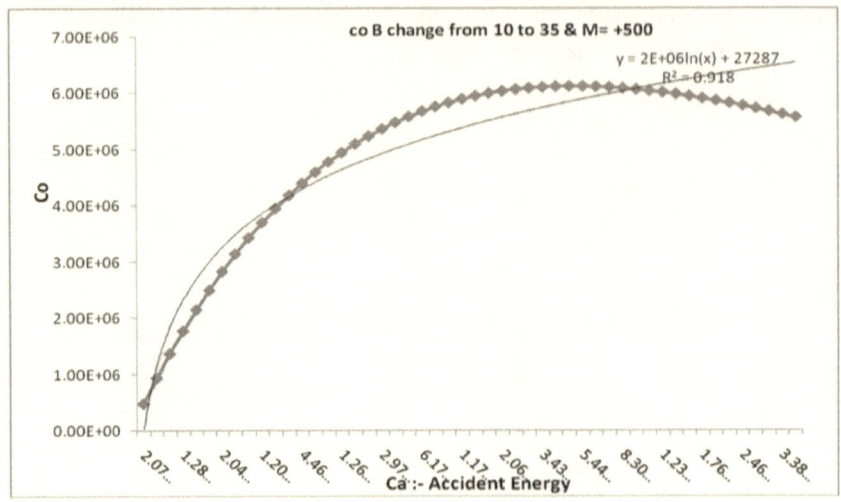

Figure 6. 20: Present value of cost

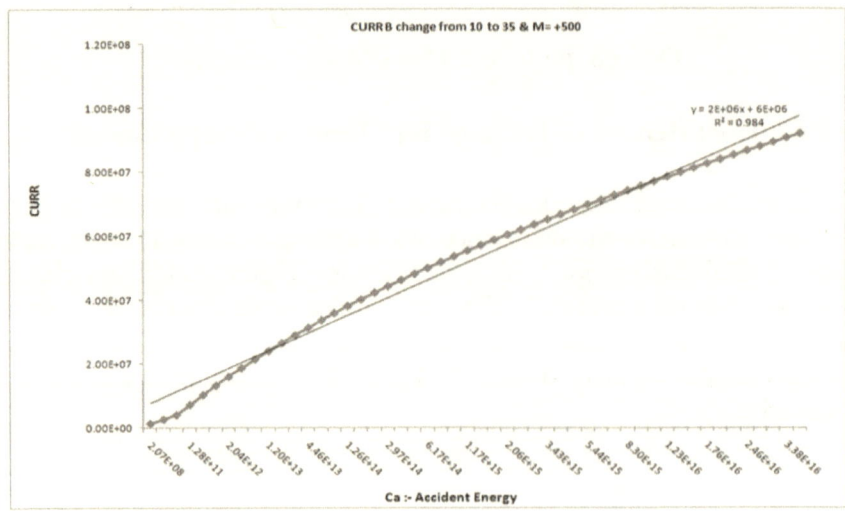

Figure 6. 21: Cost of Unit Risk Reduction

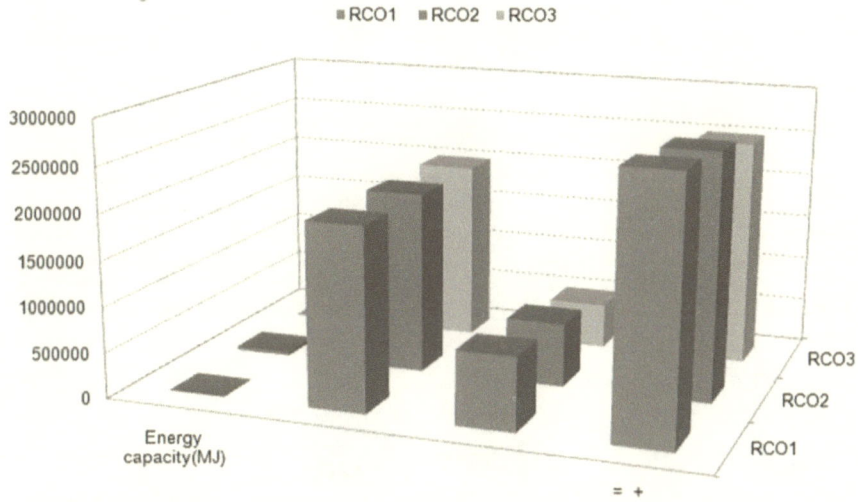

Figure 6. 22: RCO's analysis for total cost of damage

Figure 6.22 show risk cost benefit and sustainability analysis According to recent discussion with Langat River, a decision is already made no pass the bridge over the river. Therefore for Langat River that need to be included in analysis, but benefit could be quantify into cost.

6.20 Sustainability

Figure 6.23 show it cost much more to implement navigation and machineries failure system. The maximum cost is indicated by the point where the total cost (Ct), the present value of loss, and NPV coincide, about RM30 million, where the cost of unit risk reduction still stand at about RM2000, 000. Figure 6.23 shows cross plot of the risk level and optimal cost require for the channel maintenance. From this Figure it is observed by spending more than 50Million, high speed craft or freighter of 35 knot will be able to navigate on Langat River in future. Table 6.4 shows the regression equation derived from the model.

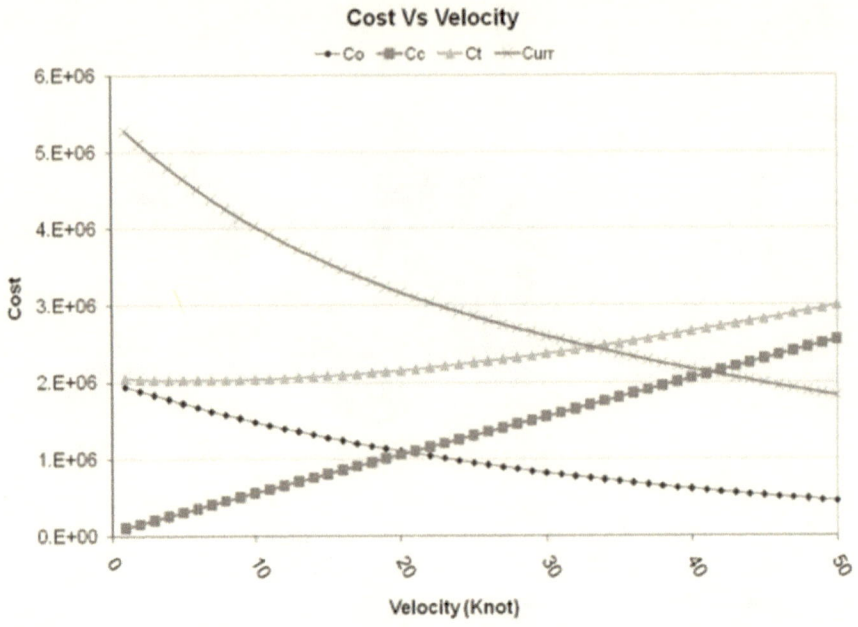

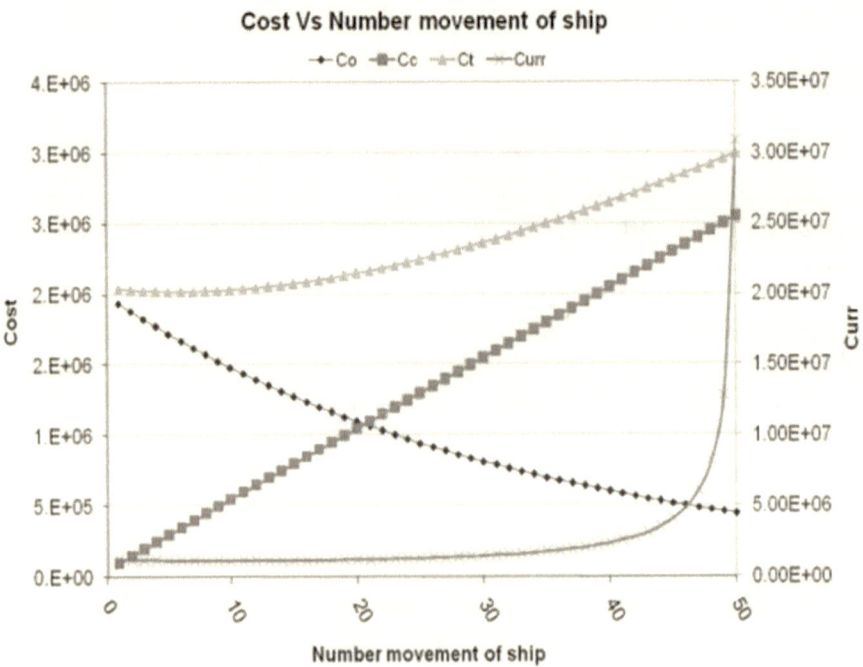

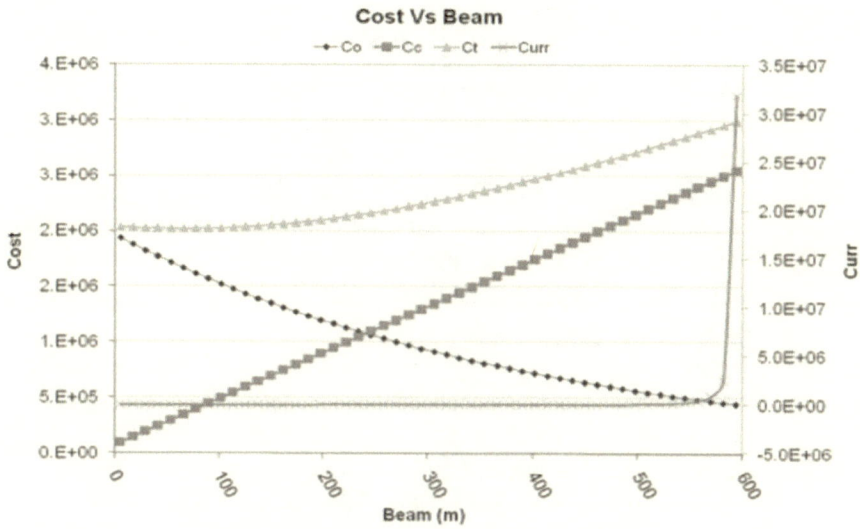

Figure 6. 23: Risk cost benefit analysis

Table 6. 4: Regression equation derived from the model.

Cost Benefit and Sustainability				
CURR	Ca	y = 3E+07ln(x)—3E+07	R² = 0.896	Logarithm
NCAF	Ca	y = 1E+07ln(x)—2E+07	R² = 0.896	Logarithm
ICAFT	Ca	y = 2E+06x + 6E+06	R² = 0.984	Logarithm
GCAF	Ca	y = 2E+07ln(x)—3E+07R² = 0.832	R² = 0.832	Logarithm

References

1. DnV. *Formal Safety Assessment of cruise navigation.* DNV Report No. 2003-0277, Det Norske Veritas, Høvik, Norway. Norway, 2005

2. Kite, Powell, H. L., D. Jin N. M. Patrikalis, J. Jebsen, V. Papakonstantinou. *Formulation of a Model for Ship Transit Risk.* MIT Sea Grant Technical Report. Cambridge, MA. 1996. 96-19.

3. Skjong, R., Vanem, E., Endresen, Ø. Risk. *Evaluation Criteria.* SAFEDOR report D. 2006.

4. Lempert, R. J., S. W. Popper and S. C. Bankes. *Shaping the Next One Hundred Years: New Methods for Quantitative Long-Term Policy Analysis.* RAND:Santa Monica, CA. 2003. pp. 187.

5. Roach, P.J. *Verification and Validation in Computational Science and Engineering.* Hermosa Publishers. Albuquerque. NM, 1998.

6. Coleman, H.W., W.G. Steele, Jr. *Experimentation and Uncertainty Analysis for Engineers.* John Wiley & Sons. 1989.

7. Kitamura, O. FEM approach to the simulation of collision and grounding damage. In Proceedings of 2nd International Conference on *Collision and Grounding of Ships* July 2001, pp. 125-136 (Maritime Engineering, Department of Mechanical Engineering, Technical University of Denmark).

8. Axtell, R., R. Axelrod, J. Epstein and M. D. Cohen. *Aligning Simulation Models: A Case Study and Results.* Computational and Mathematical Organization Theory. 1996. pp123-141.

9. Yacov T. Haimes. *Risk Modeling, Assessment and Management.* John Wiley & Sons, INC. Canada. 1998. pp. 159—187.

10. N. Soares, C. A. P. Teixeira. *Risk Assessment in Maritime Transportation. Reliability Engineering and System Safety.* 74:3, 2001, 299-309.

11. Fujii, Y. and Mizuki, N. Design of VTS system for water with bridges. In *Proceedings of Ship Collision Analysis* (Eds H. Gluver and D. Olsen), 1998, pp. 177-190 (Balkema, Rotterdam).

12. Camm, Jeffrey D. & Evans, James R. *Management Science &Decision Technology.* South-Western College Publishing, 2000.

CHAPTER 7

Human Reliability Analysis (HRA) Emanating from Use of Technology for Ships Navigating within Coastal Area

"Everything that is really great and inspiring is created by the
individual who can labor in freedom"
Albert Einstein

Summary

The traditional approach to the study of human factors in the maritime field involves the analysis of accidents without considering human factors reliability analysis. Main approach being use to analyze human errors are statistical approach, probability theory approach. Another suitable approach to the study of human factors in the maritime industry is the quasi-experimental field study where variations in performance can be observed as a function of natural variations in performance shaping factors. This chapter discus methods of analyzing of modeling human error and human reliability emanating from the use of technology on board ship navigation in coastal water area by using qualitative and quantitative tools. Accident reports from marine department are used as empirical material for quantitative analysis.

The literature on safety is based on common themes of accidents, the influence of human error resulting from technology usage design, accident report from MAIB and interventions information are use for qualitative assessment. Human reliability assessment involves analysis of accident in waterways, emanating from human-technology factors interface. The chapter present enhancement requirement of the methodological issues with previous research study, monitoring and deduce recommendations for technology modification of the human factors necessary to improve maritime safety performance. The practical cased discussed can contribute to rule making, and safety management leading for development of guideline and standards for human reliability risk management for ship navigating within inland and coastal waters.

> "Shipping is perhaps the most international of all the world's great industries and one of the most dangerous."(IMO, 2000)

7.1 Introduction

Humans have relied on oceans, lakes, and rivers to ship goods from one end to another throughout the recorded history. Today, over 90% of the world's cargo is transported by merchant ships due to various reasons; including the fact that it is the cheapest form of transportation. The shipping industry has a fairly good safety record, however maritime accidents have a high potential for catastrophes. Past experiences research report indicate that in the shipping industry around 80% of all accidents are rooted in human error (Fortland, 1996). Safety has been an immense public concern, especially caused in operations risk like: nuclear power generation, nuclear weapons, aviation, chemical/petroleum processing, and marine transportation (Robb et al, 1996).

There are several basic aspects of maritime activity that make it unique. Ships are complex, confined and isolated systems. They are sufficient on energy supply, they have a limited manpower and resources, and they have a limited response capacity to face emergencies. These particular characteristics made maritime trade a risky activity, where a fault in navigation or in usual port operations can leads to injuries or lost of life, to damage of property and sometimes irreparable damage to maritime environment (Portela, 2005).

The main purpose of navigation is a safe and efficient sailing of the ship between diverse points which require steering the ship movement on a planned trajectory. The accident occurrence of factors affecting the ship's movement causes its limitation. The ship's sailing should be safe, so that it does not cause navigation accident. A navigational accident is an unwanted occurrence which can cause loss of life or health, loss or damage to ship or cargo, the pollution of natural environment, damage to the hydro-technological structure, and economical loss due to delay in the port and associated activities.

This chapter presents the result of application of quantitative and empirical approach to analyze human factors for reliability assessment of ship navigating in coastal water. The approach is based upon a theoretical framework of well-known models. It is possible to benefit from the causal connection between human errors and accidents. It is possible to get a fast and easy access to empirical material from historical data that are analised, are compared to field studies or laboratory studies. The use of modest approach to standard developments through qualitative and quantitative risk assessment and analysis methods is necessary for HRA is performed. Quantitative risk assessment and analysis for HRA are best analysed using Failure Modes and Effects Analysis (FMEA), Fault Tree Analysis (FTA), Event Tree Analysis (ETA). Computer Relex reliability software is use for quantitative risk. While for the qualitative risk assessment and analysis method, some checklists and safety or review audits is emphasized.

7.2 Past Work

Human factors deal with human abilities and limitations in relation to the design of systems, organizations, tools etc. Important parameters are safety, efficiency and comfort. Human errors and human factors are often studied separately; therefore, the relationship between them is often overlooked. According to Gordon et al (1998), proposes a framework for describing the relationships between underlying human factors and more immediately evident human errors. Gordon categorizes human factors as individual, group, or organizational, following the Rasmussen model "Perceptions on the Concept of Human Error," that categorize human errors as skills-based, rule-based, or knowledge-based.

System-induced errors reflect deficiencies in the way the total system is designed. They include mistakes in designating the numbers and types of personnel, in training, in data resources, in logistics, and in maintenance requirements and support. Design-induced errors result from inadequacies in the design of individual items of equipment. The new equipment characteristics create special difficulties for the operator which substantially increases the potential for error. Operator-induced errors can be traced directly to an incompetency on the part of the individual who makes that error. They include errors resulting from lack of capability, training, skill, motivation, or from fatigue.

Several studies and case reviews have found that organizational factors may be the most critical in considering human factors contributions to marine accidents. At the organizational level, various factors may contribute to an increase in incidents and accidents, including cost-cutting programs and the level of communication between work-sites (Gordon, 1998).

According to the US Coast Guard's (USCG) risk-based decision-making guidelines categorize human error into four categories, which form a matrix: intentional errors, unintentional errors, errors of omission, and errors of commission. An unintentional error is an act committed or omitted accidentally, with no prior thought; therefore, intentional errors have also been referred to as "routine violations". An error of omission occurs when an operator fails to perform a step or task. An error of commission occurs when an operator performs a step or task incorrectly (USCG, 2006). Maritime transportation is a complex socio-technical system formed by four interdependent factors as technology, environment, people and organizational structures. Each of these dimensions has direct or indirect effects on maritime casualties, but failures of human action and judgment have often been seen as an important part of the causes. Growing number of accidents main cause has been attributed specifically to "human error".

Nivolianitou et al (2006), pointed out those technical factors that are more readily resolved than human factors through technological and regulatory "fixes" leaving human-related errors and breakdowns as the most probable cause of industrial accidents. (Hee et al, 1999). Support this theory, noting that structural or technological failures are generally responsible for less than 20% of accidents involving complex systems, and noting that this is "a tribute to technology". By comparison, more than 80% of accidents can be attributed to the "unanticipated actions of people" leading to undesirable outcomes. Hee et al (1999) concluded that human inputs to technological and engineering processes may actually contribute to accident risks from the beginning stages of equipment design.

There are many methods and techniques that have been developed to perform various types of analysis, in areas such as reliability and safety. Several different accident forecasting models and analytical tools have been developed in an attempt to identify root cause errors in human systems and develop preventative preventive measures that intervene at the appropriate level, although the proper categorization of human and organizational errors is critical to this process (Nivolianitou et al, 2006). Quantitative analysis relies on statistical methods and databases that identify the probability and consequence.

—

This objective approach examines the system in greater detail for risk (Robb et al, 1996). Quantitative risk analysis generally provides a more uniform understanding among different individuals, but requires quality data for accurate results. Qualitative risk analysis uses expert opinion to evaluate the probability and consequence. This subjective approach may be sufficient to assess the risk of a marine system (Robb et al, 1996). The qualitative method for risk assessment or analysis is designed for the purpose of enhancing one's awareness of potential problems and can assist one in analysing these risks. A combination of both qualitative and quantitative risk analysis can be used depending on the situation.

Failure Mode and Effect Analysis (FMEA) is another powerful tool used by system safety and reliability engineers/analysts to identify critical parts, functions and components whose failure will lead to undesirable outcomes such as production loss, injury or even an accident. The tool was first proposed by NASA in year 1963 for their obvious reliability requirements. Since then, it has been extensively used as a powerful technique for system safety and reliability analysis of products and processes in wide range of industries—particularly aerospace, nuclear, automotive and medical

The concept of Fault Tree Analysis was originated by Bell Telephone Laboratories as a technique with which to perform a safety evaluation of the Minuteman Launch Control System. Bell engineers discovered that the method used to describe the flow of "correct" logic in data processing equipment could also be used for analyzing the "false" logic which results from component failures. A FTA is useful for understanding the mode of occurrence of an accident logically. Furthermore, given the failure probabilities of system components, the occurrence probability of the top event (TE) can be obtained. Traditionally, it is usually assumed that the basic events within a fault tree are independent of each other and could be represented in terms of probabilistic numbers. With this assumption, quantitative analyses of fault trees are usually performed by considering two cases: (1) fault trees without repeated event, and (2) fault trees with repeated events.

Event tree analysis is a binary form of a decision tree for evaluating the various multiple decision paths in a given problem. ETA appears to have been developed during the WASH-1400 nuclear power plant safety study. The WASH-1400 team realized that a nuclear power plant PRA could be achieved by FTA; however, the resulting fault trees (FTs) would be very large and cumbersome, and they therefore established ETA to condense the analysis into a more manageable picture, while still utilizing FTA.

7.3 Human error and accident waterways

The 21st century shipping industry faces new challenges in term of accident and its consequence. For instance, 25 years ago the average cargo ship would have been manned with a crew of between 40 and 50 (Grech et al, 2002) Today technological advances have contributed to decrease manning, in some cases to just 22 seafarers on a Very Large Crude Carrier (VLCC).

There are two sides to the technological advances. Improvements in ship design and navigation aids have reduced the frequency and severity of shipping incidents. In turn, the reduction of failures in technology has revealed the underlying level of influence of human error in accident causation (Catherine, 2006). The fact that human factors contribute to accidents is generally accepted, but there is no consensus on the importance of this factor. Suggestions regarding the proportion of marine accidents caused by human errors vary from 50% to 90% of the total (Kletz, 1991). The main causes of accidents is shown in Figure 7.1 where first 60% of the total number of claims recorded that human error was the direct cause and further 30% human error is from indirect contributory cause.

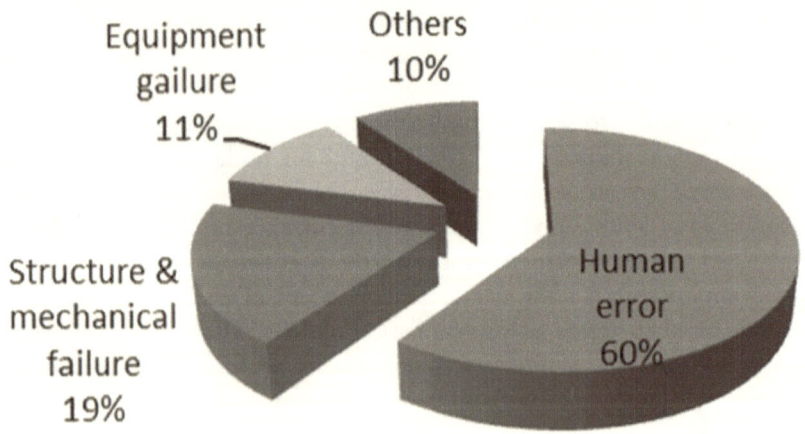

Figure 7. 1: Main Causes of Accidents (Marine department, Malaysia, 2009)

Human factors are based on the acknowledgement that human characteristics and behaviors are intrinsically linked with the functioning of the technology, people, design, building, maintains area, and operation. The human-technology relationship works in both directions. Not only do humans impact the functioning of our technology, but technology can also influence human decisions and actions. Since human factors is triggered by human errors are the main source of risk in maritime activities, it seems interesting to develop methodologies that allow evaluating quantitatively and qualitatively the real incidence of several human factors over maritime accidents happening with the aim of taking human factors into account in properly developed risk management plans (Portela, (2005).

Human reliability assessment required system based approach analysis that makes it easy to determine human factors risk levels through statistical analysis of maritime accidents. The risk of disasters cannot be eliminated, but risks can be reduced by establishing better safety criteria prior to an accident. This chapter present the finding of human reliability analysis that can reduce the probability risk in ship navigation in coastal area by considering the relationship between the human factors and technologies, the cause of the accident from technological influence to human error and deduce solution from the analysis (Ayyub, 2002).

7.4 Methodology

There are a few of qualitatively and quantitatively approach methods available to in order to analyse perform the relationship between human factors and technologies follow analyzing the causes of accidents from technological influence to associated human error and deduce solution from the analysis. The methods include application of the checklists, Failure Modes and Effects Analysis (FMEA), Fault Tree Analysis (FTA) and Event Tree Analysis (ETA).

7.4.1 Checklist

This is a qualitative approach to ensure that the organizations are complying with standard practices. The checklists can be used as a preparation for a port call to avoid unnecessary problems and delays. The checklists may be included in the International Safety Management (ISM) procedures as documentation to checks for maintenance etc. The list can be filled in manually or printout or electronically. The list is qualitatively assessed in correlation between human and technology, and management for operation of ships. Checklist is observed to capture the gaps in the system.

7.4.2 Failure Modes and Effects Analysis (FMEA)

FMEA is a systematic tool for identifying the effects or consequences of a potential product or process failure; and the methods to eliminate or

—

reduce the chance of a failure of occurring. It involves identification of the process functions that has been clearly articulated. It requires preparation of a failure mode analysis and preparation of worksheets by using reliability analysis software, like Relex or isograph. Then identification of the failure modes and describe the effects of those failure modes. This is followed by establishment of a numerical ranking for the severity of the effect in order to identification causes of each failure mode. The occurrence factor and the likelihood of detection are determined. The Risk Priority Numbers (RPN) is determined by a product of the numerical values of Severity, Occurrence and Detection ratings:

RPN = (Severity) x (Probability) x (Detection) Eq.   7.01

Finally, recommendation of action(s) to address potential failures that have a high RPN can be made.

7.4.3 Fault Tree Analysis (FTA)

Fault tree analysis are generally performed graphically using a logical structure of AND and OR gates. There should be only one Top Event and all risk contributing factors must tree down from it. Actual number of failure probabilities, area assigned to the contributing factor. Traditionally, it is usually assumed that the basic events within a fault tree are independent of each other and could be represented in terms of probabilistic numbers. There are five steps involved for the basic Relex FTA, which are to define the undesired event to study. Next, obtain an understanding of the system. Then only construct the fault tree and evaluate it. Lastly, controls for the hazards are identified.

7.4.3 Event Tree Analysis (ETA)

Event tree analysis is based on discrete binary logic, in which an event is in either ON or OFF state that indicate that failure did not happen, failure happened or a component of it has not failed. It is valuable in analyzing the consequences arising from a failure or undesired event. An event tree begins with an initiating event of interest. The consequences of the event are followed through a series of possible paths. Each path is assigned a probability of occurrence and the probability of the various possible outcomes can be calculated from them. THERP (Technique of human error probability) is another recommended alternative approach for HRA

7.5 Accident data

7.5.1 Casualty Statistics

The total losses of all ships and boats during the years 2000—2009 are 289 in number, according to the Malaysian Marine Department data that reported to Port Klang's Vessel Traffic Management System (VTMS) for annual casualty statistics in Peninsular Malaysia which include coastal areas of Peninsular Malaysia and the Straits of Malacca. Total losses in number during the years of 2000—2009 are presented in Figure 7.3 and 7.4 respectively. Sinking or foundering accounts for total losses of almost 50 % of accident compared to collision which accounts for total losses amounting to only 26.30 % as shown in Figure7.2.

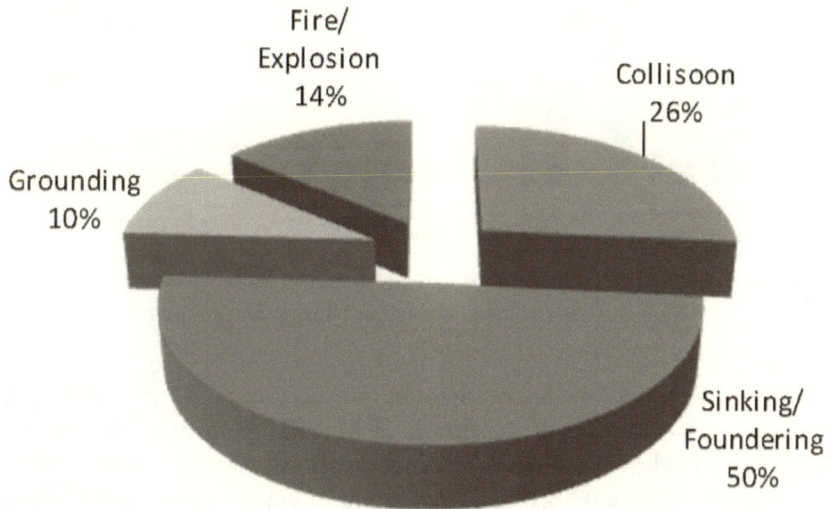

Figure 7. 2: Total Losses of Ships in Number during the Years of 2000—2009

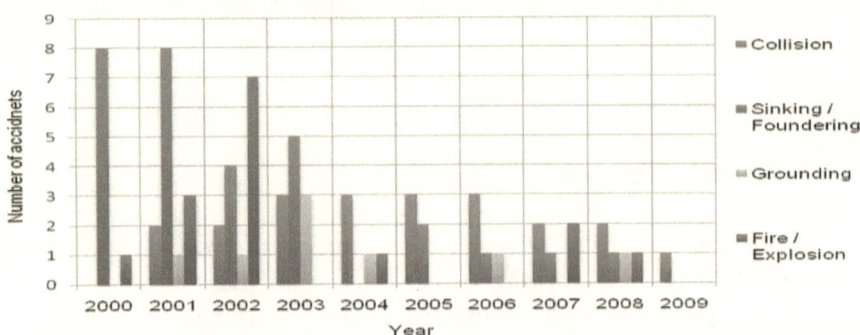

Figure 7. 3: Total of Accident of Ships and Boats
at Malaysia Coastal Areas from Years 2000—2009

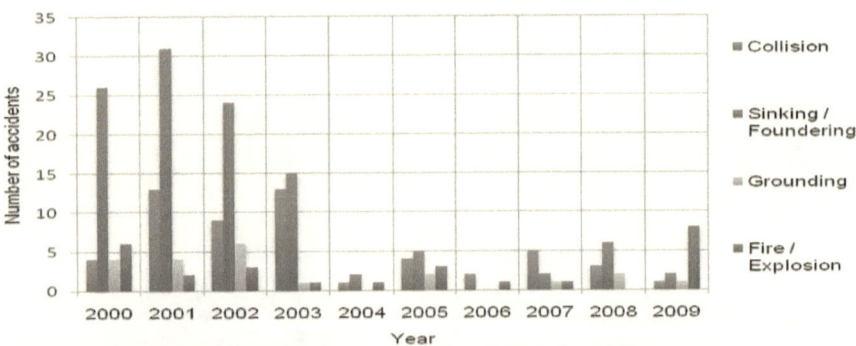

Figure 7. 4: Total of Accident of Ships and Boats at Malacca Straits from Years
2000—2009, (Source: Malaysian Marine Department)

Figure 7.3 and 7.4 4 shows that the total accident of ships and boats in
Malaysia coastal areas and the Straits of Malacca from year 2000 till year
2009 respectively. From Figure 7.3, analysis shows that the highest ranking
of accident which only occurred in average of eight cases at the Malaysia
coastal areas in ten years. From 2000 until 2009, a total of 215 accidents
including sinking or foundering, collision, fire or explosion and grounding
occurred in the Straits of Malacca with an average of 30 accidents per year.
About 353 vessels of all types passes through the Straits of Malacca each
year and 35% of them are oil tankers and this has potential to increase the
discharges of oil in the sea including ballast water, oil, sewage and others
solid wastes. As shown from the bar chart, the probability of an accident to

occur at the Straits of Malacca is high if compared to the Malaysia coastal areas.

This is because Malacca Straits is a golden heritage of the littoral states such as Malaysia, Singapore and Indonesia. It is not only rich in marine resources but is also one of the oldest and busiest shipping lanes in the world. As the years come by, from year 2004 till 2009, there were reducing numbers in maritime casualties which in average of ten accidents per year. In accordance with the International Maritime Organization's (IMO) Rules for vessels navigating through the Straits of Malacca, an under keel clearance of 3.5m is required for shipping safety of navigation and reduction in the risk for an accident to occur.

7.6 Qualitative Risk Assessment and Analysis Method

7.6.1 Checklists

Prior to arrival and departure of vessel in port, checks on operation procedure should be carried out after long ocean passages and before entering restricted coastal areas. The Incident Command System (ICS) Emergency checklist for collision is carried out, actions include switch the VHF to Chanel 16, check for the watch alarm system etc. In the navigational watch checklist, the primary duties of the Officer of the Watch (OOW) are watch keeping, navigation and GMDSS radio watch keeping complying all times with the COLREGS and STCW95. The Officer of the Watch is not allowed to leave the bridge until properly relieved. While in the sole lookout checklist, a sole lookout is allowed only during day light according to the STCW Code. The qualitative exercise is matched closely with guideline given by reference IMO (2002) and FortLand (2004).

The Master, before allowing has to ensure that it is safe to operate with a sole lookout and carefully assessment of the situation, made the state of weather, visibility and traffic density. In addition, basically radar also should be kept running and fully operational at all times. The log books checklist which includes a correct record of the movements and activities of the vessel should be kept in the appropriate log book during the watch. Instructions for the completion of log books should be strictly observed as per respective national regulations and rules. Navigational and radio equipment has to be checked periodically to ensure satisfactory and safe operation.

7.6.2 Failure Modes and Effects Analysis (FMEA)

Analysis is carried out by using Relex FMEA. The following is the Relex FMEA worksheet (Figure 7.5) based on the process: water leaking into vessel. The potential failure modes for the sinking or foundering process includes the failure system of bilge alarm, inadequate of watertight bulkhead, failure of seawater pipe work and the perforation of hull plating. One of the causes to the failure of the bilge alarm system is the electrical failure where the terminal connector block is not suitable for used and the screw terminals provide an opportunity for corrosion resulting in the connection failing. Other than that, as previously mentioned, human factor play a main role, it is also possible that a bilge level alarm that activates every time, vessel motion will be ignored or turned off.

Relex

POTENTIAL
FAILURE MODE AND EFFECTS ANALYSIS
(PROCESS FMEA)

Name:
Process Responsibility:
Key Date:
Core Team:

FMEA Identifier: FMEA2
Page 1 of 1
Prepared By:
FMEA Date (Orig.)　　(Rev.) 3/1/2010

Process Function / Requirements	Potential Failure Mode	Potential Effect(s) of Failure	Sev	Class	Potential Cause(s)/ Mechanism of Failure	Occur	Current Design Controls Prevention	Current Design Controls Detection	Detec	R.P.N.	Recommended Actions	Responsibility & Target Completion Date	Action Results				
													Actions Taken	Sev	Occ	Det	R.P.N.
Sinking / Foundering Water leaking into vessel	Perforation of the hull plating	Hull plating cracked	8			6			4	192	Periodically using ultrasonic thickness measurements			8	5	4	160
	Bilge alarm system failure	Undetectable	9		Electrical failure	6			4	216	Regular manual checks			9	5	4	180
			9		Lack of inspection and maintenance	6			4	216	It should be tested before sailing			9	5	4	180
	Seawater pipework failure	Fatigue cracks	9		Loose fittings	7			4	252	Periodically inspected internally of pipework			9	6	4	216
			9		Pipework corrosion / erosion	7			4	252				9	6	4	216
	Lack of watertight bulkhead	Watertight integrity inadequate	7			6			4	168	Maintenance and inspection of watertight integrity carry out frequently			7	5	4	140

Page #: 1

Print Date:　3/1/2010
Print Time:　4:55 PM

Figure 7. 5: A Relex FMEA Worksheet which Based On the Process: Water Leaking into Vessel

As a result, the Risk Priority Number (RPN) is established as well. The graph of FMEA Risk Level (Figure 7.6) is to view a graph of the failure modes with the highest RPN values. In the graph that is obtained, the highest RPN is 252 for seawater pipe work failure. The pipe failure which led to the flooding was caused by simple sea water corrosion.

—

Piping failures are so far the main cause of all fishing vessel flooding and foundering where the cause has been identified. While for the potential failure modes like bilge alarm system failure and perforation of hull plating has RPN, 216 and 192 respectively. Steel hull plating failures occur frequently. Failure mode of lack of watertight bulkhead has its RPN at 168 where the severity ranks at 7 and occurrence ranking at 6. To have a watertight machinery compartment, the maintenance and inspection of the watertight integrity must be adequate. The outcome of risk outcome is measure using guideline provided by reference (USCG, 2006).

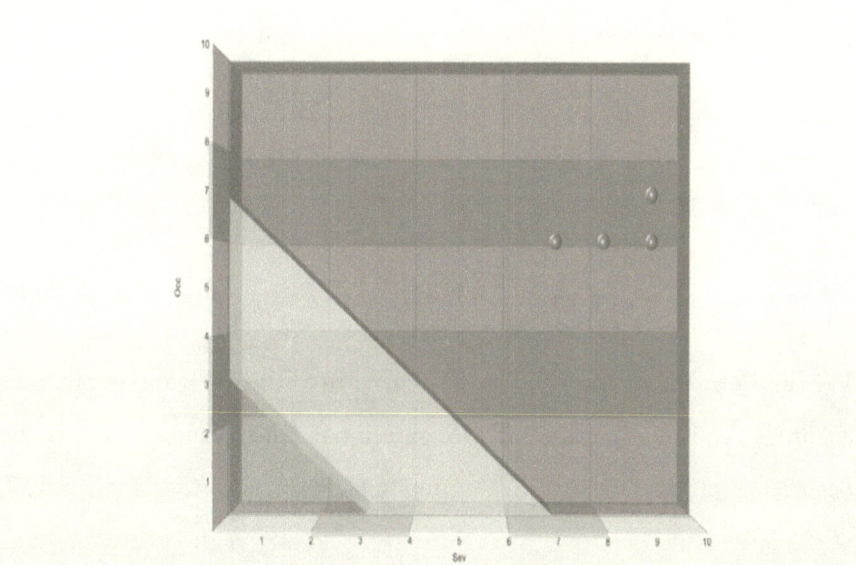

Figure 7. 6: A FMEA Risk Level Based On the Process:
Water Leaking into Vessel

7.6.3 Fault Tree Analysis (FTA)

From accident data it has been identified that sinking or foundering accident has the highest ranking among the other accidents. Thus, it means that a systematic approach must be undertaken to identify all possible causes and their consequences, so that risk can be reduced to a minimum level through appropriate safety measures. By using the fault tree symbols, a simple fault tree for a top event, water leaking into vessel, is shown in Figure 7.7. The occurrence probability of the top event of a fault tree can be calculated when the probabilities of the occurrence of basic fault events are known. This can only be obtained by first calculating the occurrence probability of the resultant (i.e., output) fault events of intermediate by using lower logic gates such as AND and OR. Thus, the probability of occurrence of the number of AND gate output fault event is expressed by:

$$P(x_o) = \Pi_{i=1}^n P(x_i)$$

7.01

Where: $P(\chi_0)$ is the probability of occurrence of the AND gate output fault event, x_0, n is the number of AND gate input fault events. $P(x_i)$ is the occurrence probability of AND gate input fault event x_i; for i = 1, 2, 3, . . ., n

Similarly, the probability of occurrence of the OR gate output fault event is given by:

$$P(y_o) = 1 - \Pi_{i=1}^k \{1 - P(y_i)\}$$

7.02

Where: $P(y_0)$ is the probability of occurrence of the OR gate output fault event, y_0. k is the number of OR gate input fault events. $P(y_i)$ is the occurrence probability of OR gate input fault tree event y_i; for i = 1, 2, 3, . . ., k

By substituting the given probabilities and calculated values of the top event: Water leaking into vessel into the equations. Thus, the probability of occurrence of the top event: water leaking into vessel is 0.1759. In this regard, better training and procedures can help to promote better communications and coordination on and between vessels. Poor equipment design was a causal factor in one-third of major marine casualties. A proper consideration by equipment designers to factors such as how a given piece of equipment will support the mariner's tasks and how it can be integrate into the entire equipment "suite" used by the mariner can be a very helpful step.

The issue of "inexperience, lack of knowledge and training" is concerned with poor understanding of mariners of how automation works or under what conditions it was designed to work effectively.

—

Consequently, mariners sometimes commit errors in using the equipment. Poor inspection and maintenance is another important issue because poor maintenance of navigational equipments can lead to dangerous work environments and lack of backup systems that need to carry out emergency repairs.

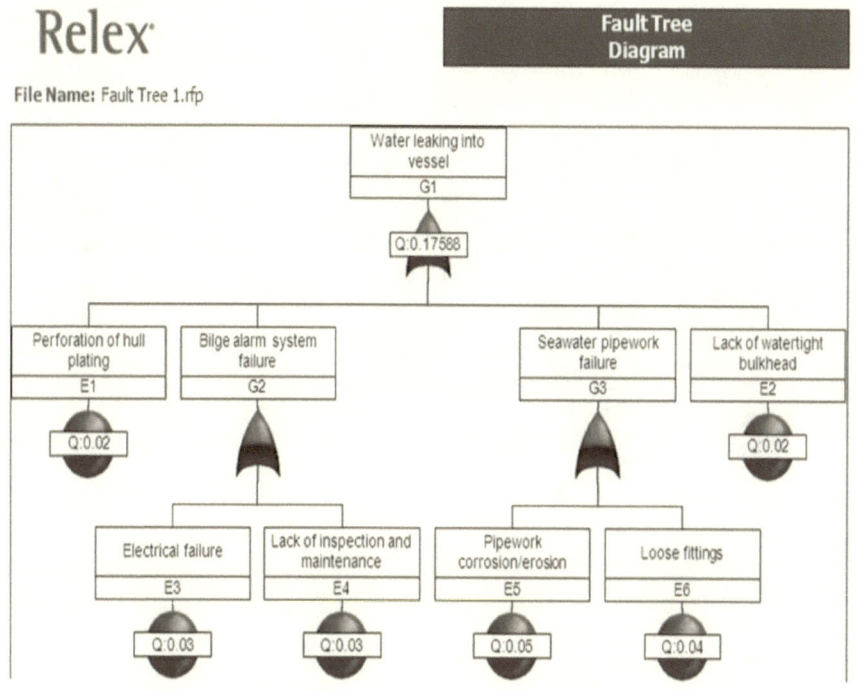

Figure 7. 7: A Fault Tree Diagram for the Top Event: Water Leaking into Vessel

After the first top event, fault tree analysis is repeated for the second top event: Collision. Fault tree for a top event analysis by RELEX Reliability software, collision, is shown in Figure 7.8. Also by substituting with manual calculation the given probabilities and calculated values of the top event: collision into the equations. Thus, the probability of occurrence of the top event: collision is 0.1121 (ETA, 2001).

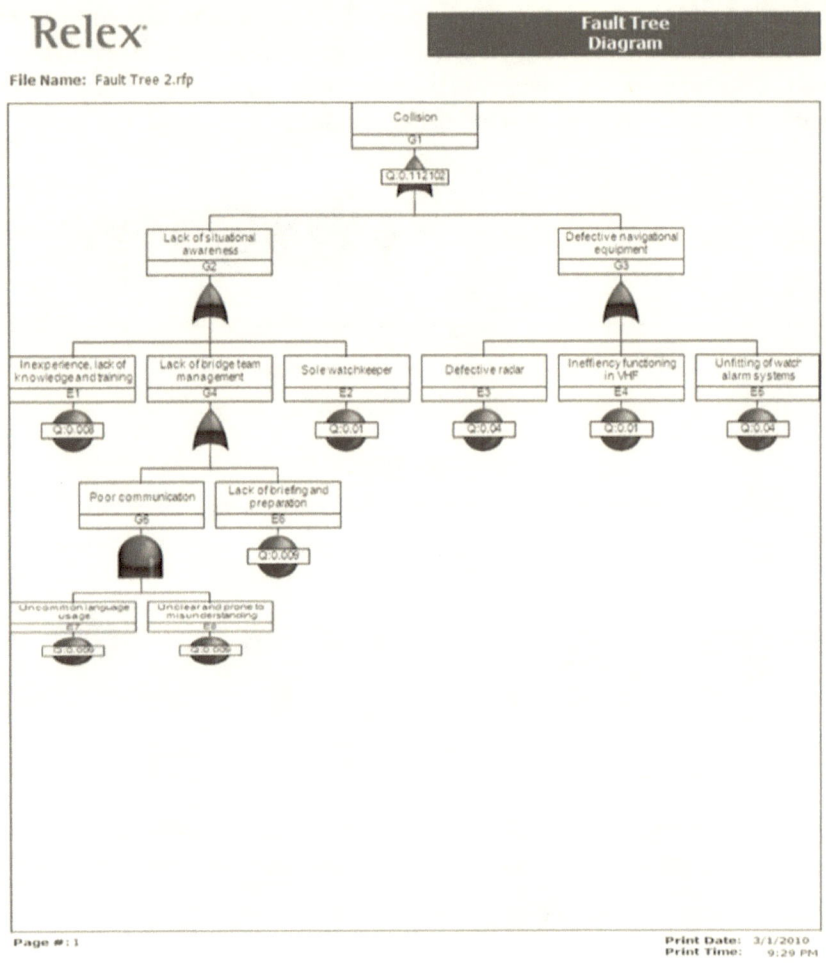

Figure 7. 8: A Fault Tree Diagram for the Top Event: Collision

7.6.4 Event Tree Analysis (ETA)

The goal of ETA is to determine the probability of all the possible outcomes resulting from the occurrence of an IE. By analyzing all possible outcomes, it is possible to determine the percentage of outcomes that lead to the desired result and the percentage of outcomes that lead to the undesired result. Each safety design method is evaluated for the contributing event: Operates successfully and Fails to operate.

Each success/failure event is assigned a probability of occurrence, and the final outcome probability is the product of the event probabilities

—

254

along a particular path. When computing the success/fail probability for each contributing PE that the PE states must always sum to 1.0, based on the reliability formula that:

$$P_{Success} + P_{Failure} = 1$$

7.03

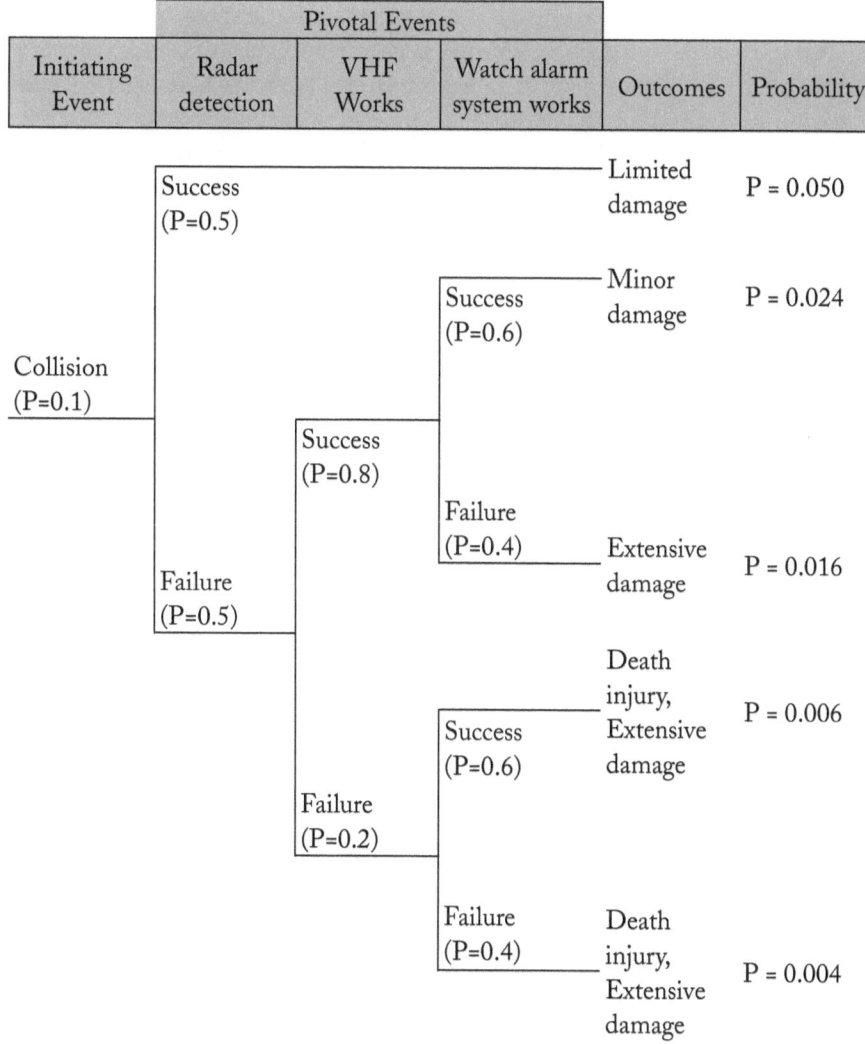

Figure 7.9:Event Three Analysis

There are errors (both human and technology) common to accidents in waterways. Employing methods through which these can be moderated and reduced could potentially enhance shipping safety the practical application of human reliability analysis is clear. It requires obtaining the cause parameters, both direct and indirect parameters, from the studied factor. The parameters that help to better understand the root of the presence of such a factor, and help to take punctual, specific and direct corrective actions to try to minimize the accident risk. The use of systemic HRA with right data, assessment and analysis can be is one mainstay to accident reduce, control and prevent maritime accident and sub sequential improvement of maritime safety Nowadays, it has becomes a very important tool to identify the problems related to human factor. Technological and engineering improvements in the marine sector have shown, in some cases, to increase the risk of an accident occurring due to human factors. The human factor accidents are mostly caused by lack of skill or knowledge, or other risk compensation.

Reference

1. IMO. (2002). Introduction to IMO. UK, London
2. Fotland, H. (2004). Human Error: A Fragile Chain of Contributing Elements.The International Maritime Human Element Bulletin, No. 3, pp. 2–3. Nautical Institute. 202 Lambeth Road, London, U.K.
3. Robb C. Wilcox, Zbigniew J. Karaszewski and Bilal M. Ayyub. (1996). Methodology for Risk-Based
4. Technology Applications to Marine System Safety. The Society of Naval Architects and Marine Engineers and the Ship Structure Committee. *http://www.shipstructure.org/pdf/1996 symp07*
5. De La Campa Portela. R. (2005). Maritime Casualties Analysis as a Tool to Improve Research about
6. Human Factors on Maritime Environment. Journal of Maritime Research. Vol. II.No. 2, pp. 3-18. http://www.jmr.unican.es/pub/00202/0020201.pdf [12 August 2009].
7. Grech, M., et. al (2002). Human error in maritime operations: Situation awareness and accident reports. Paper presented at the 5th International Workshop on Human Error, Safety and Systems Development, Newcastle, Australia.
8. Catherine H. et al (2006). Safety in shipping: The human element. Journal of Safety Research The Industrial Psychology Research Centre. .University of Aberdeen, Kings College, Old Aberdeen. 37, 401-411.
9. Ayyub, B.M., et al (2002). Risk Analysis and Management for Marine Systems. Naval Engineers Journal. Vol. 114, No. 2, pp. 181–206.
10. Hee, D.D.et al. (1999). Safety Management Assessment System (SMAS): a process for identifying and evaluating human and organization factors in marine system operations with field test results. Reliability Engineering and System Safety. Vol 65: pp. 125–140.
11. Kletz, T.A. (1991), Plant Design for Safety: A User-Friendly Approach, Hemisphere, New York.
12. Nivolianitou, Z. et al (2006). Development of a database for accidents and incidents in the Greek petrochemical industry. Journal of Loss Prevention in the Process Industries. Vol. 19: pp. 630-638.

13. United States Coast Guard. (2006). Risk Based Decision Making Guideline. Retrieved on 26 August 2009, from *http://www.uscg.mil/hq/gm/risk/e-guidelines/rbdm.htm*
14. Gordon, R.P.E. (1998). "The contribution of human factors to accidents in the offshore oil industry." Reliability Engineering and Systems Safety. Vol. 61, pp. 95-108.
15. Event Tree Analysis—ETA. (2001). Relex software Corporation. Retrieved on 19 October 2009, from *http://www.event-tree.com/*

CHAPTER 8

Risk and Reliability for Offshore Aquaculture Oceanic System

"Everything should be made as simple as possible,
but not one bit simpler"
Albert Einstein

Summary

Complex system design is increasingly being based on risk and reliability analysis. Aquaculture is the fastest growing sector for seafood production and other bio-base technological processes. Considerable interest exists in developing open ocean aquaculture in response to a shortage of aquaculture product, suitable system, sheltered inshore locations and possible husbandry advantages of oceanic system. Concept of very large floating structure is adapted in aquaculture farming in ocean to produce more aquaculture product like seaweed. The risk analysis study of offshore aquaculture ocean plantation system is very important to determine the system functionality and capability that meet sustainable and reliability requirement. This chapter describes process required to qualitatively assess system risk and quantify mooring failure probability, maximum force and required number of mooring as well as associated cost with the system.

8.1 Introduction

Since the early 1970s the technology, for very large floating structures has developed continually, while changing societal needs have resulted in many different applications of the of the technology for floating structure. Very large floating structure for offshore aquaculture of seaweed could be adapted offshore aquaculture ocean plantation system for oceanic farming of fish, prawn, squid and many more.

Seaweed encompass *macroscopic, multicellular, polyphyletic benthic marine algae* that includes some members of the *red, brown* and *green algae*. Different type of seaweeds are available, they are classified by use for example as food, medicine, fertilizer, industrial, biomass and others. In addition, some tuft-forming blue green algae (*Cyanobacteria*) are sometimes considered as seaweeds. The usual type of seaweed that is used in ocean plantation is Cottonii seaweed or also known as Kappaphycus (Eucheuma spp).

The design of very large floating structure for offshore aquaculture ocean plantation system required a reliable and risk free system with robust mathematical and simulation, risk and reliability of the hydroelastic structure, mooring system, structure, and material. Hence, the study of risk and reliability for the mooring system of offshore aquaculture ocean plantation system is required to make sure the system can function well, be monitored, and accessed safety and efficiency. Typical mooring structure for offshore aquaculture include piers, docks, floats and buoys and their associated pilings, ramps, lifts and railways. Generally, mooring structure is required to follow local and international requirements for offshore standards, materials, installation timing and surveys. The mooring structures should; be able to withstand in critical saltwater and freshwater habitats when the standards, overwater structures shall be constructed to the minimum size necessary to meet the needs of ocean resources exploration use.

Mooring system for VLFS need risk and reliability analysis of the associated criticality. Risk analysis of offshore aquaculture ocean plantation system focus on analyzing mooring structure with hope to help determine safe, reliability and efficiency of the system. Qualitative assessment and quantitative risk assessment analysis methods are explored towards reliable decision support for VLFS. Qualitative assessment analysis employed qualitative tools like checklist, and HAZOP (Hazard and Operability Study) that define the system while quantitative risk analysis, the methods employed include Failure Modes and Effects Analysis (FMEA), Fault Tree Analysis (FTA), Risk Control Option based on HAZID (Hazard

—

Identification) process. The risk of disaster cannot be eliminated, but risk can be reduced by employing better safety detection technique and establishing safety criteria prior to an accident occurrence. This section describe development of simplified but holistic methodology that determine risk based decision support for reliable design and development of VLFS system, the risk analysis focus on mooring structure failure and reliability through employment of risk tools like FMEA, FTA, RCO and HAZID.

8.2 Background

The seaweed extract, (Carrageenan) is an important hydrocolloids product for food additive ingredient and it is highly demanded in the world market. Seaweed is also used for biomass energy production as well as pharmaceutical and medicinal product The demand for seaweed has created huge market for this raw material, especially, the Cottonii seaweed also known as Kappaphycus (Euchema spp). For example, under the Malaysian Government NKEA, there is need to produce 1 million tones seaweed every year. Unfortunately, currently there is no proper system or platform to deliver this demand.

The mooring system failure analysis is very important part in the development offshore aquaculture ocean plantation system; risk analysis is required to determine the system function duty and performance. Besides that, there will be increasing demand for concept of floating technology worldwide, so the concept of offshore aquaculture ocean plantation system can be applied for the technology platform required. There is currently no systematic and formal proactive methodology for offshore aquaculture floating structure design. Offshore floating structure is required to be reliable in order to to withstand harsh environment. A risk and reliability studies of offshore aquaculture system for mooring structure will contribute to sustainable development of the seaweed farming industry as well as improvement of technology platform for other aquaculture farming in open seas.

The section describes conduct and determination of the reliability analysis that can reduce the probability of accident risk occurrence and impact in offshore aquaculture system for ocean plantation. Especially mooring structure system integrity and reduction of consequence of failure. The study accesses the risk, system functionality and capability of offshore aquaculture seaweed plantation for mooring structure. The study also estimate the risk in design of mooring structure for deployment of very large floating structure for oceanic aquaculture seaweed plantation

—

and decision recommendation will be offered for level integrity of oceanic aquaculture seaweed plantation for mooring structure.

8.3 Important of Risk Based Design VLFS

The study VLFS, fall under complex and new system, unlike system like ship, offshore structure, that have of the shelf guideline, new method is required for design and development of reliable VLFS for aquaculture seaweed farming. The risk and reliability analysis is one best approach to use for offshore aquaculture ocean plantation system of mooring system and VLF structure that withstands the aspects that has been tested in the system design and development. The risk approach investigate more detailed the risk, specification and requirement that the system needs to make sure that it is reliability for deployment of capability. The significant of this using risk method for VLFS are:

i.   To avoid system failure according recommendation from quantifying and deduction of improvement measures
ii. Identify inadequate mooring strength due to poor material quality of fatigue in order to determine required mitigation.
iii. Identified excessive environmental forces for example under estimated or freak environmental condition and determine solution for system additional uncertainty.
iv. Predicted incorrectly mooring tension based on the reviews and analysis of the system.
v.   Performed risk and reliability leads to recommend the best safety level integrity of oceanic aquaculture seaweed plantation for mooring structure to alert the risk and improve reliability of this system.

8.4 Data Collection and Choice of Case Study Area

The study of risk analysis should consider region with harsh metaocean condition like the North Sea in Europe and South China Sea in Asia. The area identified and applied in the water body of water that is known for harsh weather, thus inheritably buffer area that provide and right way for sea traffic and water depth is prevalent location consideration of for seaweed farming. Other location ideal for seaweed cultivation are areas with absence of larger feeder river systems in its vicinity and reasonable deep waters that will ensure stability in its salinity and some degree of shelter protected by the chain of island off coast of the area.

Such location required to fulfils the prime factors for seaweed culture which are suitable water condition and good exchange of seawater. Besides these two factors, the relatively developed infrastructure and logistic network are also some of the important supporting factors for this system development. The data collection should be from specific source and method. The right source of data should be chosen to make sure the data are true and valid for this study analysis. The data are also obtained from model test, Meteorology Department, Department of Environment, Offshore Company, Aquaculture Company.

8.5 Very Large Floating Structures (VLFSs)

Very Large Floating Structures (VLFSs) OR very large floating platforms (VLFP) can be constructed to create floating airports, bridges, breakwaters, piers and docks, storage facilities (for oil and natural gas), wind and solar power plants, for military purposes, to create industrial space, emergency bases, entertainment facilities such as casinos, recreation parks, mobile offshore structures and even for habitation. VLFS for habitation could become reality sooner than one may expect. Currently, different concepts have been proposed for building floating cities or huge living complexes (A.l. Andrianov, 2005). The system is constituted by vertical tethers.

This characteristic makes the structure very rigid in the vertical direction and very flexible in the horizontal plane. Both these features results particularly attractive. The vertical rigidity helps tie in wells for production, while, the horizontal compliance makes the platform intensive primary effect of waves, (Li, Y. and Kareem, A., 1993). Pontoon type VLFS are mat-like VLFSs because of their small draft in relation to the length dimensions. Very large pontoon type floating structure is often called Mega Floats. As a rule, Mega Floats is floating having at least one length dimension greater than 60 meters. Horizontally large floating structures can be from 500 to 5000 meters in length and 100 to 1000 meters in width, while their thickness can be of the order of about 2-10 meters (Jan Van Kessel's, 2000).

8.5.1 Analysis and Design of Very Large Floating Structures

Clauss state that in year 1992, the analysis and design of floating structures need to account for some special characteristics. That statement is valid when comparing to land-based structures. In a floating structure, the static vertical self-weight and payloads are carried by buoyancy. If a

—

floating structure has got a compliant mooring system, consisting for instance of catenaries chain mooring lines, the horizontal wave forces are balanced by inertia forces. That shows that if the horizontal size of the structure is larger than the wavelength, the resultant horizontal forces will be reduced due to the fact that wave forces on different structural parts will have different phase which is direction and size (Clauss, 1992).

The forces in the mooring system will then be small relative to the total wave forces. The main purpose of the mooring system is then to prevent drift-off due to steady current and wind forces as well as possible steady and slow-drift wave forces which are usually more than an order of magnitude less than the first order wave forces. Sizing of the floating structure and its mooring system depends on its function and also on the environmental conditions in terms of waves, current and wind. (Moan, 2004) The design may be dominated either by peak loading due to permanent and variable loads or by fatigue strength due to cyclic wave loading. Moreover, it is important to consider possible accidental events and ensure that the overall safety is not threatened by a possible progressive failure induced by such damage.

Clauss, (1992) explain that, unlike land-based constructions with their associated foundations poured in place, very large floating structures are usually constructed at shore-based building sites remote from the deepwater installation area and without extensive preparation of the foundation. Each module must be capable of floating so that they can be floated to the site and assembled in the sea. Owing to the corrosive sea environment, floating structures have to be provided with a good corrosion protection system and also possible degradation due to corrosion or crack growth (fatigue) requires a proper system for inspection, monitoring, maintenance and repair during use.

8.6 Design Considerations for Mooring Structure

The mooring system must be well designed to ensure that the very large floating structure is kept in position so that the facilities installed on the floating structure can be reliably operated and to prevent the structure from drifting away under critical sea conditions and storms (E. Watanabe, 2004). There are a number of mooring systems such as the dolphin-guide, frame system, mooring by cable and chain, tension leg method and pier/quay wall method. The design procedure for a mooring system may take the following steps. First select the mooring method, the shock absorbing material, the quantity and layout of devices to meet the environmental conditions and the operating conditions and requirements. The layout of the mooring dolphins for example is such that the horizontal displacement of the floating structure is adequately controlled and the mooring forces are appropriately distributed. In role of reliability analysis, the behavior of the floating structure under various loading conditions is examined. The layout and quantity of the devices are adjusted so that the displacement of floating structure and the mooring forces do not exceed the allowable values. Finally, the floating structure is designed by applying the design load based on the calculated mooring forces.

The materials for the mooring system shall be selected according to the purpose, environment, durability and economy. (Maeda et al.2000), (Shimada& Miyajima, 2002), (Ookubo et al. 2002) and (Shiraishi et al.2002). According to C.W.Lee, (2005), the deformation of the mooring line for floating structures can happen due to current act. The shapes of the structures change to a considerable degree by the water flow and the large tension working on the mooring line of the upper side. The degree of deformation of the structures and the tension of the mooring line depend on the size of flow speed and materials. The tension of the mooring line on the upper side of the tide and waves changes regularly and become higher by a maximum of 250% than when only tides works on it.

8.7 The Seaweed farming

Harvesting seaweed from wild population is an ancient practices dating back to the fourth and sixth centuries in Japan and China, respectively, but it was not until the mid-twentieth century that methods for major seaweed cultivation were developed (McHugh, 2003). Since that time, seaweed farming or marine agronomy has grown rapidly due to demand that has outpaced the productivity of natural populations. Today almost 90% of seaweed for human use comes from cultivation, rather than wild harvests (Zemke-White & Ohno, 1999). Seaweed has traditionally been grown in near shore coastal waters, with some smaller operations on land. An offshore system which is the focus of this study are an emerging seaweed culture technology. .he global production of all aquaculture products in 2004 was 59.4 million metric ton with a total value of $70.3 billion (Chen, 2006). Of this almost a quarter by weight, but only a tenth by value ($6.8 billion) were aquatic plants; 99.8% of which were farmed in Asia and the Pacific Region. Seaweed farms worldwide are estimated to produce 13.9 million metric ton wet weight per year.

Seaweed farming has become an economically important natural resource for Malaysia since 1978, when it was first introduced to Semporna, east coast of Sabah on a commercial scale. It has develop the

—

aquaculture activities in Sabah as a second largest contributor from marine aquaculture which produce 60% from total value of exported fisheries product at MR$114 million (1994-1997) (Sade, 2006). It has wide application potentials similar to other commodities such as palm oil and cocoa.

This has been approved during the Ninth Malaysia Plan (2006-2010) and the Third National Policy (1998-2010) with seaweed being mentioned specifically as one of the most important aquaculture product of food farming commodities for the country. Although the sector of seaweed industry has developed enormously over the past few years (111,298 tonnes wet weight in 2008), seaweed production and national target on 2010 of 250,000 tonnes (wet weight) is however yet to be achieved (Kaur, 2009). With the nature elements of the deep and open water environment, seaweed farming is hard to be applied in this area. However, Marine System Engineering can deliver system that could solve this problem

8.7.1 Seaweed growth of life cycle

Seaweed life cycles are complex in many species, with annual and perennial species and sexual and asexual reproductive modes, resulting in isomorphic or heteromorphic life history forms, commonly referred to as alternation of generations. Understanding the complex and diverse life cycles of different seaweeds is of practical significance in controlling growth and reproduction for optimal plant husbandry. An example in which our increased understanding of life cycle had clear economic impact was the identification of the conchocelis, originally considered a separate organism, as a one of the diploid stages of Porphyra spp. (Drew, 1949; Drew, 1954). This recovery revolutionized the culture of commercially cultivated seaweed in Japan, China, and Korea. The conchocelis become the seed stock source for artificial propagation of this seaweed (Choi et al., 2002).

8.8 Risk Analysis

Risk analysis are best used for assessing and evaluating uncertainties associated with an event, risk is defined as the potential for loss as a result of a system failure, and can be measured as a pair of factors, one being the probability of occurrences of an event, also called a failure scenario, and the other being the potential outcome or consequence associated with the event's occurrence (Bilal M. Ayyub, 2002). A risk assessment is the process used to determine the risk based on the likelihood and impact of an event.

—

Failure history through experience (qualitative) and data (quantitative) may be used to perform a risk assessment (Glickman and Gough, 1993).

Risk analysis is concerned with using available data to determine risk posed by safety hazards and usually consists of steps such as scope definition, hazard identification and risk determination. The phase in which the decision process is inundated with metrics and judgments is called the risk evaluation. The purpose of analysis is to determine the contributory causes and circumstances of the accident as a basis for making recommendation, if any, with the aim of preventing similar accidents occurring again.

8.8.1 The Concept of Risk

Risk is defined as an objective, systematic, standardized and defensible method of assessing the likelihood of negative consequences occurring due to a proposed action or activity and the likely magnitude of those consequences, or, simply put; it is "science-based decision-making". "Risk" is the potential for realization of unwanted, adverse consequences to human life, health, property or the environment. Its estimation involves both the likelihood (probability) of a negative event occurring as the result of a proposed action and the consequences that will result if it does happen.

For example, in some sector, "Risk—means the likelihood of the occurrence and the likely magnitude of the consequences of an adverse event to public, aquatic animal or terrestrial animal health in the importing country during a specified time period." While some sectors incorporate consideration of potential benefits that may result from a "risk" being realized (e.g. financial risk analysis), others specifically exclude benefits from being taken into account. The definition of "risk" varies somewhat depending on the sector. Most definitions incorporate the concepts of:

i. Uncertainty of outcome (of an action or situation)
ii. Probability or likelihood (of an unwanted event occurring)
iii. Consequence or impact (if the unwanted event happens)

"Risk analysis" is also defined either by its components and/or its processes. The Society for Risk Analysis www.sera.org offers the following definitions of "risk analysis":

—

i. A detailed examination including risk assessment, risk evaluation and risk management alternatives, performed to understand the nature of unwanted, negative consequences to human life, health, property or the environment
ii. An analytical process to provide information regarding undesirable events
iii. The process of quantification of the probabilities and expected consequences for identified risks

There are many sources of risk to marine systems including human error, external events, equipment failure, and installation error (Ayyub *et. al.*, 2002). Risk is defined as the product of likelihood of occurrence and consequences of an accident (Glickman and Gough, 1993). Risk analysis or assessment helps to answer basically three questions, as shown in Figure 8.1

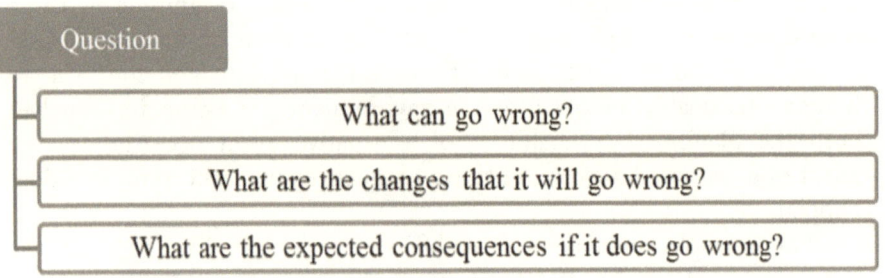

Figure 8. 1: Questions Answered by Risk Assessment or Analysis (Glickman and Gough, 1993)

8.8.2 The concept of "hazard"

All risk analysis sectors involve the assessment of risk posed by a threat or "hazard". The definition of "hazard" depends on the sector and the perspective from which risk is viewed (e.g. risks to aquaculture or risks from aquaculture). A hazard thus can be:

i. A physical agent having the potential to cause harm, for example:
   a) A biological pathogen (pathogen risk analysis);
   b) An aquatic organism that is being introduced or transferred (genetic risk analysis, ecological risk analysis, invasive alien species risk analysis);

—

    c) A chemical, heavy metal or biological contaminant (human health and food safety risk analysis, environmental risk analysis); or

ii. The inherent capacity or property of a physical agent or situation to cause adverse affects, as in

iii. Social risk analysis

iv. Financial risk analysis

v. Environmental risk analysis

8.8.3 General Principles Risk

Principles that are common to all types of risk analysis are presented below. These involve the broader concepts of common sense, uncertainty, precaution, objectivity, transparency, consistency, scientific validation, stakeholder consultation, stringency, minimal risk management, unacceptable risk and equivalence.

i. *The Principle of Common Sense*—In assessing risks, the use of "common sense" should prevail. In many cases, the outcomes of a risk analysis are obvious and uncontroversial, and a decision can be made without resulting to a full risk analysis, which can be a lengthy and expensive process.

ii. *The Principle of Uncertainty*—All risk analyses contain an element of uncertainty. A good risk analysis will seek to reduce uncertainty to the extent possible.

iii. *The Principle of Precaution*—Those involved in the aquaculture sector have a responsibility to err on the side of caution, particularly if the outcomes of a given action may be irreversible. If the level of uncertainty is high, the Precautionary Principle can be applied to delay a decision until key information is obtained. However, steps must be taken to obtain the information in a timely manner.

iv. *The Principle of Objectivity*—Risk analyses should be conducted in the most objective way possible. However, due to uncertainty and human nature, a high degree of subjectivity may be present in some risk analyses. A risk analysis should clearly indicate where subjective decisions have been made.

v. *The Principle of Transparency*—Risk analyses, particularly those conducted by public sector agencies, should be fully transparent, so that all stakeholders can see how decisions were reached. This includes full documentation of all data, sources of information, assumptions, methods, results, constraints, discussions and conclusions.

vi. *The Principle of Consistency*—Although risk analysis methodology continues to evolve, it is important that decisions, particularly those made by government, are reached via standardized methods and procedures. In theory, two risk analysts independently conducting the same risk analysis should reach roughly similar conclusions.

vii. *The Principle of Scientific Validation*—The scientific basis of a risk analysis and the conclusions drawn should be validated by independent expert review.

viii. *The Principle of Stakeholder Consultation*—If the results of a risk analysis are likely to be of interest to, or impact upon others, then stakeholder consultations should be held. This is accomplished by risk communication, the interactive exchange of information on risk among risk assessors, risk managers and other interested parties. Ideally, stakeholders should be informed/ involved throughout the entire risk analysis process, particularly for potentially contentious risk analyses (e.g. ecological, genetic and pathogen risk analyses for the introduction of new aquatic species).

ix. *The Principle of Stringency*—The stringency of the risk management measures to be applied should be in direct proportion to the risk involved.

x. *The Principle of Minimal Risk Management*—Risk management measures that impinge on the legitimate activities of others should be applied only to the extent necessary to reduce risk to an acceptable level.

xi. *The Principle of Unacceptable Risk*—If the level of risk is unacceptable and no effective or acceptable risk management measures are possible, then the activity should not take place.

xii. *The Principle of Equivalence*—Risk management measures proposed by trading partners that meet the acceptable level of risk should be accepted by the importing country.

8.9 Qualitative and Quantitative Analysis

Qualitative analysis relies on statistical methods and databases that identify the probability and consequence. This objective approach examines the system in greater detail for risk (Robb *et. al.,*

—

1996). Quantitative risk analysis generally provides a more uniform understanding among different individuals, but requires quality data for accurate results. Qualitative risk analysis uses expert opinion to evaluate the probability and consequence. This subjective approach may be sufficient to access the risk of a marine system (Robb *et. al.*, 1996).

The qualitative method for risk assessment or analysis is designed for the purpose of enhancing one's awareness of potential problems and can assist one in analyzing the risks. There are many methods and technique that have been developed to perform various types of analysis, in areas such as reliability and safety. In order to perform risk assessment and analysis method, this can be determined by quantitative and qualitative risk analysis tools presented in Table 8.1 below. A combination of both qualitative and quantitative risk analysis can be used depending on the situation.

Table 8. 1: Quantitative and qualitative risk analysis.

QUANTITATIVE METHODS
Failure Modes and Effects Analysis (FMEA) Identifies the components (equipment) failure modes and the impact on the surrounding components and the system.
Fault Tree Analysis (FTA) Identify combinations of equipment failure and human errors that can result in an accident.
Event Tree Analysis (ETA) Identify various consequences of events, both failures and successes that can lead to an accident.
QUALITATIVE METHODS
ALARP Possible to demonstrate that the cost involved in reducing the risk further would be grossly disproportionate to the benefit gained.
Checklist Ensures that organizations are complying with standard practice.
Safety/Review Audit Identify equipment conditions or operating procedures that could lead to a casualty or result in property damage or environment impacts.
What-If Identify hazards, hazardous situations, or specific accident events that could lead to undesirable consequences.

| Hazard and Operability Study (HAZOP) |
| Identify system deviations and their causes that can lead to undesirable consequences and determine recommended actions to reduce the frequency and/or consequences of the deviations. |
| Preliminary Hazard Analysis (PrHA) |
| Identify and prioritize hazards leading to undesirable consequences early in the life of a system. |
| Determine recommended actions to reduce the frequency and/or consequences of prioritized hazards. |

8.9.1 Failure Modes and Effects Analysis (FMEA)

Failure Modes and Effects Analysis (FMEA) is a powerful tool used by the system safety and reliability engineers/analysts to identify critical parts functions and components whose failure will lead to undesirable outcome such as production loss, injury or even an accidents. The tool was first proposed by NASA in year 1963 for their obvious reliability requirements (NASA, 1999). Since then, it has been extensively used as a powerful technique for safety and reliability analysis of products and process in wide range of industries that are particularly aerospace, nuclear, automotive and medical (Rezaie *et al.*, Amalnik, Gereie *et al.*, Ostadi *et al.*, & Shakheseniaee *et al.*, 2007; Rezaei *et al.*, Gereie *et al.*, Ostadi *et al.*, & Shakhseniaee *et al.*, 2008).

Failure Modes and Effects Analysis (FMEA) is a method to analyze potential reliability problems in the development cycle of the project, making it easier to take actions to overcome such issues, enhancing the reliability through design (Sharma *et al.*, Kumar *et al.*, & Kumar *et al.*, 2008). FMEA is used to determine actions to mitigate the analyzed potential failure modes and their effect on the operations. Expected failure modes, being the central step in the analysis, needs to be carried on extensively, in order to prepare a list of the maximum potential failure modes. FMEA is also a procedure for evaluating the various aspects of a system in order to identify all catastrophic and critical failure possibilities so that they can be eliminated or minimized through design correction at the earliest possible time (MIL-STD-1629A, 1980).

8.9.2 Fault Tree Analysis (FTA)

Fault Tree Analysis (FTA) is a tool for analyzing, visually displaying and evaluating failure paths in a system, thereby providing mechanism

—

for effective system level risk evaluation. Many people and corporation are already familiar with this tool and use it on a regular basis for safety and reliability evaluations, (Clifton A. Ericson II, 1999). Clifton A. Ericson II, (1999), FTA has become an important tool in system design and development, and history related to the basic should be recorded and appropriate people duty recognized. FTA is based on Reliability theory, Boolean algebra and probability theory.

8.10 Reliability Analysis

Reliability analysis methods have been proposed in several studies as the primary tool to handle various categories of risks (Billinton 2004; Janjic and Popovic 2007). Traditionally, the research and the development of reliability analysis methods have focused on generation and transmission (Kwok 1988). However, several studies have shown that most of the customer outrages depend on failures at the distribution level (Billinton and Allan 1996; Billinton and Sankarakrishnan 1994; Bertling, 2002). Furthermore, there is an international tendency towards adopt new performance based tariff regulation methods (Billinton 2004; Mielczarski 2006; Mielczarski 2005).

Reliability of a system can be defined as the system's ability to fulfil its design functions for a specified time. This ability is commonly measured using probabilities. Reliability is, represent the probability that the complementary event that will occur will leads to failure. Based on this definition, reliability is one of the components of risk. Safety can be defined as the judgment of a risk's acceptability for the system safety, making it a component of risk management (Bilal M. Ayyub, 2002).

8.11 Risk Analysis in Maritime Industry

International Maritime Organization state that, Formal Safety Assessment (FSA) is a structured and systematic methodology, aimed at enhancing maritime safety, including protection of life, health, the marine environment and property, by using risk analysis and cost benefit assessment. FSA can be used as a tool to help in the evaluation of new regulations for maritime safety and protection of the marine environment or in making a comparison between existing and possibly improved regulations, with a view to achieving a balance between the various technical and operational issues, including the human element, and benefit between maritime safety or protection of the marine environment and costs. FSA consists of five steps which are, firstly is identification of

—

273

hazards that means a list of all relevant accident scenarios with potential causes and outcomes, secondly is assessment of risks means that the evaluation of risk factors, thirdly is risk control options that is devising regulatory measures to control and reduce the identified risks, fourthly is cost benefit assessment which determining cost effectiveness of each risk control option and lastly recommendations for decision-making conclusion from the information about the hazards, their associated risks and the cost effectiveness of alternative risk control options.

8.12 The ALARP principle

ALARP (As Low As Reasonably Practicable), is a used in the analysis of safety-critical and high-integrity systems. The ALARP principle define residual risk that shall be as low as reasonably practicable, it has been used for decision support for Nuclear Safety Justification, is derived from legal requirements in the UK's Health & Safety at Work Act 1974 and is explicitly defined in the Ionizing Radiation Regulations, 1999. The ALARP principle is part of a safety culture philosophy and means that a risk is low enough that attempting to make it lower would actually be more costly than cost likely to come from the risk itself. This is called a tolerable risk.

The ALARP principle arises from the fact that it would be possible to spend infinite time, effort and money attempting to reduce a risk to zero. It should not be understood as simply a quantities measure of benefit against detriment. It is more a best common practice of judgment of the balance of a risk and societal benefits (Marvin Rausand, 2005). The meaning and value of the ALARP tolerability risk presented in Figure 8.2 the triangle represents increasing levels of 'risk' for a particular hazardous activity, as we move from the bottom of the triangle towards the top".

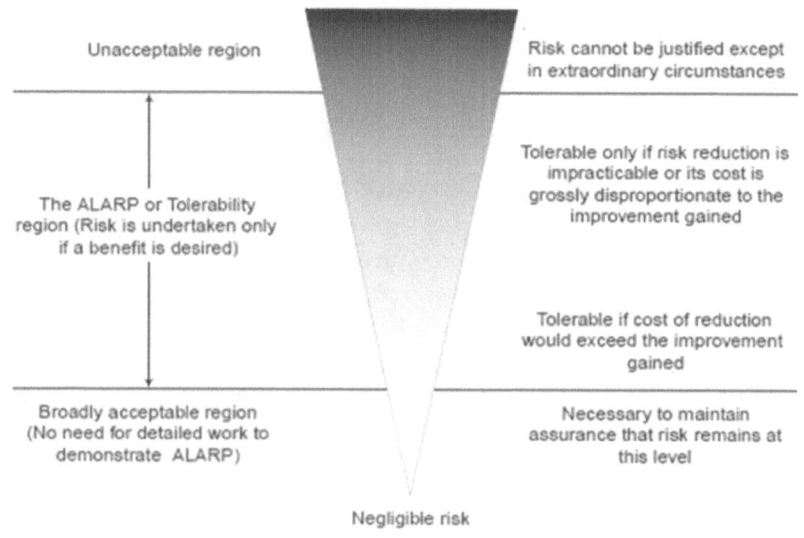

Figure 8. 2: Levels of Risk and As Low As Is Reasonably Practicable (ALARP)

8.13 Offshore Industry Risk Analysis

Traditionally, offshore quantitative risk analyses (QRAs) have had a rather crude analysis of barrier performance, emphasizing technical aspects related to consequence reducing systems. However, recently the Petroleum Safety Authority Norway (PSA) has been focusing on safety barriers and their performance, both in regulations concerning health, safety and environment (PSA, 2001) and in their supervisory activities. The development of offshore Quantitative Risk Assessment (QRA) has been lead by the mutual influence and interaction between the regulatory authorities for the UK and Norwegian waters as well as the oil companies operating in the work sea. Also, other countries have participated in this development, but to some extent this has often been based on the British and Norwegian initiatives according to DNV Consulting Support, GI 291, Det Norske Veitas AS, 1322 Hovik, Norway.

In more recent times, efforts to protect citizens and natural resources, has make governments to be more involved, requiring corporations to employ risk-reducing measures, secure certain types of insurance and even, in some cases, demonstrate that they can operate with an acceptable level of risk. During the 1980's and 1990's, more and more governmental agencies have required industry to apply risk assessment techniques. For instance, the U.S. Environmental Protection Agency requires new facilities to describe "worst case" and "expected" environmental release scenarios as part of the permitting process. Also, the United Kingdom requires submittal of "Safety Cases" which are intended to demonstrate the level of risk associated with each offshore oil and gas production facility (ABS Guidance Notes on Risk Assessment, 2000)

8.14 Offshore Rule for Offshore Structure

The variety of offshore structures concerning the function, size, geometrical configuration and material selection as well as the variability of the environmental factors complicate the development of a unique design procedure (Research Centre Asia Classification Society, 2003). Therefore, the separate investigation of the interaction between the actual structure and the environment is necessary.

For mooring system offshore rules (Bureau Veritas, 2010) use reference documents NI 493 "Classification of Mooring System for Permanent Offshore Units". The design and specification of mooring structure for offshore aquaculture ocean plantation system must be based on all requirements had listed and mention in NI 493 document.

8.15 Safety and Risk of Offshore Aquaculture

The EC–JRC International Workshop on "Promotion of Technical Harmonization on Risk-Based Decision Making" (Stresa/Ispra, May 2000) investigated the use of risk-based decision making across different industries and countries. Under the UK safety case regulations (UK Health and Safety Executive, 1992), each operator in the UK Sector is required to prepare a Safety Case for each of its installations, fixed or mobile, to demonstrate that;

i. The management system adequately covers all statutory requirements;.
ii. There are proper arrangements for independent audit of the system;
iii. The risks of major accidents have been identified and assessed;
iv. Measures to reduce risks to people to the lowest level reasonably practicable have been taken;
v. Proper systems for emergency arrangements on evacuation, escape and rescue are in place.

Before an installation is allowed to operate, the Safety Case must be formally accepted by the Health and Safety Executive (HSE). Like any aquaculture industry, offshore aquaculture will benefit from thoughtful site selection. Offshore enterprises should be sited in areas that meet optimal biological criteria for species grow-out and minimize user conflicts with other established groups. Careful site selection may also ensure the development of offshore aquaculture zones or parks to expedite industry development.

8.16 Failure of Mooring System

It is clearly identified that mooring systems on Floating Production Systems are category 1 safety critical systems (Noble Denton Europe Limited, 2006). Multiple mooring line failure is required to put lives at risk both on the drifting unit and on surrounding installations. There is also a potential pollution risk. Research to date indicates that there is an imbalance between the critical nature of mooring systems and the attention which they receive. The mooring system failure probability is considerably reduced with increases safety factor in particular for system with several parallel loads sharing element. For system with low overall safety factor, the mooring system failure probability is expected to increase with increasing in number of lines, whereas for high safety factors, the system failure probability is expected to reduce with the increasing number of lines. While for the same load distribution and number of lines, a wire system is in general more reliable than a chain system with the same overall safety factor.

8.17 Methods OF Risk Modeling

8.17.1 The risk analysis process

    i.   Risk analysis typically seeks to answer 4 questions:
    ii. What can go wrong?
    iii. How likely is it to go wrong?
    iv. What would be the consequences of its going wrong?
    v.   What can be done to reduce either the likelihood or the consequences of its going wrong? (MacDiarmid, 1997; Rodgers, 2004; Arthur et al., 2004).

8.17.2 Risk Framework

The general framework for risk analysis typically consists of four major components:

    i.   *Hazard identification*—the process of identifying hazards that could potentially produce consequences

ii. *Risk assessment*—the process of evaluating the likelihood that a potential hazard will be realized and estimating the biological, social, economic, environmental and failure consequences

iii. *Risk management*—the seeking of means to reduce either the likelihood or the consequences of it going wrong

iv. *Risk communication*—the process by which stakeholders are consulted, information and opinions gathered and risk analysis results and management measures communicated

8.17.3 Risk analysis process

The risk analysis process is a flexible process, its structure and components vary and depend on the sector (e.g. technical, social or financial), the user (e.g. government, company or individual), the scale (e.g. international, local or entity-level) and the purpose (e.g. to gain understanding of the processes that determine risk or to form the basis for legal measures). It can be qualitative (probabilities of events happening expressed, for example, as high, medium or low) or quantitative (numerical probabilities).

General idea of the risk and reliability analysis study of offshore aquaculture ocean plantation system focus on mooring structure of offshore aquaculture systems well as investigation of the problem, goal and objectives, advantage, disadvantage, limitation, design for environment, data reliability. Analysis of historical information from various sources play important role in the outcome of system identification. Flow chart, tables and mathematical governing equation are used to present detail of the process and procedure. The outcome of risk leads to recommendation for system reliability of future work. This study process followed three tier, preliminary system identification, qualitative risk assessment that involve HAZID process and quantitative risk. The process of the approach is more elaborated as followed.

i. Preliminary system assessment and involve the review of past work data collection and general requirement for mooring structure. Data of analyses of offshore aquaculture ocean plantation mooring system and structure are collected in order to define system, deduce system risk areas and reliability areas.

ii. (HAZID) Hazard Identification qualitative process involves clarification risk. For risk analysis had two processes which are qualitative analysis and quantitative analysis. Qualitative assessment use HAZOP and checklist, Fault Modes and Effect Analysis (FMEA), Fault Tree Analysis (FTA).

–

iii. Quantitative analysis involves Analytical process that employed hybrid of deterministic, statistical, reliability and probabilistic method to redefine system behavior in the past, present and future. These use of law physics, help to strength the analysis and support the study of the risk and reliability of this system.

iv. In result of each of the tier can lead to risk matrix, ALARP graph, Risk Control Option (RCO) and cost Effectiveness Analysis.

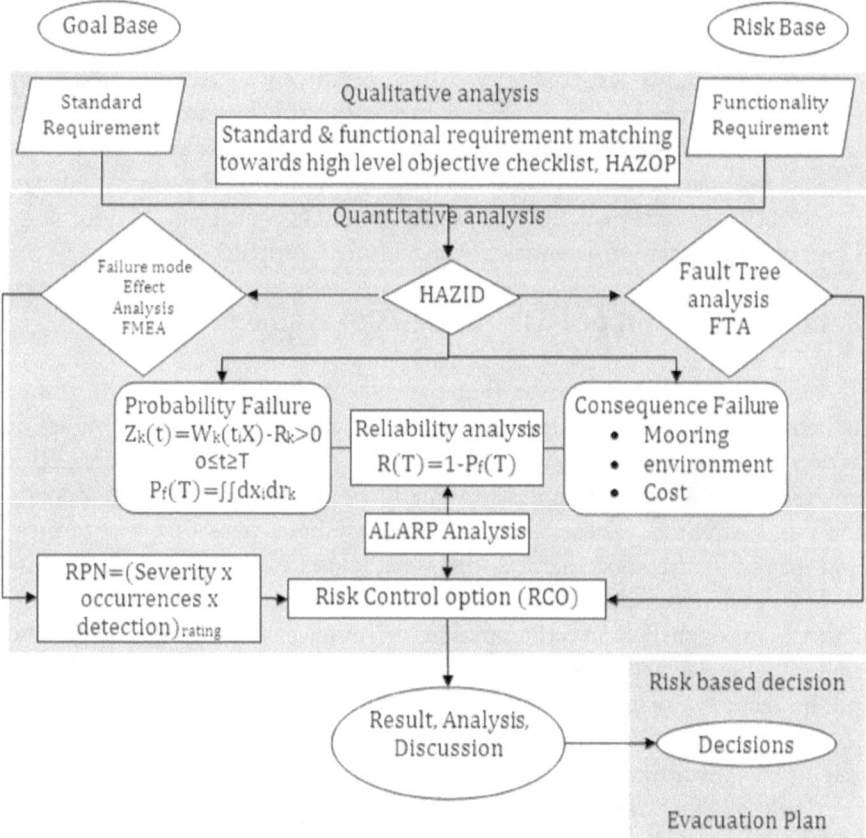

Figure 8. 3: Quantitatively accidents frequency and consequences VLFS

Since the design of VLFS for seaweed farming is required new methodology based on risk, guideline systems for solving a problem with specific components such as phases, tasks, methods, technique and tools that are incorporated are (Irny, S.I. and Rose, A.A, 2005). It can define as follows:

i. "The analysis of the principles of method, rules, and postulates employed by a discipline"
ii. "The systematic study of methods that are, can be, or have been applied within a discipline"
iii. "The study of description of methods"

8.18 Safety and Environmental Risk Model (SERM)

SERM methology adapted from O.O.Sulaiman (2010) intends to address risk over the entire life of the complex system. SERM address qualitative aspect as well quantitatively accidents frequency and consequences VLFS, as shown in Figure 8.3. Relevant data can be obtained from specific places and method. The right sources should be chosen to make sure the data is reliable and valid for the study analysis. Some of the data will be obtained from model test, Meteorology Department, Offshore Company, Aquaculture Company.

8.19 Application of Risk Analysis to Aquaculture

Risk analysis is a process that provides a flexible framework within which the risks of adverse consequences resulting from a course of action can be evaluated in a systematic, science-based manner. Risk analysis is now widely applied in many fields that touch human daily lives and activities. These include decisions about risks due to chemical and physical stressors (natural disasters, climate change, contaminants in food and water, pollution etc.), biological stressors (human, plant and animal pathogens; plant and animal pests; invasive species, invasive genetic material), social and economic stressors (unemployment, financial losses, public security, including risk of terrorism), construction and engineering (building safety, fire safety, military applications) and business (project operations, insurance, litigation, credit, cost risk maintenance etc.).

Risk analysis has wide applicability to aquaculture. So far, it has mainly been applied in assessing risks to society and the environment posed by hazards created by or associated with aquaculture development depending on aquaculture farming in question. The risks include risks of environmental degradation; introduction and spread of pathogens, pests and invasive species; genetic impacts; unsafe foods; and negative social and economic impacts. Governments and the private sector often make decisions based on incomplete knowledge and a high degree of uncertainty. Such decisions may have far-reaching social, environmental and economic consequences. The use of risk analysis can provide insights and assist in making decisions that will help to avoid such negative impacts, thus helping aquaculture development to proceed in a more socially and environmentally responsible manner.

8.19.1 System Definition

The risk analysis approach permits a defendable decision to be made on whether the risk posed by a particular action or "hazard" is acceptable or not, and provides the means to evaluate possible ways to reduce the risk from an unacceptable level to one that is acceptable. Risk analysis is thus a pervasive but often unnoticed component of modern society that is used by governments, private sector and individuals in the political, scientific, business, financial, social sciences and other communities. The risk analysis is start with the system definition, Figure 8.5 shows example for offshore aquaculture for seaweed biomass farming).

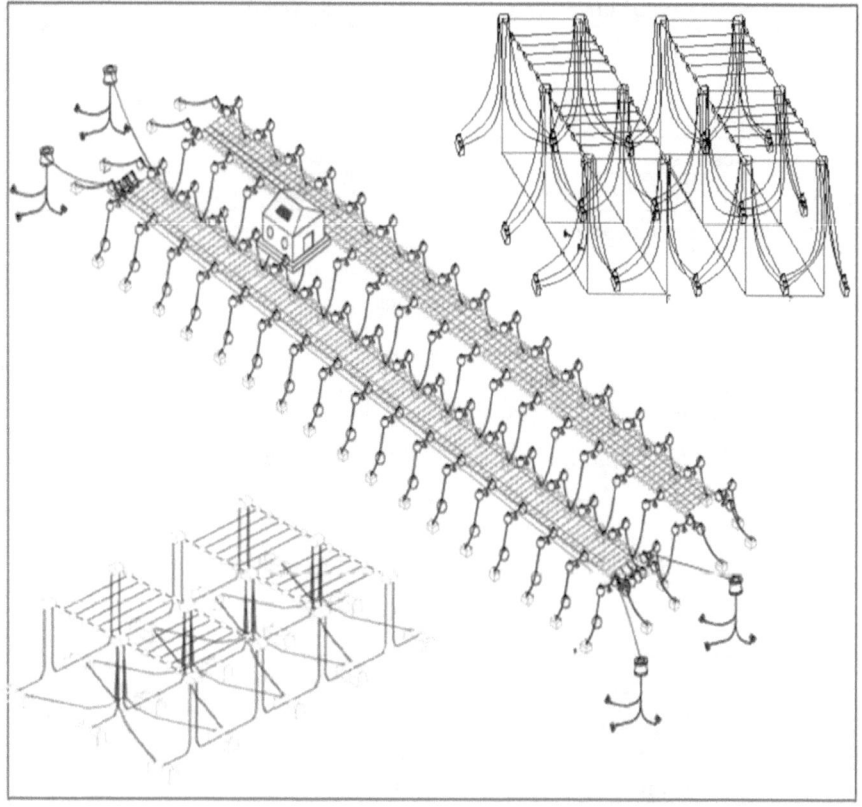

Figure 8.5: Offshore aquaculture system for seaweed oceanic plantation

8.19.2 Qualitative Assessment

The process is followed by qualitative risk as assessment. Risk analysis is less commonly used to achieve successful and sustainable aquaculture by assessing the risks to aquaculture posed by the physical, social and economic environment in which it takes place Table 8.2. These include reduction of environmental risks (e.g. due to poor sitting or severe weather events), biological risks (infection by pathogens via transfer from native stocks, predation by seals and sharks; red tides etc.), operational risks (poor planning, work-related injuries), financial risks (e.g. market changes, currency fluctuations, emergence of new competitors, etc.) and social risks (negative image and resulting product boycott, lack of skilled manpower, competition from other sectors). There exists, therefore, considerable scope to develop and expand the use of risk analysis for the benefit of aquaculture and the social and physical environments in which it takes place.

Table 8. 2: Societal risk from aquaculture

Environmental risks
—pollution from feeds, drugs, chemicals, wastes
—alteration of water currents & flow patterns
Biological risks
—introduction of invasive alien species, exotic pests & pathogens
—genetic impacts on native stocks
—destruction/modification of ecosystems and agricultural lands (mangrove deforestation, salivation of rice lands)
Financial risks
—failure of farming operations
—collapse of local industry/sector Social risks
—displacement of artisanal fishers Human health risks
—food safety issues

Table 8. 3: Risks to aquaculture from society and the environment

Environmental risks
• severe weather patterns
• pollution (e.g. agricultural chemicals, oil spills)
Biological risks
• pathogen transfer from wild stocks
• local predators (seals, sharks etc.)
• toxic algal blooms, red tide

—

Operational risks
• poor planning
• poor design
• workplace injuries
Financial risks
• market changes
• inadequate financing
• currency fluctuations
• emergence of new competitors
Social risk
• negative image/press
• lack of skilled manpower
• competition for key resources from other
Sectors
• theft, vandalism

8.19.3 System Functionality and Standard Analysis

System functionality and standard analysis should is best composed using block diagram of simplifies system with the input and output and feedback that represent sensing and detection point.

8.20 Qualitative Risk Assessment

Tools employed for qualitative assessment are described below:

8.20.1 Checklist

This is qualitative approach to insure the organization is complying with standard practice. The checklist can be used as a preparation for system design, deployment, maintenance and monitoring to avoid unnecessary problems and delays. The checklist included in the International Safety Management (ISM) procedures as documentation about checks for maintenance can be adopted for this study. The list can be filled in manually or printout electronically.

Checklist analysis is a systematic evaluation against pre-established criteria in the form of one or more checklists. It is applicable for high-level or detailed-level analysis and is used primarily to provide structure for interviews, documentation reviews and field inspections of the system being analyzed. The technique generates qualitative lists of conformance

—

and non-conformance determinations with recommendations for correcting non-conformances. Checklist analysis is frequently used as a supplement to or integral part of another method especially what-if analysis to address specific requirements. The quality of evaluation is determined primarily by the experience of people creating the checklists and the training of the checklist users.

The checklist analysis used most often to guide inspection of critical systems. It is also used as a supplement to or integrates part of another method, especially what-if analysis, to address specific requirements. Procedures for Checklist Analysis

i.   Define the activity or system of interest
ii. Define the problems of interest for the analysis
iii. Subdivide the activity or system for analysis
iv. Create relevant checklists
v.   Respond to the checklist questions
vi. Further subdivide the elements of the activity or system (if necessary or otherwise useful)
vii. Use the results in decision making

Table 8.3 shows checklist for development of offshore aquaculture system.

Table 8. 4: Risk to the system

Potential Risk			L-M-H	
			Likelihood	Impact
System	Configuration	The shape can it hold the system	L	L
	Buoy	Failure	H	H
	Rope (horizontal mooring/vertical)	Fatigue	M	M
	Mooring	Collapsed	H	H
	Anchor	Corrosion	H	H
	Material	Unsuitable material	M	M
	Structure		H	H
	Mooring failure			
	Disaster			
	Structural integrity			

—

Environment	Wind	Including normal to extreme wind	H	H
	Wave	The wave height	H	H
	Current	The maximum speed of the current	H	H
	Speed direction	The direction of speed came from	H	H
	Type of soil	The highest tide and the low tide	L	L
	Tide level	Type of soil underneath the sea	L	L
	Depth of sand	The maximum depth of sand layer	M	M
GHG	Global warming	Release the greenhouse gas	L	L
Cost	Theft	Treat from human being want to steal the seaweed	M	M
	Predator	Treat from turtle or any animals that eat seaweed	M	M
Location	Setiu, Terengganu			
Design	Inappropriate design	Poorly designed, constructed and maintained farms are more likely to pose a hazard to navigational safety.	H	H
Natural disaster	Tsunami	The natural disaster is unpredictable, when occurs may collapsed all system	M	M
	Swirl			
	Hurricane			
	Heat Wave			
Pollution	Water pollution	May the system effect /harm the sea water	L	L

Seaweed	long lasting	Can stand long time in sea water	M	M
	suitable type	The right type of seaweed that have many functions and give benefits		
Human	Error	Installation	H	H
		Procedure to farm the seaweed	L	L
Manual	System	Always take a look the system, site visit	L	L
Operation	The system cannot function	Make sure the system is function	H	H

Table 8. 5: Risk from the system to the environment

Potential Risk			L-M-H	
			Likelihood	Impact
Ecology	Habitat	May affect the ecosystem of living organism under sea water.	H	H
	Organism			
Passing vessel/ Navigation	Ships			
	Ferries	Disturb the sea traffic.	H	H
	Fishing boats			
Health	Medicine	More seaweed we farm	M	M
Human	Systematic system for human	Easy to farm seaweed in a proper way.	L	L

8.20.2 Hazard and Operability Study (HAZOP)

A hazard and Operability (HAZOP) study is a qualitative risk analysis technique that is used to identify weaknesses and hazards in a processing facility or system; it is normally used in the planning phase (design). The HAZOP technique was originally developed for chemical processing facilities, but it can also be used for other facilities and systems. For example, it is widely used in Norway in the oil and gas industry.

A HAZOP study is a systematic analysis of how deviation from the design specifications in a system can arise, and an analysis of the risk

—

potential of these deviations. Based on a set of guidewords, scenarios that may result in a hazard or an operational problem are identified. The following guidewords are commonly used: no/not, more of/less of, as well as, part of, reverse and other than. The guidewords are related to process conditions, activities, materials, time and place. The question would be:

i. What must happen to ensure the occurrence of the deviation "no throughput" (cause)?
ii. Is such an event possible (relevance/probability)?
iii. What are the consequences of no throughput (consequence)?

As a support in the work of formulating meaningful questions based on the guidewords, special forms have been developed. The principle that is used in a HAZOP study can be illustrated in the following way:

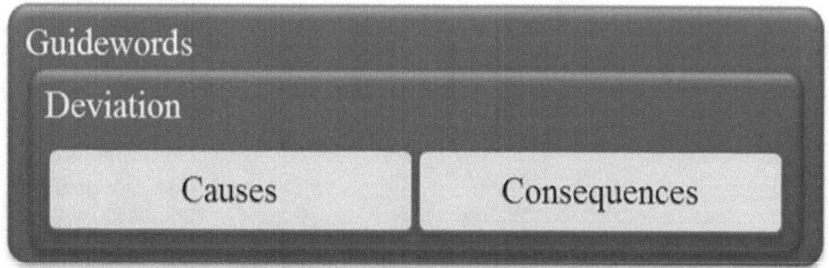

Figure 8. 4: HAZOP process

Table 8. 6: HAZOP assessment

No	Guideword	Description	Causes	Safety measure
1	No Pitch		operation, control mechanism, alignment failure	address by 2, 3, 4, 5
2	No blade	No rotational energy is transformed	Object in the water break the blade	implementation of propeller protection such as grating jet, sail in ice free water, +7& 9

3	No control bar	All blade on random pitch, loss of operational control	Material weakness	Improve design and construction
4	No crank wheel	On all blade have independent pitch		
5	NOT enough material strength	part of propeller breakdown	Wrong design, corrosion or cavitations, alignment different pitch, extra load on bearing	Validate propeller design, catholic protection, appropriate propeller material, test the propeller against cavitations, periodic alignment adjustment
6	MORE pitch than optimal	Too heavy load on propulsion system. Cavitations	Operation failure	Surveillance, increase operator competency
7	LESS pitch than optimal	Too little load on propulsion system. Cavitations		
8	LESS draft than allowed	Propeller I not sufficiently submerged. Loss of Thrust		
9	LESS depth than necessary	Propeller hit the ground and it is damaged		Technical equipment, surveillance, increase operator competency

In HAZOP study, critical aspects of the design can be identified, which requires further analysis. Detailed, quantitative reliability and risk analyses will often be generated after that. A HAZOP study of a planned plant or system will, in the same way as an FMEA, normally be most useful if the analysis is undertaken after the System Operation and Monitoring have been worked out. It is at this point in time that sufficient information about the way the plant is to be operated is available. A HAZOP study is a time and resource demanding method. Nevertheless, the method has been widely used in connection/, with the review of the

—

design of process system for a safer, more effective and reliable system. Figure 8.4 shows typical analysis for Hazard Operability (HAZOP). Table 8.7 shows typical analysis of PrHA and highlight risk contributing factor to collision.

Table 8. 7: PrHA for collision event

Hazards/ Events	Causes	Probabilities analysis	Consequence analysis	Risk	Possible measure to reduce risk
Powered vessel to Structure collision	Visual restriction at night	Probable (4)	Death / serious injuries (D)	4D	Night navigational aids made available
Moored vessel collision to structure	Unaware	Remote (2)	Death / disability (D)	2D	Installation of proper signage. Site identified with buoys.
Passing vessel to vessel collision	Too close/ Net caught under dredger		Probably major injury (C)	4C	Restrict navigation at nearby site. Install buoys.
Propulsion failure	Break down	Probable (4)	Probably major injury (C)	4C	Regular maintenance
Mooring failure	Weather damage		Probably major injury (C)	4C	Make assessment on weather conditions.

8.21 Quantitative risk assessment and analysis method.

8.21.1 Hazard Identification

Hazard identification (HAZID) and risk assessment involves a critical sequence of information gathering and the application of a decision-making process. These assist in discovering what could possibly cause a major accident (hazard identification), how likely it is that a major accident would occur and the potential consequences (risk assessment) and what options there are for preventing and mitigating a major accident (control measures). These activities should also assist in improving

—

operations and productivity and reduce the occurrence of incidents and near misses. The flowchart below summaries all the steps needed in a HAZID process. Major accidents by their nature are rare events, which may be beyond the experience of many employers. These accidents tend to be low frequency, high consequence events as illustrated in Figure 1 below. However, the circumstances or conditions that could lead to a major accident may already be present, and the risks of such incidents should be proactively identified and managed. Figure 8.5 shows the HAZID flowchart for typical rare events.

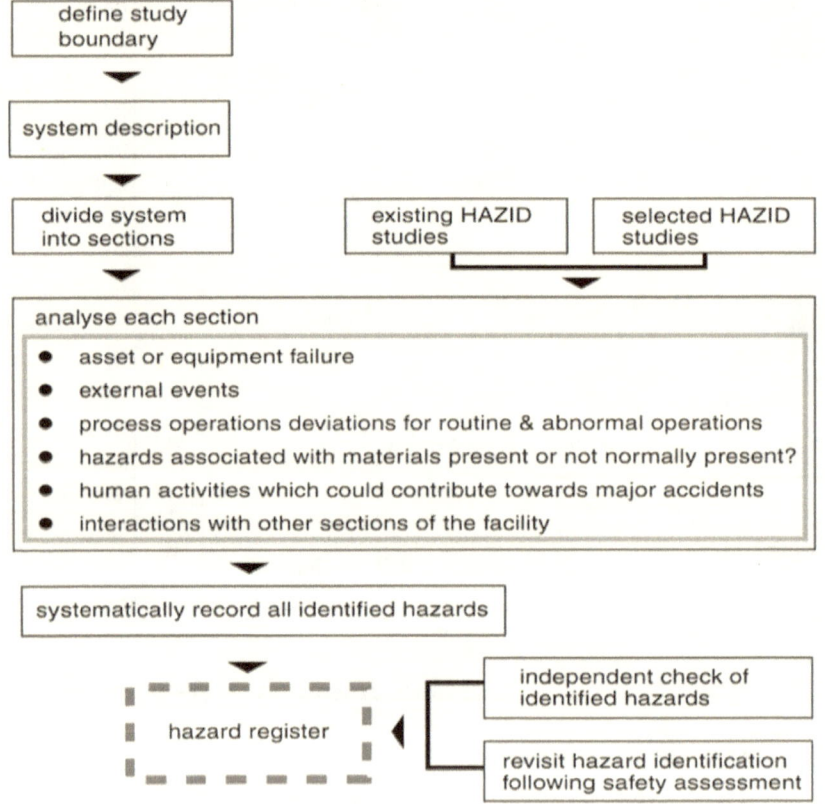

Figure 8. 5: HAZID rare events process and HAZID events process.

In assessing safety systems, towards deduction of option to mitigate the effects of external hazards, the assessor should have due regard to Reliability, redundancy, diversity and segregation. External hazards may particularly give rise to common mode or common cause failures.

8.21.2 Failure Modes and Effect Analysis (FMEA)

A failure modes and effects analysis (FMEA) is a *procedure* in *product development* and *operations management* for analysis of potential failure modes within a system for classification risk i2by the severity and likelihood of the failures. A successful FMEA activity helps a team to identify potential failure modes based on past experience with similar products or processes, enabling the team to design those failures out of the system with the minimum of effort and resource expenditure, thereby reducing development time and costs. It is widely used in manufacturing industries in various phases of the product life cycle and is now increasingly finding use in the service industry. Failure modes are any errors or defects in a process, design, or item, especially those that affect the intended function of the product and or process, and can be potential or actual. Effects analysis refers to studying the consequences of those failures. The Figure 8.6 below shows FMEA process toward determining the Risk Priority Number (RPN).

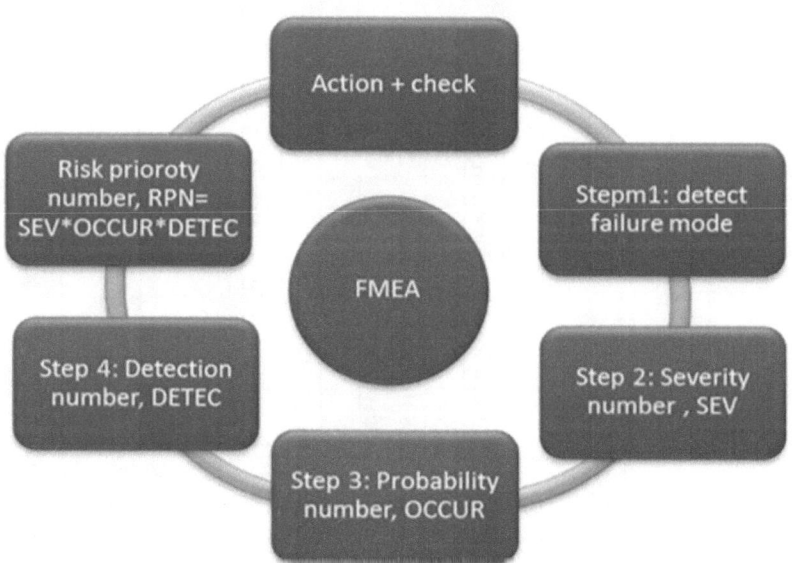

Figure 8. 6: Risk1 Priority Number.

The RPN (Risk Priority Number) is the product of Severity, Occurrence and Detection (RPN = S x O x D), and is often used to determine the relative risk of a FMEA line item. In the past, RPN has been used to determine when to take action. RPN should not be used this way. RPN is a technique for analyzing the risk associated with potential problems identified during a Failure Mode and Effects Analysis. RPN = Severity Rating x Occurrences Rating x Detection Rating, is the formula used in FMEA.

Table 8. 8: FMEA

Requirement	Potential failure mode	Potential effect of failure	Severity	Potential causes of failure	Occurrence	Current control	Detection	Risk priority number	Recommended action	responsibilities & target completion date	Revise			
											Severity	Occurrence	Detection	Risk Priority
Rope (Main)- 11 Unit	Degradation	Corrosion, wear, crushing, and wear on winch, torsion effect, chasing and fatigue	6	the salinity of sea water, weather	2	use the high quality of rope (PE rope, UV resistant)	2	24	Use the high quality of rope		5	5	5	125
Rope (Load line)- 300 unit								24	select the best rope					125

Typical FMEA for Offshore aquaculture farming is shown in Table 8.8 FMEA procedures are:

i.   Define the system and its performance requirements
ii. State all assumptions and ground rules that will be used in the analysis
iii. Develop block diagrams of the system and identify possible failure modes for example breaking, cracking, snap weather and others
iv. Identify cause of each failure mode
v.   Determine impact of every possible failure mode on the operation of affected items, items of subsequent assemblies, and the total system.

—

vi. List the possible symptoms of all failures and the means used to detect the failure.

vii. Assign a severity ranking to each failure mode.

viii. Assign an occurrence ranking to each failure mode for example estimate of the probability of the failure based on actual event occurrence.

ix. For each potential failure mode, perform a criticality analysis.

x. Evaluate and recommend any corrective actions and improvements to the design.

8.21.3 Fault Tree Analysis (FTA)

Fault tree analysis (FTA) is a top down, deductive failure analysis in which an undesired state of a system is analyzed using *boolean logic* to combine a series of lower-level events. This analysis method is mainly used in the field of *safety engineering* and *Reliability engineering* to determine the probability of a safety or accident or a particular system level (functional) failure. In Aerospace the more general term "system Failure Condition" is used for the "undesired state" v Top event of the fault tree. These conditions are classified by the severity of their effects. The most severe conditions require the most extensive fault tree analysis. These "system Failure Conditions" and their classification are often previously determined in the functional *Hazard analysis*. FTA can be used to:

i. Understand the logic leading to the top event / undesired state.

ii. Show compliance with the (input) system safety / reliability requirements.

iii. Prioritize the contributors leading to the top event—Creating the Critical Equipment/Parts/Events lists for different importance measures.

iv. Monitor and control the safety performance of the complex system (e.g. is it still safe to fly an Aircraft if fuel valve x is not "working"? For how long is it allowed to fly with this valve stuck closed?).

v. Minimize and optimize resources. Assist in designing a system. The FTA can be used as a design tool that helps to create (output / lower level) requirements.

vi. Function as a diagnostic tool to identify and correct causes of the top event. It can help with the creation of diagnostic manuals / processes.

—

Many different approaches can be used to model a FTA, but the most common and popular way can be summarized in a few steps. Remember that a fault tree is used to analyze a single fault event and that one and only one event can be analyzed during a single fault tree. Even though the "fault" may vary dramatically, a FTA follows the same procedure for an event, be it a delay of 0.25 msec for the generation of electrical power, or the random, unintended launch of an (Intercontinental Ballistic Missiles) *ICBM*. FTA analysis involves five steps:

i. Define the undesired event to study; Definition of the undesired event can be very hard to catch, although some of the events are very easy and obvious to observe. An engineer with a wide knowledge of the design of the system or a system analyst with an engineering background is the best person who can help define and number the undesired events. Undesired events are used then to make the FTA, one event for one FTA; no two events will be used to make one FTA.

ii. Obtain an understanding of the system; Once the undesired event is selected, all causes with probabilities of affecting the undesired event of 0 or more are studied and analyzed. Getting exact numbers for the probabilities leading to the event is usually impossible for the reason that it may be very costly and time consuming to do so. Computer software is used to study probabilities; this may lead to less costly system analysis. System analysts can help with understanding the overall system. System designers have full knowledge of the system and this knowledge is very important for not missing any cause affecting the undesired event. For the selected event all causes are then numbered and sequenced in the order of occurrence and then are used for the next step which is drawing or constructing the fault tree.

iii. Construct the fault tree; After selecting the undesired event and having analyzed the system to know all the causing effects (and if possible their probabilities) we can now construct the fault tree. Fault tree is based on AND and OR gates which define the major characteristics of the fault tree.

iv. Evaluate the fault tree; After the fault tree has been assembled for a specific undesired event, it is evaluated and analyzed for any possible improvement. This step is as an introduction for the final step which will be to control the hazards identified. In short, in this step it is not required to identify all possible hazards affecting the system in a direct or indirect way the system.

—

v. Control of the hazards identified; This step is very specific and differs largely from one system to another, but the main point will always be that after identifying the hazards all possible methods are pursued to decrease the probability of occurrence. Fault trees are developed using gate and events symbols. A gate may have only one input and one or more outputs. Dhillon and Kapur have defined the following common gate and event symbols for use in FTA (Figure 8.7).

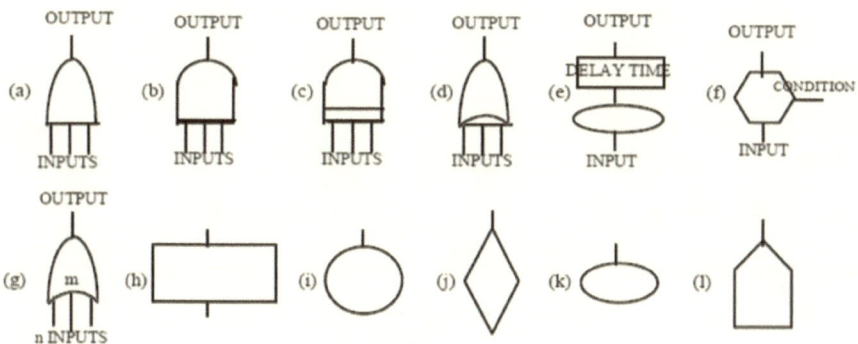

Figure 8. 7: Logic symbol for FTA

a. Gate: used when output event occurs when one or more input events occur.
b. AND Gate: used when output event occurs when all input events occur.
c. Priority AND Gate: like AND gate but input events occur in a specified order.
d. Exclusive OR Gate: used when output occurs when one, and only one of the input events occur.
e. Delay Gate: used when output event occurs after a specified time delay.
f. Inhibit Gate: used when output event occurs based on a conditional event occurring.
g. M-out-of-N Gate: used when output event occurs based on a m out of n input events occurring.
h. Resultant Event: used to represent an event resulting from some combination of preceding fault events.
i. Basic Fault Event: used to represent failure of component or subsystem.

j. Incomplete Event: used to represent a fault event whose cause has not yet been determined.
k. Conditional Event: used to represent the condition associated with an Inhibit gate.
l. Trigger or Switch Event: used to examine special cases by forcing events to occur, or by forcing them not to occur.

Equation for FTA:

Ab	=	ba	(cumulative law)
a + b	=	b + a	(cumulative law)
(a + b) + c	=	a + (b + c) = a + b + c	(associative law)
(ab) c	=	a (bc) = abc	(associative law)
a (b + c)	=	ab + ac	(distributive law)

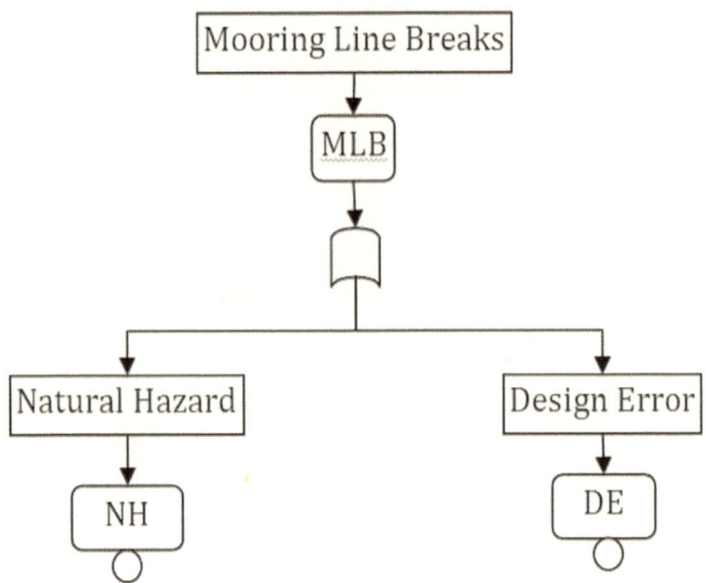

Figure 8. 8: A FTA diagram for MLB

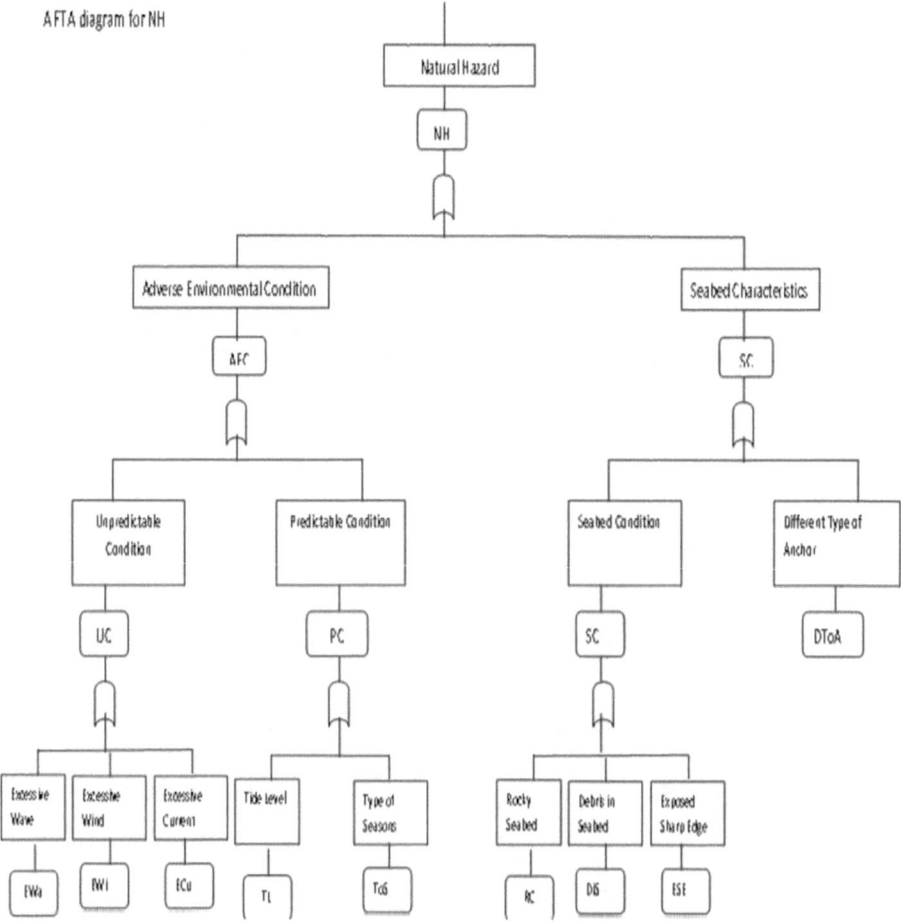

Figure 8. 9: FTA for Natural hazard

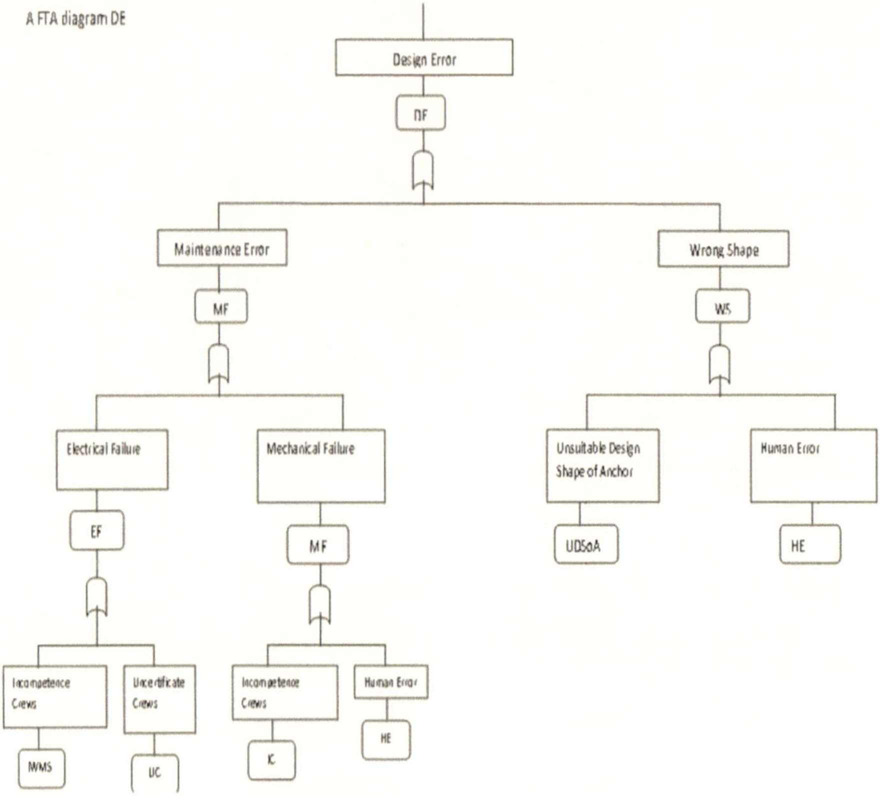

Design Error

DF

Maintenance Error

MF

Wrong Shape

WS

Electrical Failure

EF

Mechanical Failure

MF

Unsuitable Design Shape of Anchor

UDSoA

Human Error

HE

Incompetence Crews

MMS

Uncertificate Crews

UC

Incompetence Crews

IC

Human Error

HE

Figure 8. 10: FTA for design error

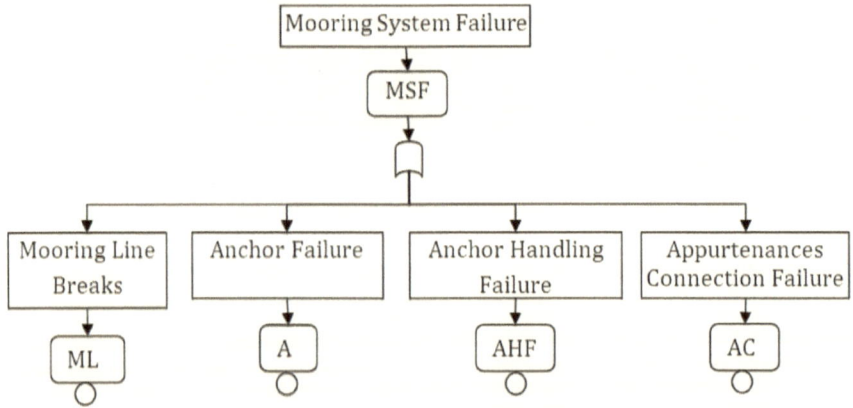

Figure 8. 11: A FTA diagram for MSF

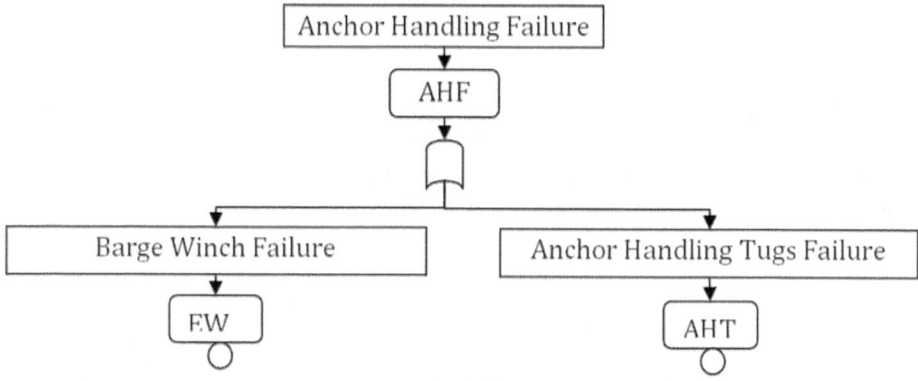

Figure 8. 12: A FTA diagram for AHF

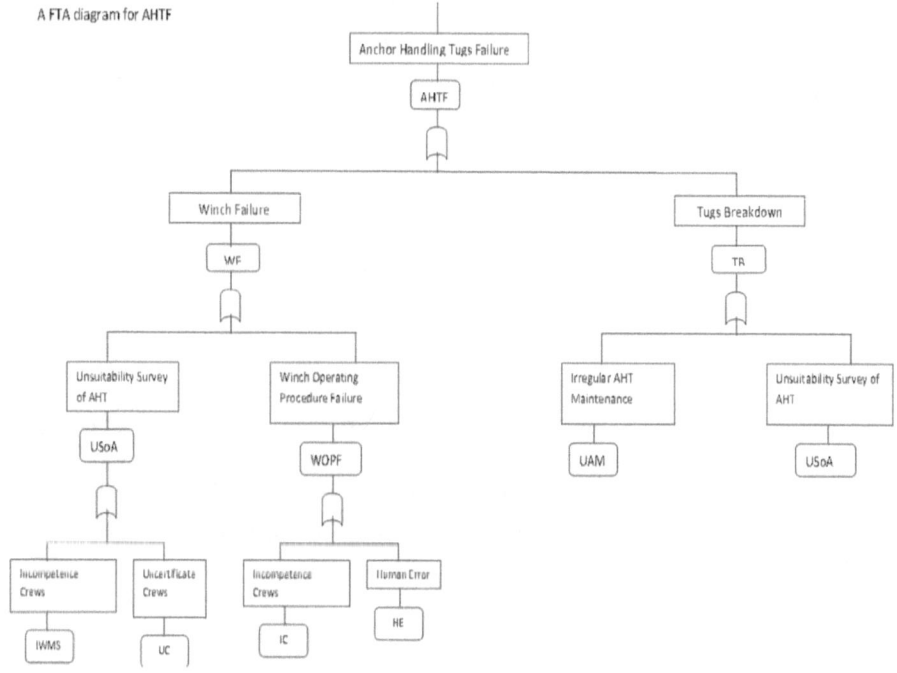

Figure 8. 13: FTA for anchor handling tug failure

8.22 Quantitative failure probability and consequence risk analysis for very large floating structure for offshore aquaculture structure

8.22.1 Failure probability

A mooring device is failed when the mooring reaction force W, due to oscillation of the floating structure, exceeds the yield strength R. The floating structure drifts when all its mooring devices are failed. Failure of a mooring device indicates presence of an event satisfying the following condition:

$$0 \leq t \leq TZ_k(t) = W_k(t;X) - R_k > 0 \qquad 8.01$$

Where X is natural condition parameters, T duration of the natural condition parameters, and R_k the random variable for the final yield strength of mooring device k, X and R_k are independent of each other.

The probability of a multi-point mooring system being failed by strong wind and waves in specified service life is given by the following equations:

$$P_f(T) = \iint dx_i dr_k$$
$$\text{Prob}\left[\bigcup_{k=1}^{m} Z_k(t) > 0; 0 \leq t \leq T \mid X = x_k, R_k = r_k\right] f_X(x_i) f_R(r_k) \qquad 8.02$$

Where Prob [A| B = C] is the probability of under the condition of B = C, and $f_x(x)$ and $f_R(r)$ are probability density functions of natural condition parameters and final yield strength of mooring device, respectively. Using the extreme-value distribution of the annual maximum values as the distribution of natural condition parameters, we define the annual reliability as follows:

$$R(T) = 1 - P_f(T) \qquad 8.03$$

The total reliability for years of service life is approximated by the following equation:

$$R_n (T) = (1 - P_f (T))^N \qquad\qquad 8.04$$

Estimation of failure probability

The governing equation for oscillation of the floating structure is defined as follows:

$$[M_{ij} + m_{ij}(\infty)]X(t)'' + F_v(\dot{X}) + \sum_{j=1}^{n} \int_{-\infty}^{t} \dot{x}_j(\tau)L_{ij}(t-\tau)d\tau + F_M(X,\dot{X}) = F_{wind}(t) + F_1(t) + F_2(t)$$

$$\qquad\qquad 8.05$$

Where $\overline{X}$: displacement vector of horizontal plane response of the floating structure; M_{ij} : inertia matrix of the floating structure; $M_{ij}(\infty)$: added mass matrix at the infinite frequency; F_v: viscous damping coefficient vector; L_{ij} : Memory influence function; F_M: Mooring reaction force vector; F_{wind} : Wind load vector, F_1 and F_2: first and second wave force vectors respectively.

Estimation wave force

Wave force vector is generally expressed as the sum of linear wave force proportional to wave height and the slowly varying drift force proportional to the square of the wave height. See the equation below.

$$F(t) = F_1(t) + F_2(t)$$
$$= \int h_1(\tau)\zeta(t-\tau)d\tau + \iint h_2(\tau_1,\tau_2)\zeta(t-\tau_1)\zeta(t-\tau_2)d\tau_1 d\tau_2 \qquad\qquad 8.06$$

Where $h_1(\tau)$ $h_2(\tau_1, \tau_2)$ are the vectors of impulse response function of wave force. $\zeta(t)$ is the time series of surface elevation of incident waves.

Estimation of risk of current load

Floating structure and the pressure drag for the lateral walls. Average wind velocity distribution on the horizontal plane is assumed uniform. The velocity profile in the perpendicular direction expressed using the logarithmic rule. For the fluctuating wind velocity, the mainstream direction (average wind velocity direction) is the sole element of consideration. The power spectrum of fluctuating current load is given in the following equation that considers spatial correlation:

—

$$S_{FF}(f) = \rho_a^2 \iint_A C_{di}(f)C_{dj}(f)U_iU_j \, Re[R_{ij}(f)]\sqrt{S_i(f)S_j(f)} \, dA_i dA_j \qquad 8.07$$

The spatial correlation is defined as follows:

$$R_{ij}(f) = \exp\left(-\frac{k_1 f|y_i - y_j|}{\sqrt{U_iU_j}}\right)\exp\left(i\frac{k_2 f(x_i - x_j)}{\sqrt{U_iU_j}}\right) \qquad 8.08$$

where ρ_a: air density; U: average wind velocity; dA: area element of the floating structure surface; C_d: drag coefficient; $S(f)$: power spectrum of fluctuating wind; x: coordinate in the mainstream direction of plane element; y: coordinate at right angles to the mainstream direction of plane element; k_1: spatial correlation coefficient at right angles to the mainstream; and k_2: spatial correlation coefficient in the direction of mainstream.

Estimation natural environmental condition

Assuming yield strength R is a deterministic value and wave height and others are a function of wind velocity, in equation as given below. This enables us to calculate annual initial failure probability from the distribution of the conditional failure probability and the distribution of the probability of annual maximum current velocity.

$$P_f = \int_B^\infty P[T \mid U_{10}]f(U_{10})dU_{10} \qquad 8.09$$

Where P[T|.] is conditional initial failure probability during duration time T and $f(U_{10})$ probability density function for annual maximum wind velocity. The lowest value B for integration varies with which extreme-value distribution the conditional failure probability is approximated by the equation.

8.22.1 Aessment of Functional and Serviceability

Modern safety criteria for marine structures are expressed by limit states as indicate in the Table 8.9 below and are briefly outlined in the following. This will be applied to stages of risk and reliability assessment and analyzing the system required.

Table 8. 9: Safety Criteria (e.g. ISO 19900 1994, Moan 2004)

Limit State	Description	Remarks
Ultimate (ULS)	Overall structure stability. Ultimate strength of structure. Ultimate strength of mooring system.	(Not relevant for VLFS) Component design check
Fatigue	Failure of joint-normal welded joins in hull and mooring system.	Component design check depending on residual system strength after fatigue failure.
Accidental collapse (ALS)	Ultimate capacity of damaged structure (due to fabrication defects or accident loads) or operational error.	System design check
Serviceability (SLS)	Structure fails its serviceability if the criteria of the (SLS) are not met during the specified service life and with the required reliability	Disruption of normal use due to excessive deflection, deformation, motion or vibration.

8.20.2 Typical Result from Quantitative Analysis

The expected results of this study analysis are declared from based on the methodologies applied to the study analysis. In Figure 8.15 it is expected to obtain the probability of exceedance of the mooring reaction relative to average current velocity.

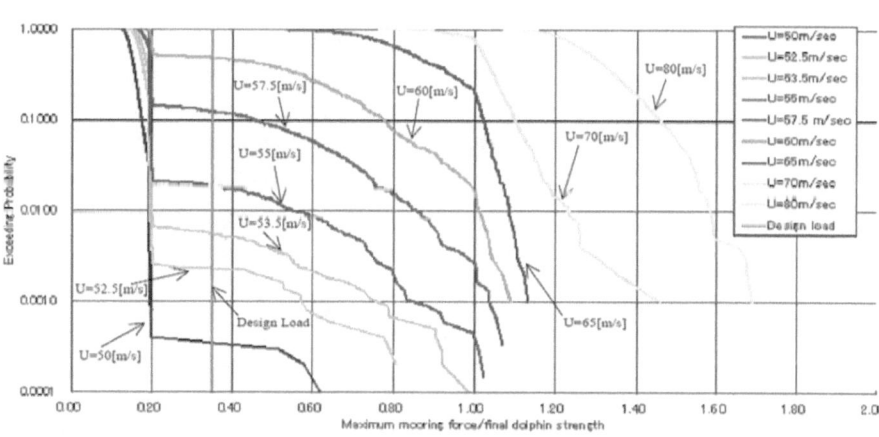

Figure 8. 14: Graph of quantitative analysis

—

303

The Figure 8.16 shows example of the extreme value distribution of annual maximum current velocities graph based on natural environmental condition that shows below. The probability of failure and reliability of the system is very low compared to structure of very large floating structure.

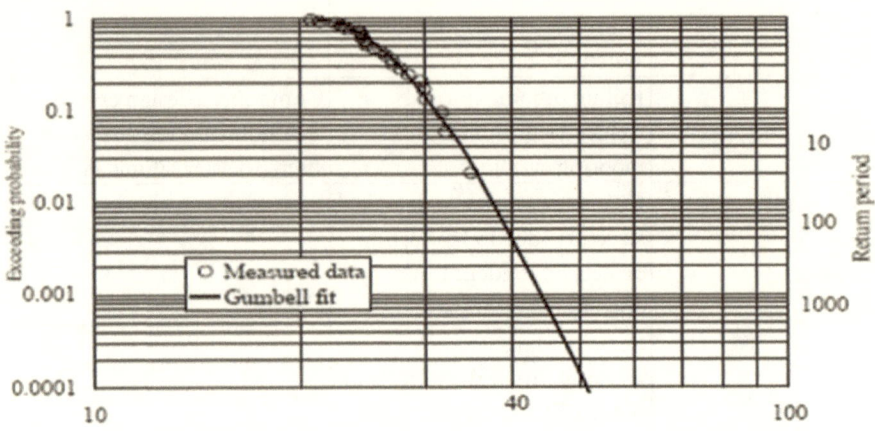

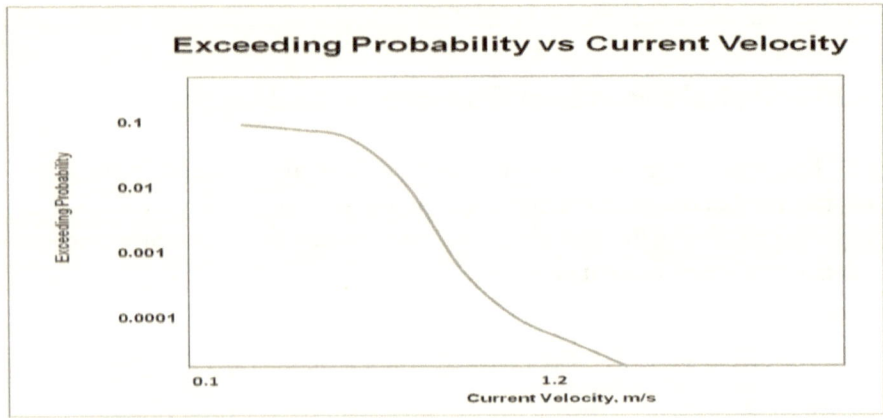

Figure 8. 15: The extreme value distribution of annual maximum current velocities graph

In Figure 8.17 expected require conditional failure probabilities and mean current reliability could be obtained.

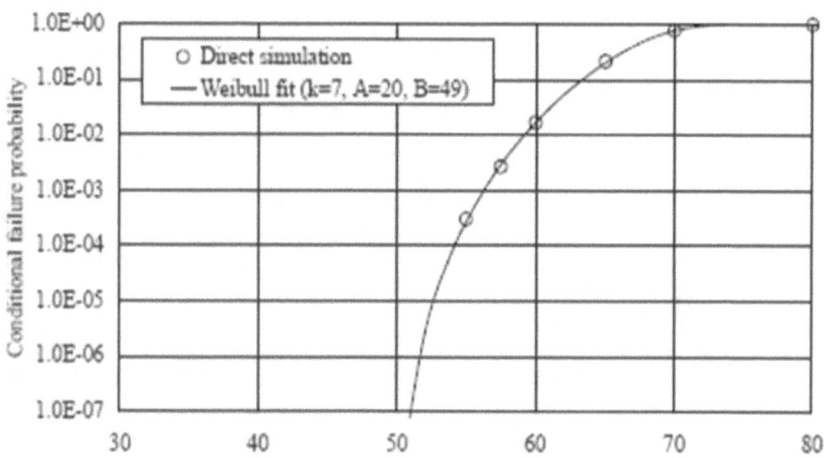

Figure 8. 16: The conditional failure probabilities and mean current reliability

Figure 8.18 shows expected result for variation of failure probability to a number of mooring on the system. The system can be able to withstand the harsh condition. However, a lower reliability for accidental loads may be accepted, as society understands that the engineer is less able to predict these events and there may be less of an outcry if failure occurs due to an accidental load.

An integrated approach to risk analysis will assist the aquaculture sector in reducing risks to successful operations from both internal and external hazards and can similarly help to protect the environment, society and other resource users from adverse and often unpredicted impacts. This could lead to improved profitability and sustainability of the sector, while at the same time improving the public's perception of aquaculture as a responsible, sustainable and environmentally friendly activity.

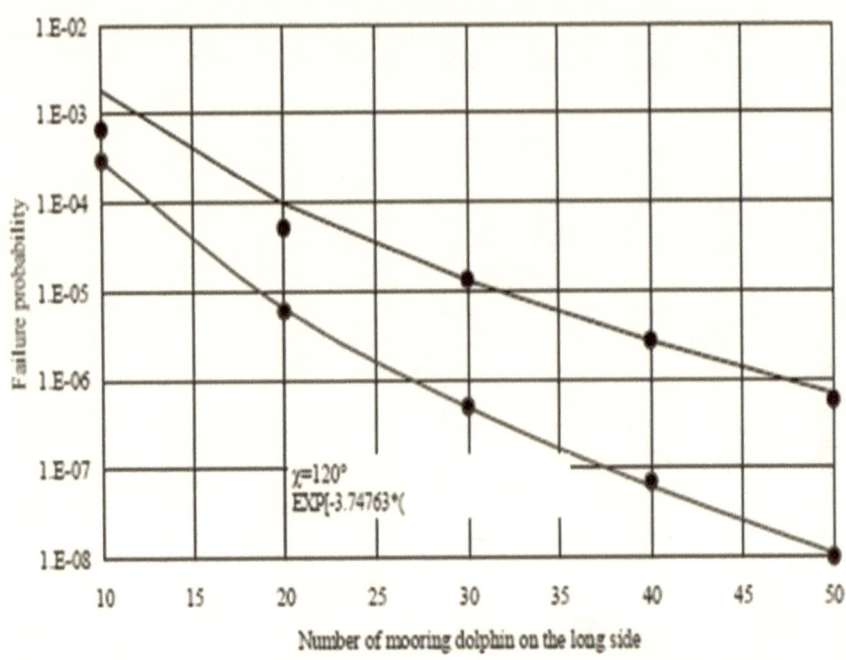

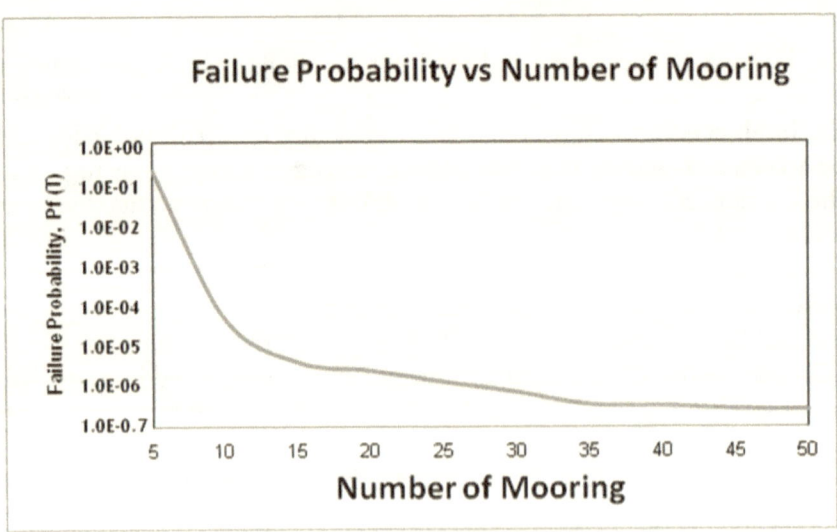

Figure 8. 17: The variation of failure probability
to a number of mooring on the system.

Reference

1. Bercha, FG, Cervosek, M, and Abel W.(2004). Assessment of the Reliability of Marine Installation Escape, Evacuation, and Rescue Systems and Procedures, in Proceedings of the 14th International offshore and Polar Engineering Conference (ISOPE), Toulon, France.
2. Sade, A. (2006). "Seaweed Industri in Sabah, East Malaysia". In A. T. Phang Siew-Moi, *Advances In Seaweed Cultivation And Utilization In Asia* (pp. 41-52). Kota Kinabalu, Sabah: University of Malaya Maritime Research Centre.
3. Sulaiman Oladokun Olenwanju (2012). "*Safety and Environmental Risk Model for Inland Water Transportation*". University Technology Malaysia.
4. Stamatis, D.H (2003). Failure Mode and Effect Analysis: FMEA from Theory to Excecution, 2nd Ed. United States of American Society for Quality.
5. Li, Y. and Kareem, A. 1993. Multivariate Hermite expansion of hydrodynamic drag loads on tension leg platforms. J. Engrg. Mech. ASCE, 119 (1), 91-112.
6. Ayyub, B.M., Beach, J.E., Sarkani, S., Assakkaf, I.A. (2002). *Risk Analysis anad Management for Marine System*. Naval Engineers Journal, Vol. 114, No.2, pp.181-206.
7. E. Watanabe, C. W. (2004). *Very Large Floating Structures: Application, Analysis & Design*. Singapore: Centre for Offshore Research and Engineering National University of Singapore.
8. Huse, E. (1996). Workshop on Model Testing of Deep Sea Offshore Structures. *ITTC1996, 21st International Towing Tank Conference* (pp. 161-174). Trondheim, Norway,: NTNU, Norwegian University of Science and Technology, 1996.
9. ISSC2006. (2006). ISSC Commitee VI.2 "Very large Floating Structures". *16th International Ship & Offshore Structures Congress 2*, (pp. 391-442). Southampton, UK.
10. Koichiro Yoshida, K. K.-S. (1993, May). Model Tests on Multi-Unit Floating Structures In Waves. (N. Saxena, Ed.) *Recent Advances In Marine Science and Technology, 92*, 317-332.
11. Moan, T. (2004). "Safety of floating offshore structures" *Proc. 9th PRADS Conference*, Keynote lecture, PRADS Conference, Luebeck-Travemuende, Germany, September 12—17, 2004.
12. Snell, R., Ahilan, R. B. and Versavel, T. (1999), "Reliability of Mooring Systems: Application to Polyester Moorings," *Proceedings*

—

of Annual Offshore Technology Conference, OTC 10777, Houston, TX, USA, 125-130.

13. Tang, W. H., and Gilbert, R. B. (1993). "Case study of offshore pile system reliability." *Proceedings of Annual Offshore Technology Conference*, OTC 7196, Houston, TX, USA, 677-683.

14. D.Dessi, A. G. (2004, March 28-31). *Experiment and Numerical Analysis of a Moored Floating Strucuture Response to Waves*. Retrieved April 2012, from International Workshop on Water Waves and Floating Bodies: *http://www.iwwwfb.org/Abstracts/ iwwwfb19/iwwwfb19_10.pdf*

15. *http://assakkaf.com/Papers/Journals/Risk_Analysis_and_ Management_for_Marine_Systems.pdf*

16. Kaur, C. R. (2009, November 17). *Developing Malaysia's Seaweed Aquaculture Sector*. Retrieved November 1, 2011, from Baird Maritime: http://www.bairdmaritime.com

17. Wikipedia. (2012, 10 19). *Failure Mode and Effect Analysis*. Retrieved 11 14, 2012, from Wikipedia:

18. *http://en.wikipedia.org/wiki/Failure_mode_and_effects_analysis*

19. J. Richard Arthur, UN Food and Agriculture Organisation (FAO) consultant explains the general principles of the risk analysis process and its application to aquaculture.

CHAPTER 9

Determination of Hydrodynamic Loads on Seaweed for Use in Design of Aquaculture Mooring System

"If the fact does not fit the theory, change the fact"
Albert Einstein

Summary

Seaweed farming has potential economic benefits in Malaysia as a source of sustainable supplementary income for fishermen in remote areas. Prototype plantations have been deployed in coastal areas and therefore are competing for limited space with other shore side activities such as recreation and industry. If seaweed can be cultivated in relatively deeper waters away from populated areas, production may be expanded using larger areas not presently exploited for other purposes. Prototype systems aquaculture plantations have so far relied on ad hoc, field developed mooring systems, rather than engineering design principles. Such systems may not be sufficiently reliable for deepwater deployments, where repairs are more costly and resources are scarce. Nevertheless, considerable experience and hands-on knowledge has been obtained. So that the basic design concepts can be used as a basis for further improvements.

9.1 Introduction

In order to design the mooring system for a typical offshore installation, dynamic and hydrodynamic design loads are needed. In the case of a jacket, the loading can readily be calculated using Morison's formulations based on empirical data or for a floating structure like a Spar or FPSO existing software like WAMIT and dynamic simulation tools can be used. However, for a deep water aquaculture facility no similar tools are available to calculate the hydrodynamic forces on such flexible structures. It was decided that the quickest way to obtain design loadings for seaweed would be to carry out model tests on some available seaweed samples. The main difficulty is that the flow kinematics (fluid velocities/accelerations) that can be produced in the available model basin, at the Marine lab of the Universiti Teknologi Malaysia in Johor Bahru, where the tests are to be carried out, are suitable only at smaller scale (in a range typically around 1/50), whereas the seaweed samples are actually fullscale (1/1) size and mass. So, there is a mismatch that has to be overcome by suitable simplifying assumptions. This section describes the model tests and analysis used to develop the loads needed so that the mooring system for the aquaculture system can be designed based on methods like those used for offshore structures.

9.2 Offshore Aquaculture for Oceanic Seaweed Marine Algae Farming

The ocean represent one big frontier for humanity to explore, much of it biological species remain untap. The case of seaweed or marine algae stands significant because of its potential to that benefit solution to human activities that end with unwanted effect. For example, the green algae (Chlamydomonas reinhardtii) seaweed is used as biomass, hydrogen fuel source, as it can produce superior amount of vegetable oil and heat compare with terrestrial crop. Marine algae are used as food supplement that provide the body with additional fuel and immune body system regulatory response. It contains extensive fatty acid profile including Omega 3 and Omega 6 and it contain abundant vitamins, mineral, and trace elements. Chondrus crispus know as carrageen is an excellent stabilizer in milk, gelatin and cosmetic product. Algae are used as fertiliser for livestock and soil; Algae are also used for waste water pollution control. Figure 9.1 shows seaweed under test in towing tank.

—

Figure 9. 1: Seaweed under test

Components of the Floating Structure is as followed (Figure 9.2)

i. *Frame Line:* This line support the planting line for each block of the structure. It was made from synthetic fibre rope.

ii. *Planting Line:* Planting line is the main part of the structure which contains the planted seaweed. This line must be able to withstand the mature growth of seaweed weight.

iii. *Separator Line:* When the nature hit floating structure, the planting line tends to tangle. Therefore, this line will act as the separator between each planting line.

iv. *Mooring Line:* This line holds the whole structure at the surface to be in place with connection to the anchor. Since mooring system is a crucial part for floating structure. It must be designed to withstand the natural force and achieve stability through the use of mooring line tension.

v. *Buoy:* Buoy is to provide a convenient means for connection of the floating structure on the surface to the mooring. Through the use of distributed buoyancy for each buoy, floating structures can achieve stability. The buoy also has to withstand the structural weight and additional load from the seaweed.

vi. *Sinker:* A sinker is made of concrete and placed on a catenary mooring line to ensure horizontal mooring at anchor, enhance mooring line energy absorption and affect mooring line pretension in a way that can be useful in controlling structural stability.

vii. *Anchor Block:* Anchor is designed in a large mass of concrete to keeps the floating structure at the place also to resists both horizontal and vertical movement.

—

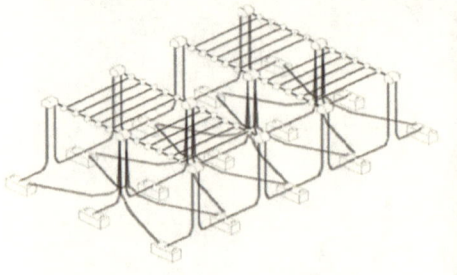 

a. Physical system deployed at sea   b. Mooring configuration

Figure 9. 2: Floating Structure for Ocean Farming System

The study focuses to produce a design for offshore floating structure of seaweed farming. This design is required to meet the operating conditions, strength and serviceability requirements, safety requirements, durability, visually pleasing to the environment and cost-effective. An appropriate design service life is prescribed depending on the importance of the structure and the return period of natural loads. Its service life is generally expected to be as long as 50 to 100 years with preferably a low maintenance cost. The structure will be operating 200 meters from the shore as a result; the structure is likely to experience more energizes wave action and stronger wind associated with deep water region. This design also considered 1-2 boat lanes within the structure blocks which is about 5 meters wide at the original size. In the structural design of floating offshore structures, the external load and major load effect, such as cross sectional forces, are determined from the rigid body motions.

The dimensions of structural members and arrangement are subsequently determined so that the structure has sufficient strength and stiffness against the given loads and loads effects. The hydrodynamic loads measured will be used to build approximate scale models of the seaweed. The model seaweed will mimic the Froude-scaled properties (mass, dimensions, added mass and damping) of the seaweed measured previously. Suitable material such as plastic ribbon, rubber tubing or even young seaweed seedlings will be used to build a sufficiently quantity of scaled seaweed.

9.3 Environmental Load Consideration

The weather in Malaysia is mainly influenced by two monsoon regimes, namely, the Southwest Monsoon from late May to September, and the Northeast Monsoon from November to March. (K.C.Low, 2006).

However the east coast of peninsular Malaysia is the area that exposed directly to the strong sea currents and periodic monsoon season which is prevalent off the east coast. Furthermore, with the existence of nature elements of the deep and open water environment, seaweed farming is hard to be applied in this area.

Regular waves were considered and generated by the wave maker for a few tests. Random waves spectrum was based on the Jonswap spectrum for less than 1Yr or for 1Yr or greater, respectively. Froude scaling was applied to establish the relationship between full scale wave height (Hp) and period (Tp) and the corresponding model scale wave height (Hm) and period (Tm), where Hm = Hp/50 and Tm = Tp/Ö50 (Table 9.1). Incident waves will be measured and analyzed prior to the tests. Two wave probes will be installed for calibrations: one in front of the carriage at the basin centerline and one to the side of the nominal position of the model. Wave force vector is generally expressed as the sum of linear wave force proportional to wave height and the slowly varying drift force proportional to the square of the wave height. Mooring design for offshore platforms makes use of software tools which have been benchmarked against model tests, computational data and full scale measurements for their given applications. Hydrodynamic loading on the platform, risers and mooring system itself due to waves and currents are calculated using a variety of tools such as potential flow, CFD and empirical data.

Table 9. 1: Full Scale Wave

Return Period (Year)	Full Scale Wave Height (m)	Full Scale Wave Period (s)
90%	4.599	9.711
95%	4.850	10.25
1/12th	0.6	1.295
1-Yr	5.110	10.79
10-Yr	10.7	12.82
100-Yr	7.3-13.6	11.1-15.1

9.4 Model Test

The model test is required to determine the hydrodynamic loads due to waves and currents acting on seaweed and its mooring system components. The total system loads must be suitable for use in designing a seaweed culture mooring system to avoid failure with potential loss of the valuable crop and possibly requiring costly repairs or replacement of the mooring system components.

A towing test conducted at UTM marine Lab involved (Figure 9.3) one method of research to determine the hydrodynamic loading coefficients (added mass and damping) in a few different configurations [6]. The samples which are seaweed are dried, but will restore to nearly nominal properties when soaked in water for a period of time. In typical practice, the rows of seaweed are held using ropes separated by about 2.6 m between rows. For the drag measurement, the seaweed was attached to towing lines and the lines were towed from the moving carriage. A frame consisting of aluminium channel sections attached to the towing carriage is used. The seaweed clumps then in turn attached to a rope line. Tension load cells will be attached between the line and the frame and the measured forces recorded on the model basin's data acquisition system (Figure 9.3).

Component of the full-scale platform and mooring cable characteristic first need to be clarified. A typical block has an overall length of 100 meter and the breadth is 100 meter. One block of seaweed farming contains four main ropes for the frame, four main buoy and 30 load lines on which is the seaweed will be planted. Table 9.1 shows the structural properties of one block of seaweed farming (Table 9.2 and 9.3).

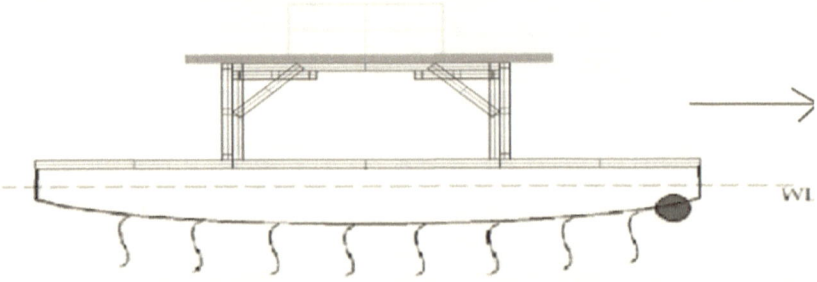

Figure 9. 3 UTM Towing Tank and Carriage

Table 9. 2: Structural properties

Item	Actual Structure	Model
Length Overall for 10 Blocks, L	1000m	2m
Breath, B	100m	2m
Dimension for Each Block	100m x 100m	2m x 2m
Mooring Depth, D	50m	2.5m

Table 9. 3: Structural dimension

Structural properties (one block)	No/Quantity	unit
Length overall	100	m
Breadth	100	m
Main buoy	4	
Main rope	4	
Load line rope	30	
Sea water density	1025	Kg/m^3

The model test is designed to investigate the modeling laws required for the system in question to be analyzed. The scaling parameters are very important in designing a model test and a few key areas of consideration in replicating a prototype structure for a physical model test. In order to achieve similitude between the model and the real structure, Froude's law is introducing as the scaling method. Froude's law is the most appropriate scaling law for the free and floating structure tests (Chakrabarti, 1998). The Froude number has a dimension corresponding to the ratio of u^2/gD, where u is the fluid velocity, g is the gravitational acceleration and D is a characteristic dimension of the structure Figure 9.4 shows the scale model constructed by UMT students. The Froude number Fr is defined as:

$$Fr^2 = gD/u \qquad\qquad 9.01$$

The subscripts p and m stand for prototype and model respectively and λ is the scale factor. Assuming a model scale of 1 and geometric similarity, the Froude model must satisfy the relationship: $u_p^2/gD_p = u_m^2/gD_m$ 9.02

Important variable quantities of importance are derived from the equation and dimensional analysis as follows:

—

Linear	$l_p = \lambda\, l_m$
Speed	$u_p = \sqrt{\lambda}\, u_m$
Mass	$m_p = \lambda^3\, m_m$
Force	$F_p = (\lambda^3 / 0.975)\, {}^*F_m$
Time	$t_p = \sqrt{\lambda}\, t_m$
Stress/Pressure	$S_p = \lambda\, S_m$

Figure 9. 4: Scale model of the physical system

The model tests involved:

i. Measure drag loads of actual seaweed by towing
ii. Dynamic tests to measure added mass and damping using PMM. Originally, it was planned to use the planar motion mechanism (PMM).
iii. Hydrodynamic tests on component tests (Buoy, ropes, float, net, . . .) considered available from coefficient of mooring components. Industry data can be substituted for component loading during initial design work.
iv. Complete system tests with scaled seaweed of 1/50 scaled model deployed at UMT

The complete system tests will be used to confirm the adequacy of the preliminary design. Instrumentation required:

i. Load cells of suitable size and range will be attached to the seaweed lines

ii. Wave probes
iii. Native carriage speed record
iv. Wave flap signal

The load cells attached to the mooring spring are water-proof aluminum ring strain gauges that measure axial tensile loads. The measured voltage outputs from the load cell strain gauges are connected by cables to the basin's native Dewetron Data Acquisition System (DAQ)) to be digitally sampled and stored. Software is used to convert the measured voltage to tension readings. The load cells are appropriately-scaled and calibrated (100N range). Other instruments used in the tests are wave probes fixed at specific locations under the carriage and accelerometers mounted on the model decks. Both are channeled to the DAQ to record measured data output. A video camera was positioned at strategic locations on the carriage for model motion recordings.

9.4.1 Tests for Seaweed Hydrodynamic Coefficients

Tests is carried out to identify hydrodynamic coefficients of an equivalent Morison model of the seaweed which will be suitable for use in typical mooring design and analysis package such as Arienne. Samples (clumps) of dried seaweed, the dried will restore to nearly nominal properties when soaked in water for a period of time (Figure 9.5). Typical size seaweed clump weigh up to 1.5kg in air, when fully grown. However, the natural buoyancy of the seaweed, make its weight in water almost insignificant. A sample clump of seaweed weight in air 4.1N and the corresponding weight in water is 0.01N in UTM lab.

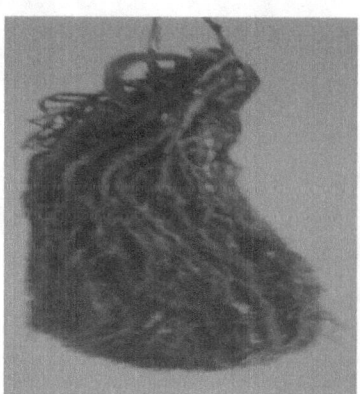 

Figure 9. 5: Seaweed

A sample row of seaweed is attached to a frame and towed from the carriage. To determine the hydrodynamic coefficient, a series of tests including towing in calm water, towing in waves and wave-only tests will be performed. As mentioned, originally, it was planned to use the PMM to perform forced oscillation tests. The resulting loads can be analysed in a straightforward manner to determine the relevant coefficients which give similar hydrodynamic loads.

However, the PMM system is not functioning at present. So, an alternative plan had to be developed. The main difficulty is that the available seaweed samples represent full-scale (prototype) clumps, whereas, the model basin is equipped to generate wave and current kinematics (wave height <0.4m 0.5s<Period<2.5s, and speed<5 m/s) typical of model scale conditions. Therefore, a mismatch exists between the typical body (clump) dimensions and the relative kinematics that can be generated. Some approximations and simplifying assumptions are needed to utilize the available facilities to determine the required coefficients. The tests of floating structure in regular and irregular waves will be carried out in the towing tank 120m x 4m x 2.5m of Marine Technology Laboratory UTM. This laboratory is equipped with the hydraulic driven and computer controlled wave generator which is capable to generate regular and irregular waves over a period range of 0.5 to 2.5 seconds. For this structural experiment, a model of 2m x 2m per block with 50 scale ration will be used (Figure 9.6).

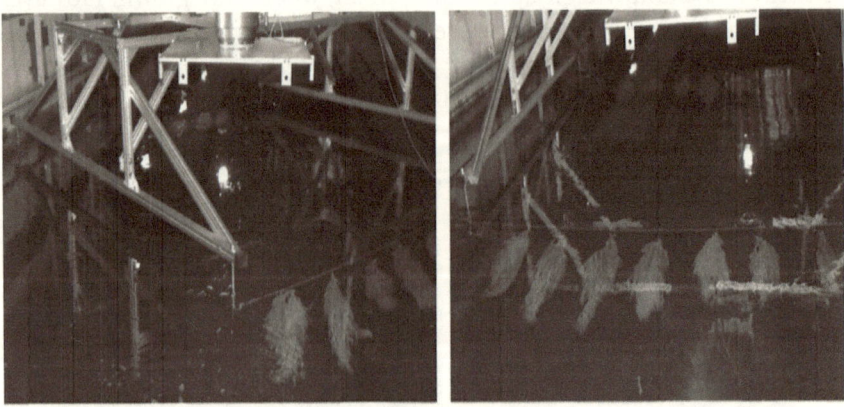

Figure 9. 6: Seaweed test in towing tank for environmental sensitivity

9.5 Environmental Sensitivity

For this research, currents are considered because of the dominant contribution of load compared with wave and wind. In addition, it is expected that drag loads will account for a large portion of the wave-driven loads as well. So, by studying the drag loads first, we hope to quickly arrive at an approximate model that is adequate for the design of major system components. Static current loads are discussed in detail below [4]. Static loads due to current are separated into longitudinal load, lateral load [5]. Flow mechanisms which influence these loads include main rope drag, main buoy drag, seaweed drag, and planting lines drag. The general equation used to determine lateral and longitudinal current load are.

$$Current\ Load = \frac{1}{2}\rho V^2 A C_d$$

9.03

Thus current loading on the system is considered at this stage, after the wave test, Morison equation will be incorporated according.

$$F = F_I + F_D \ \text{or}$$

$$F = \rho C_m V \dot{u} + \frac{1}{2}\rho C_d A u|u|$$

9.04

Where: $F(t)$ is the total inline force on the object, $\dot{u} = du\,/\,dt$ is the flow acceleration, i.e. the time derivative of the flow velocity $u(t)$, the inertia force $F = \rho C_m V \dot{u}$, is the sum of the Froude–Krylov force $\rho V \dot{u}$ and the hydrodynamic mass force $\rho C_a V \dot{u}$, the drag force $F_D = \frac{1}{2}\rho C_d A u|u|$, $C_m = 1 + C_a$ is the inertia coefficient, and C_a the added mass coefficient, A is a reference area, e.g. the cross-sectional area of the body perpendicular to the flow direction, V is volume of the body.

For instance for a circular cylinder of diameter D in oscillatory flow, the reference area per unit cylinder length is $A = D$ and the cylinder volume per unit cylinder length is $V = \frac{1}{4}\pi D^2$. As a result, $F(t)$ is the total force per unit cylinder length.

$$F = C_m \rho \frac{\pi}{4} D^2 \dot{u} + C_d \frac{1}{2}\rho D u|u|$$

9.05

The solution chosen at present is to focus on certain key non-dimensional values and try to use the use different scaling factors to apply the results at full-scale. In this way, the kinematics available in the towing tank can be used. For example, it is believed that the physical behaviour of seaweed may be similar to that of cylinders in waves and currents. Therefore, the approach closely follows that of the commentary section of the API RP-2A. For example, the effects on hydrodynamic loading of Keulegan-Carpenter number

$$KC = \frac{2\pi A}{L} \qquad\qquad 9.06$$

Where: A= amplitude of wave practical motion, L= typical length of a seaweed clump

Wave current flow reversal effects (r=ratio of current/wave orbital velocities) are expected to be similar to those for cylinders, though perhaps somewhat more complex and with different regimes for seaweed. Table 9.4 and 9.5 below shows the nondimensional parameters for a series of seastates, typical of the Southeast Asia metocean climate. Note, it is not envisaged to design the seaweed mooring for extreme environments such as rare 100-year events, typically used for offshore platforms, because the consequences of potential failure, while still undesirable, are considered much less severe. For the determination of coefficients, the KC number will be preserved.

The KC number for all the seastates is KC>12, So, relatively large wave velocities are present. Also, the effect of flow reversal can be maintained, at least in the near surface zone where the seaweed floats, by running the same waves at similar speeds. It is remarkable that the wave kinematics are very similar for all the different seastates. Therefore it is possible to greatly simplify the tests plan by appropriate choice of the scaling factors for each seastates

Table 9. 4: Full-scale Seaweed Parameters.

Exceedence Prob/ Return Period		50%	90%	99%	1-year
Hs	M	0.9	1.8	2.8	5
Hmax	M	1.9	3.6	5.4	9.6
Amax	M	0.95	1.8	2.7	4.8
Tp	S	4.6	6.4	8	10.7
Tasso	S	4.3	6	7.4	10
Uc	m/s	0.21	0.47	0.78	1.61
Mass	G	1500	1500	1500	1500
Length, L	Cm	50	50	50	50
Amax/L		1.9	3.6	5.4	9.6

—

Diameter, D	Cm	1	1	1	1
Amax/D		95	180	270	480
Uw	m/s	1.39	1.88	2.29	3.02
U	m/s	1.60	2.35	3.07	4.63
Uc/Uw		0.15	0.25	0.34	0.53
r=Uw/Uc		6.61	4.01	2.94	1.87
KC No. = 2π A/L		11.9	22.6	33.9	60.3
Re No. = UL/v		799,073	1,177,478	1,536,257	2,312,964

Table 9. 5: Proposed Model scale Parameters.

Model Scale		6	12	16	30
Exceedence Prob/ Return Period		50%	90%	99%	1-year
Hs	m	0.15	0.15	0.18	0.17
Hmax	m	0.32	0.3	0.34	0.32
Amax	m	0.16	0.15	0.17	0.16
Tp	s	1.88	1.85	2	1.95
Tasso	s	1.76	1.73	1.85	1.83
Uc	m/s	0.09	0.14	0.2	0.29
Mass	g	0.94	0.87	0.37	0.06
Length, L	cm	8	8	8	8
Amax/L		1.98	1.88	2.11	2
Diameter, D	cm	1	1	1	1
Amax/D		15.8	15	16.9	16
Uw	m/s	0.57	0.54	0.57	0.55
U	m/s	0.65	0.68	0.77	0.84
Uc/Uw		0.15	0.25	0.34	0.53
r=Uw/Uc		6.61	4.01	2.94	1.87
KC No. = 2ϖ A/L		12.4	11.8	13.3	12.6
Re No. = UL/v		52,195	54,385	61,450	67,566

In fact it is sufficient to tests a single wave to produce scaled wave kinematics for all the conditions, at least for the important parameter of wave velocity at the surface. Then what varies between seastates is actually the current velocity, and resulting ratio r, of wave/current velocities. For the present, tests, it is decided to focus on deterministic, regular waves, representing the worst design wave for each seastate. These waves are characterized by H_{max} and associated period T_{asso}. Therefore the tests can be carried out by using the same wave and changing the current velocity for each seastate (Table 4).

—

As these tests are currently ongoing, results will be presented as they become available in time for the conference.

9.6 Case Application

The test was carried out at different low direction, to simulate how the current environment will affect the seaweed. From Figure, it is clear that tow 1 line in transverse direction produce more drag. The maximum test speed at which the system get overload is 1.2m/s, at that point, the drag for 1 line is about 1.8, while for two tow line is about 0.2. Outrigger is also observed at the beginning of the speed, and nearly steady drag is observed for both cases at speed of 0.5 to 0.8 (Figure 9.7).

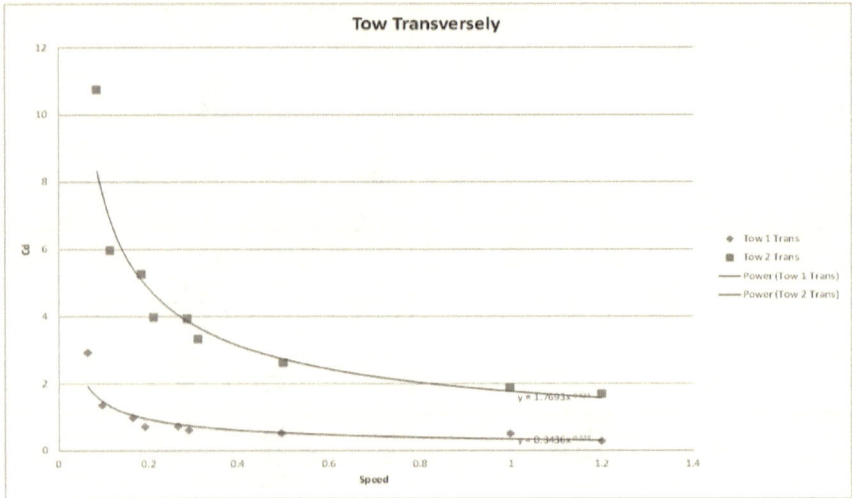

Figure 9. 7: Cd for drag in transverse direction

Towing at diagonal shows similar trend with much lower drag on the first line, however it is observed that the current speed get the chance to impact internal lines that supposed to be protected by shielding effect. This revealed that the diagonal current direction can tend to impose more added mass on the system. Steady drag is observed between 0.5-1m/s (Figure 9.8).

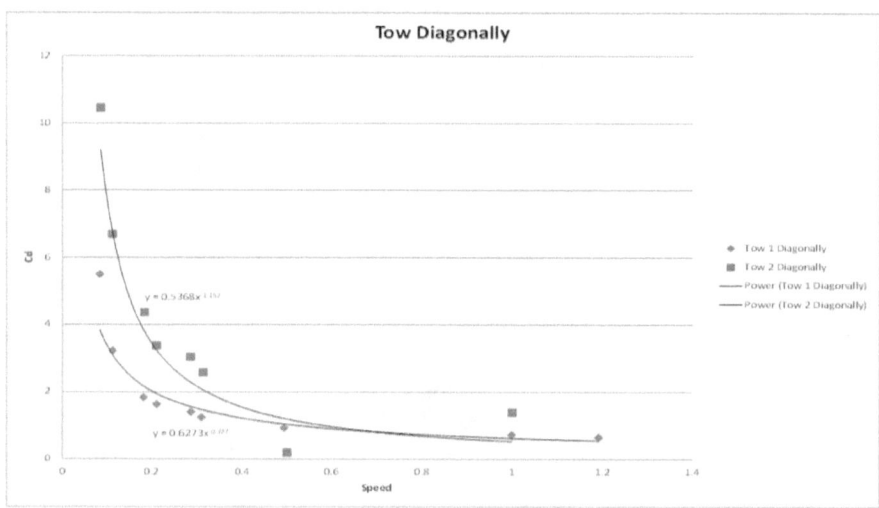

Figure 9. 8: Towing in Diagonal direction

Towing in the longitudinal direction revealed much less drag of about 0.3 at speed of 1m/s. Almost steady drag is observed between 05-0.8 (Figure 9.9).

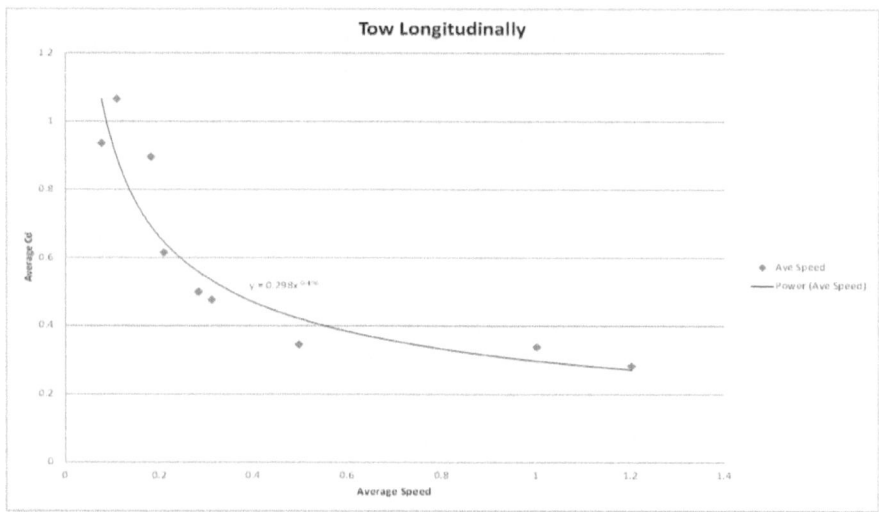

Figure 9. 9: Tow in Longitudinal direction

Figure 9.10 and 9.11 shows that strain increases with the increasing of current speed. The graph indicates that the second planting lines have a lower value compares to the first lines. This is due to the shielding

—

323

affecting which the turbulent current acting on second line resulting the lower drag. The drag current coefficient of tow two lines transverse across the basin is higher than another four test case. Furthermore, for tow one line, tow one line transverse across the basin has more drag current coefficient compared to tow one line diagonally across the basin.

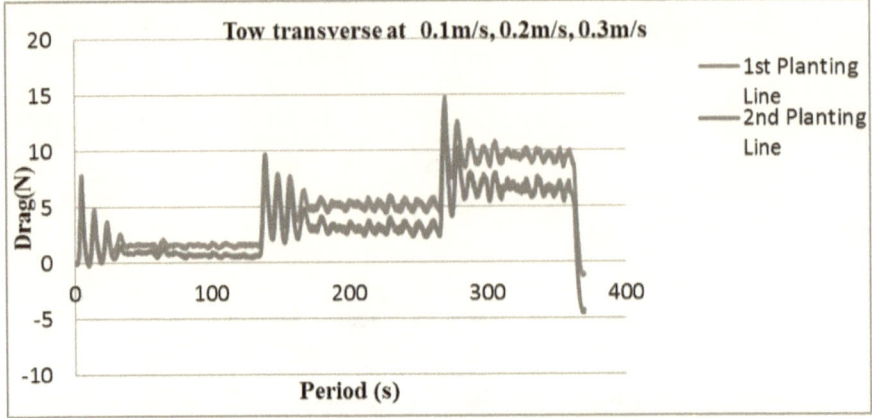

Figure 9. 10: Drag for towing two lines in transverse direction

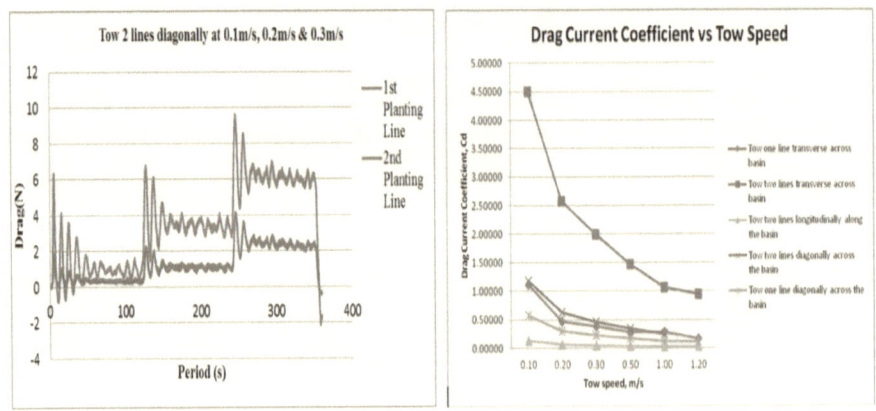

Figure 9. 11: Drag for towing two lines in diagonal direction

Aquaculture farming is widely being practice onshore or near shore, due to uncertainty about the force nature of nature, lack of established design methodology, rules and guideline. The case presented represents real life solution research to solve problem facing aquaculture industry offshore. Extreme current speed is considered for the test. The result

—

obtained represent meaningful information for the design and simulation towards reliable deployment of very large floating oceanic structure for seaweed farming and other aquaculture farming.

Reference

1. American Petroleum Institute (1997). Recommended Practice for Design and Analysis of Station Keeping Systems for Floating Structures, Recommended Practice 2SK, 2nd Edition,
2. Det Norske Veritas (2010). Offshore Standard for Position Mooring, Offshore guide
3. Buck, B. H. and Buchholz, C. M. (2004). The offshore-ring: A new system design for the open ocean aquaculture of macroalgae. Journal of Applied Phycology, 16, 355—368.
4. North, W. J. (1987). Oceanic farming of Macrocystis, the problems and non-problems. In Seaweed Cultivation for Renewable Resources (ed. Benson, K. T. B. a. P. H.), pp. 3967. Elsevier, Amsterdam
5. Chakrabarti, S. (1987). Hydrodynamics of Offshore Structures. Plainfield, Illinois: WIT Press.
6. Chakrabarti, S. (1998). Physical Model Testing of Floating Offshore Structures. Dynamic Positioning Coference.
7. Huse, E. (1996). Workshop on Model Testing of Deep Sea Offshore Structures. ITTC1996, 21st International Towing Tank Conference (pp. 161-174). Trondheim, Norway,: NTNU, Norwegian University of Science and Technology, 1996.
8. Moan, T. (2004). "Safety of floating offshore structures" Proc. 9th PRADS Conference, Keynote lecture, PRADS Conference, Luebeck-Travemuende, Germany, September 12—17, 2004.
9. O. O Sulaiman, et al (2012). Preliminary Design and Prototpe scale Model of Offshore Aquaculture Floating Structure for Seaweed Ocean Farming. International Conference Ship and Offshore Technology, ICSOT 2012 (pp. 21-34). Busan, Korea: The Royal Institution of Naval Architects.

CHAPTER 10

Mooring Analysis for Very Large Offshore Aquaculture Ocean Plantation Floating Structure

"Not everything that can be counted counts, and not everything
that counts can be counted"

Albert Einstein

Summary

Aquaculture activities are inherently done at close proximity to coastline and near shore. Issues and environmental impact concerns and challenges necessitate offshore aquaculture that required reliable structural integrity and mooring system design for ultimate state limit, fatigue state limit and accidental and progressive state limit against environmental loading and accidental loading. To avoid mooring system failure, selecting an appropriate breaking strength and limit state for mooring system components is necessary, this chapter describes mooring system design that account for forces and environmental loadings. The chapter describes evaluation of optimum mooring performance in wave, wind and current loadings on mooring components anchor, buoy and riser elements that are involved the mooring system dynamics. This section also discusses establishment of appropriate safety factors and coefficients for the design of very large offshore aquaculture floating structure.

10.1 Introduction

In Malaysia, commercial seaweed farming was started in Sabah waters in the 1970s. Sabah is still the major producer of seaweed in the country on a commercial scale, and this is mainly in Semporna, LahadDatu, Kudat, and Kunak. The recent adoption of the National Aquaculture Centre, and the latest announcement of the 2010 Budget on October 23, 2009, included seaweed specifically as one of the most important food farming commodities for the country. Although the sector has developed enormously over the past few years (111,298 tonnes wet weight in 2008), seaweed production and national target by 2010 of 250,000 tonnes (wet weight) is however yet to be achieved. There are several major issues and challenges to this, such as the unavailability of good quality seedlings, pollution in cultivation areas, diseases, shortage of raw materials, lack of capital to venture into the industry, and lack of research and development programmes[1].

It was recognized that demand for seaweed product was outstripping supply and cultivation was viewed as the best means to increase production [1]. Recently, there is renewed interest in large scale cultivation of seaweed to meet the increasing market demand and at the same time to provide alternative livelihood schemes for local populations as well as increase supply of marine biomass algae for energy production. The large farming expansion near onshore is perceived to bring so many problems. Sites as sheltered as the original marine farming culture locations are becoming limited, due to pressures from tourist development, urbanization, marine trade and the environmental related consequences. In some cases, unfavorable environmental factors such as near shore pollution from sewage have pushed marine farming sites further offshore to water depths of 50 m or greater [2]. Thus the need to move operations into more exposed sites and in totally unprotected open sea may equally face devastating natural disasters caused by tropical storms. This makes the requirement for robust mooring design and installation a necessity.

In moving offshore, wave, current and wind forces increase rapidly and these effects need to be taken into consideration in the design of mooring system. This means that maximum wave heights of 5-10 m, current speed of 2-3 knots and wind speed of 35 m/s [3] will need to be considered. A requirement is, therefore, that the farming unit may be able to withstand conditions like these. Mooring system design is a trade-off between making the system compliant in consideration to avoid excessive forces on the farming platform, and making it stiff enough to avoid difficulties, such as damage to the structure, that could be caused by excessive horizontal

—

excursions of the farming platform. This study focus uses the South China Sea, coast of Terengganu as a safety case, since the area is prone to seasonal harsh weather. Thus, the use of strategically location in the area sheltered by a chain of islands forming a good wave breaking point could be advantageous, provided the marine algae can grow there satisfactorily. This will avoid potentially even worse conditions present elsewhere.

10.2 Background

Employment of aquaculture farming facilities in offshore will face natural disasters caused by tropical storms. Without reliable design of mooring system, the farming load on the structure can easily damage during a natural disaster event. That could lead to unacceptable economic loss. Thus, ensuring the security and safe deployment of very large floating structure system is becoming one of the most important issues for industrial marine farm. Because, this comes with requirement to operate offshore aquaculture system in 50 meters of water depths, also, the suspended weight of mooring lines becomes a prohibitive factor. Several parameters need to be considered which include length/weight of mooring line, suitable of material and type of anchoring technique.

Multi-bottom total system analysis that involve use of deterministic, physical, numerical, simulation, model testing and stochastic analysis modeling of such farming platform with position keeping of multiple mooring lines is a challenge to predict the ideal performance of mooring system without ignoring cost effectiveness. The study of mooring design for VLFS of offshore aquaculture structure aims to develop a mathematical model of mooring system for offshore seaweed farming. The study focuses the investigation on the optimum mooring arrangement for offshore aquaculture marine technology for ocean farming, development of the mathematical model of mooring system offshore aquaculture marine technology for ocean farming, evaluation of the suitability of mooring components for offshore aquaculture marine technology for ocean farming and evaluation of the performance of a centenary mooring arrangement for mooring components of offshore aquaculture marine technology for ocean farming

10.3 Method application

10.3.1 VLF for Oceanic Aquaculture Farming

The system analysis is best model from deterministic approach and concludes with probabilistic and stochastic approach. First principle system consider for the design is shown in Figure 10.1. The governing equation for oscillation of the floating structure is defined as follows:

$$\left[M_{ij} + m_{ij}(\infty)\right] \ddot{X}(t) + F_v(\dot{X}) + \sum_{j=1}^{n} \int_{-\infty}^{t} \dot{x}(\tau) L_{ij}(t-\tau) \, d\tau + F_M(X, \dot{X})$$

$$= F_{wind}(t) + F_1(t) + F_2(t)$$

$$F = F_{env} + F_{moor} \qquad\qquad 10.01$$

where $\overline{X}$: displacement vector of horizontal plane response of the floating structure; M_{ij}: inertia matrix of the floating structure; $m_{ij}(\infty)$: added mass matrix at the infinite frequency; F_v: viscous damping coefficient vector; L_{ij}: Memory influence function; F_M: Mooring reaction force vector; Fenv F_{wind}: Environmental loading load force, Fmoor: mooring force.

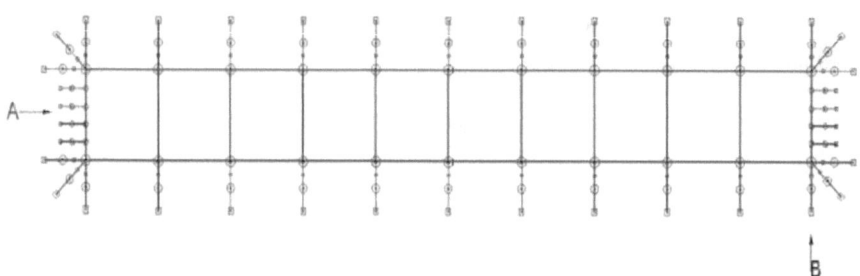

a. Offshore aquaculture System

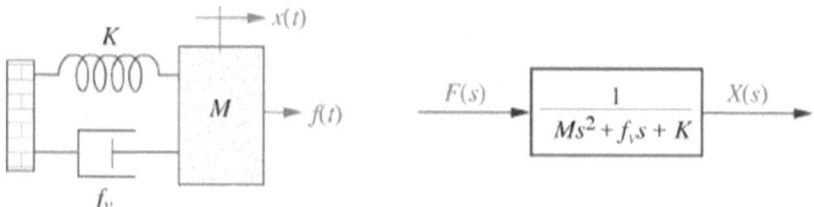

b. System conceptual free body diagram

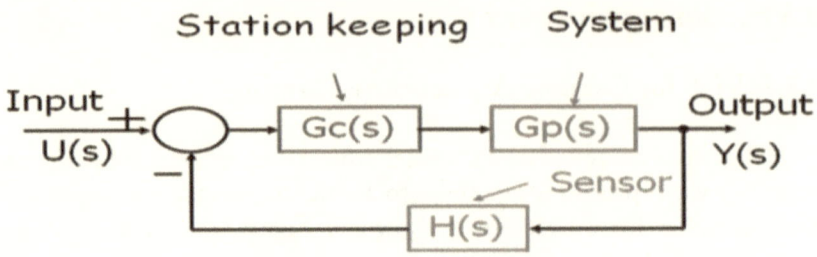

c. System control block diagram

Figure 10. 1: System description for deterministic analysis

$$(ms^2 + fs + k)\, Y(s) = F(s) \qquad\qquad 10.02$$

$$TF = G(s) = \frac{Y(s)}{F(s)} = \frac{1}{ms^2 + fs + k} \qquad\qquad 10.03$$

The elements immersed weight, WI is defined by:

$$WI = FB - W \qquad\qquad 10.04$$

WI is positive, the element is positively buoyant (polypropylene subsurface floats), and if it is negative, the element is negatively buoyant (wire rope and shackles) (Randall, 1997).

The drag Q in each direction acting on each mooring element is calculated using:

$$Q_J = \tfrac{1}{2}\, P_W C_{Di} A_J U U_J \qquad\qquad 10.05$$

Where Qj is the drag in [N] on element 'i' in water of density rw in the direction 'j' (x, y, or z), Uj is the velocity component at the present depth of the mooring element which has a drag coefficient CDi appropriate for the shape of the element, with surface area Aj perpendicular to the direction j. At the depth of the element, the drag in all three directions [j=1(x), 2(y) and 3(z)] is estimated, including the vertical component, which in most flows is likely to be very small and negligible. Figure 10.2 shows system motion analysis.

Once the drag for each mooring element and each interpolated segment of mooring wire and chain have been calculated, then the tension and the vertical angles necessary to hold that element in place (in the current) can use the equation below to estimate the three [x,y,z] component of each element:

$$Q_{xi} P_f(T) = \iint d_{xi} \, d_{rk} \qquad\qquad 10.06$$

$$Q_{xi} T_i \cos\Theta_i \sin\Psi_i = T_{i+1} \cos\Theta_{i+1} \sin\Psi_{+1} \qquad\qquad 10.07$$

$$Q_{yt} T_t \sin\Theta_t \sin\Psi_t = T_{t+1} \sin\Theta_{t+1} \sin\Psi_{+1} \qquad\qquad 10.08$$

$$B_i g + Q_{zi} + T_i \cos\Psi_i = T_{i+1} \cos\Psi_{i+1} \qquad\qquad 10.09$$

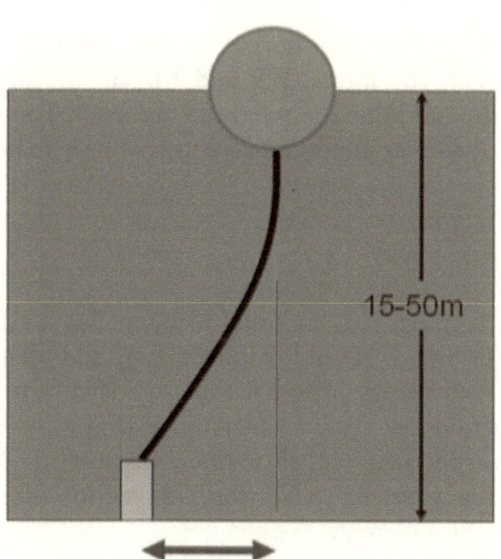

15-50m

Figure 10. 2: System motion analysis

$$X_i = X_{i+1} + L_i \cos\Theta_i \sin\Psi_i \qquad\qquad 10.10$$

$$Y_i = Y_{i+1} + L_i \sin\Theta_i \sin\Psi_i \qquad\qquad 10.11$$

$$Z_i = Z_{i+1} + L_i \cos\Psi_i \qquad\qquad 10.12$$

Drag coefficients are a function of Reynolds number, Re, as defined as:

$$Re = \frac{UD}{v}$$
10.13

Where U is the velocity of the flow, D is the characteristic buoy dimension, and v is the fluid kinematic viscosity.

By summing the forces at the attachment point, component of the tension is given by

$$T_b = F_d$$
10.14

$$T_v = F_b - W$$
10.15

Where TH and TV are the horizontal and vertical tensions in the submerged element; the forces, FD and FB are the drag and buoyant forces respectively, and W is the weight of the buoy. Total tension on the mooring lines is the other important aspects that have to be considered.

$$T_{HV} = \sqrt{T_H^2 + T_V^2}$$
10.16

For station keeping each mooring element has a time static vector force balance (in the x, y, and z directions), and that between time dependent solutions the mooring has time to adjust. The forces acting in the vertical direction are: (1) buoyancy (mass [kg]´g [acceleration due to gravity]) positive upwards (i.e. floatation), negative downwards (i.e. an anchor), (2) tension from above Newton], (3) tension from below, and (4) drag from any vertical current.

In each horizontal direction, the balances of forces are: (1) angled tension from above, (2) angled tension from below, and (3) drag from the horizontal velocity. Buoyancy is determined by the mass and displacement of the device and is assumed to be a constant (no compression effects and a constant sea water density). Other challenges in this design are system multi body analysis in respect to analysis and keeping the whole system together and system to system interaction (Figure 10.3).

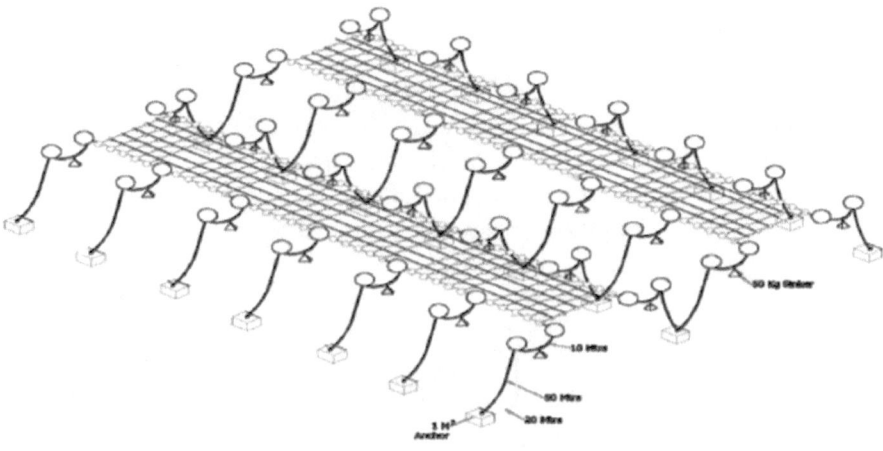

Figure 10. 3: System to system interaction

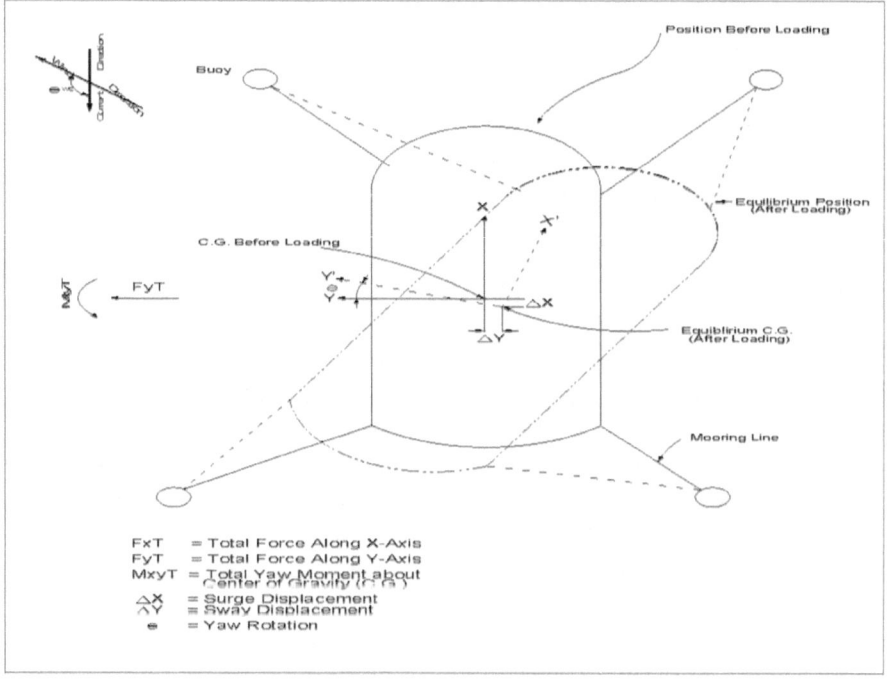

Figure 10. 4: Multi body system

10.3.2 Static Analysis

This study allows one to predict the geometry of the line between the platform and its anchoring point and the distribution of stresses from top to bottom [5]. The recurrence formulas for carrying out the computation process for the resultant forces applied at the end of any segment n [5]. The tension, the orientation angles, the stretched length and the segment coordinates can now be formulated as

$$Rx_{(n)} = Rx_{(n-1)} + Fx_n \qquad\qquad 10.17a$$

$$Rz_{(n)} = Rz_{(n-1)} + Fz_n T_n = \left[Rx_{(n)}{}^2 + Rz_{(n)}{}^2 \right]^{\frac{1}{2}} \qquad 10.17b$$

$$\Theta_n = \tan^{-1} \left[\frac{R_{z\,(n-1)}}{R_{x\,(n-1)}} \right] \qquad\qquad 10.17c$$

$$S_n = L_n + \Delta L_n \qquad\qquad 1017d$$

$$\text{Coordinate } X_{(n)} = X_{(n-1)} + S_{(n)} \cos \Theta_{(n)} \qquad 10.17d$$

$$\text{Coordinate } Z_{(n)} = Z_{(n-1)} + S_{(n)} \sin \Theta_{(n)} \qquad 10.17e$$

10.3.3 Environmental Load

For this research, currents are considered because of the dominant contribution of load compared with wave and wind. In addition, it is expected that drag loads will account for a large portion of the wave-driven loads as well. So, by studying the drag loads first, we hope to quickly arrive at an approximate model that is adequate for the design of major system components. Static current loads are discussed in detail below [4]. Static loads due to current are separated into longitudinal load, lateral load [5]. Flow mechanisms which influence these loads include main rope drag, main buoy drag, seaweed drag, and planting lines drag. The general equation used to determine lateral and longitudinal current load are

$$F = \tfrac{1}{2}\rho V^2 AC_d \qquad\qquad 10.18$$

Thus current loading on the system is considered at this stage, after the wave test, Morison equation will be incorporated accordingly.

$$F = F_I + F_D \qquad\qquad 10.19$$

$$F = \rho C_m V \dot{u} + \rho C_d A u |u| \qquad\qquad 10.20$$

where $F(t)$ is the total inline force on the object, $\dot{u} \equiv du/dt$ is the flow acceleration, i.e. the *time derivative* of the flow velocity $u(t)$, the inertia force $F_I = \rho C_m V \dot{u}$, is the sum of the *Froude–Krylov force* $\rho V u$ and the hydrodynamic mass force $\rho C_a V_u$. the drag force $F_D = \frac{1}{2} \rho C_d A u |u|$, $Cm = 1 + Ca$ is the inertia coefficient, and Ca the *added mass* coefficient, A is a reference area, e.g. the cross-sectional area of the body perpendicular to the flow direction, V is volume of the body.

For instance for a circular cylinder of diameter D in oscillatory flow, the reference area per unit cylinder length is $A = D$ and the cylinder volume per unit cylinder length is $V = \frac{1}{4} \pi D^2$. As a result, $F(t)$ is the total force per unit cylinder length:

$$F = C_m \rho \frac{\pi}{4} D^2 \dot{u} + C_d \frac{1}{2} \rho D u |u| \qquad\qquad 10.21$$

10.3.4 Model test

A towing test conducted at UTM marine Lab involved (Figure 10.7) one method of research to determine the hydrodynamic loading coefficients (added mass and damping) in a few different configurations [6]. The samples which are seaweed are dried, but will restore to nearly nominal properties when soaked in water for a period of time. In typical practice, the rows of seaweed are held using ropes separated by about 2.6 m between rows. For the drag measurement, the seaweed was attached to towing lines and the lines were towed from the moving carriage. A frame consisting of aluminum channel sections attached to the towing carriage is used. The seaweed clumps then in turn attached to a rope line. Tension load cells will be attached between the line and the frame and the measured forces recorded on the model basin's data acquisition system (Figure 10.5 and 10.6).

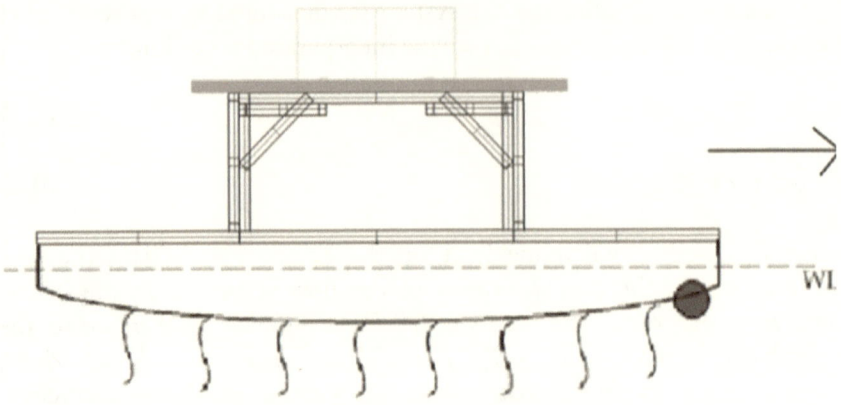

Figure 10. 5: Model test system

Figure 10. 6: Test frame construction and deployment

Figure 10. 7: UTM Towing Tank and Carriage

10.3.5 Prototype System

Component of the full-scale platform and mooring cable characteristic first need to be clarified. A typical block has an overall length of 100 meter and the breadth is 100 meter. One block of seaweed farming contains four main ropes for the frame, four main buoy and 30 load lines on which is the seaweed will be planted. Table 1shows the structural properties of one block of seaweed farming.

Table 10. 1: Structural properties

Structural properties (one block)	No/Quantity	unit
Length overall	100	m
Breadth	100	m
Main buoy	4	
Main rope	4	
Load line rope	30	
Sea water density	1025	Kg/m^3

—

10.4 Case application

10.4.1 Environmental Load

Four samples of current speeds are taken the current speeds starts with 0.5 m/s, 1 m/s, 1.5 m/s and 2 m/s. The maximum lateral current load occurs when the structure is subject to2 m/s of current speed which has the value just over 20,000N, as shown in the figure.

Current load versus current speed

Figure 10. 8: Graph current load versus current speed

Figure 3 shows that the current load increases when the current speed increases for both longitudinal and lateral load. The graph indicates that the longitudinal loads are higher than the lateral load. This is because the longitudinal current speed is moving horizontally to the x—axis of the structure (up and down in the figure) which results in the bigger load from the planting lines.

10.4.2 Spread Mooring Analysis for One Block

The mooring consists of H3, H4, H6, and H7 mooring lines, which resist longitudinal load, and four mooring lines, H1, H2, H4 and H5 placed perpendicularly to the longitudinal axis of the structure, which resist lateral load [7]. The maximum allowable working load, Tbreak is 39,865.6 N (Figure 10.8).

—

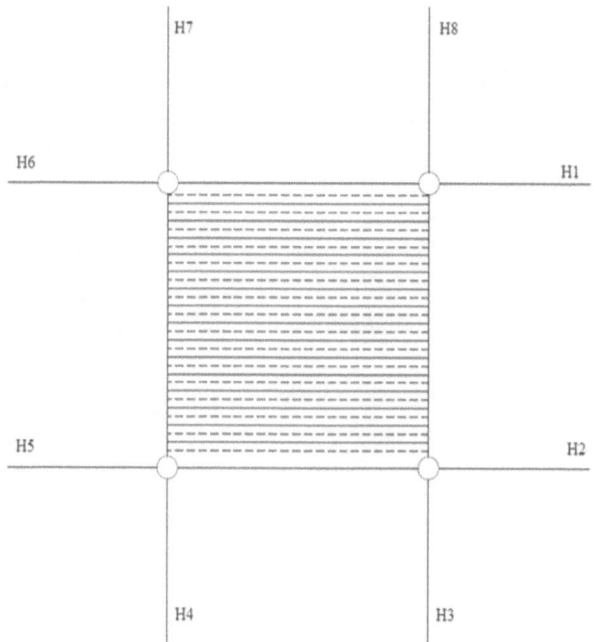

Figure 10. 9: Force diagram for the mooring system

Deadweight anchor was selected and it is a large mass of concrete or steel which relies on its own weight to resist lateral and uplift loading. Lateral capacity of a deadweight anchor will not exceed the weight of the anchor and is more often some fraction of it [8]. Deadweight-anchor construction may vary from simple concrete clumps to specially manufactured concrete and steel anchors with shear keys. Several advantages of deadweight are anchor has large vertical reaction component, permitting shorter mooring-line scope. Also, it is simple to construct using means which are readily available, even in fairly remote locations.

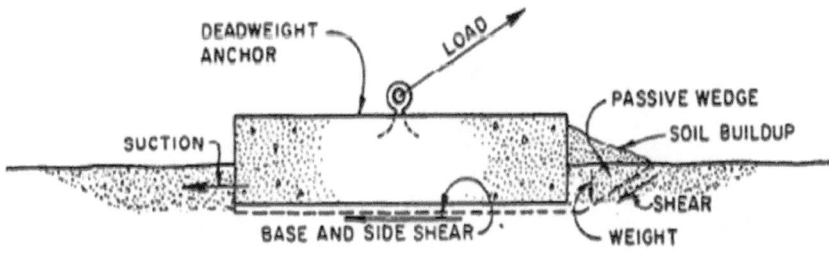

Figure 10. 10: Deadweight anchor

—

10.4.3 Model Test

The data were recorded on the basin's native data acquisition system (DAQ)at a frequency of 25Hz. The load cells attached to the mooring lines measure the tension (strain) in the line. These are water-proof aluminum ring strain gauges that measure axial tensile loads[9]. The measured voltage outputs from the load cell strain gauges are connected by cables to the basin's native Dewetron Data Acquisition System (DAQ) to be digitally sampled and stored. Figure below shows the result of several typical tests which indicate the drag value of seaweed in different current speeds.

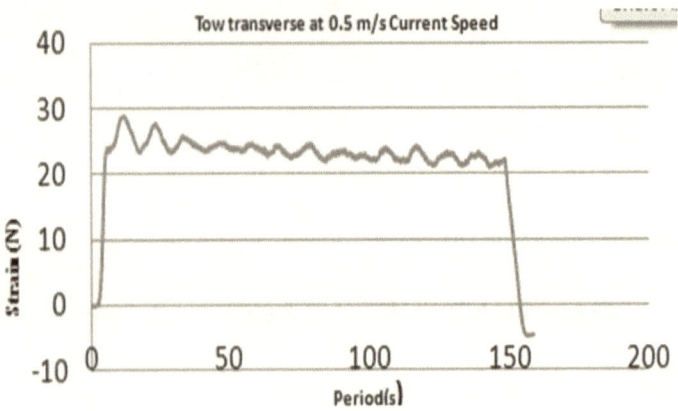

Figure 10. 11: Tow transverse at 0.5 m/s Current Speed

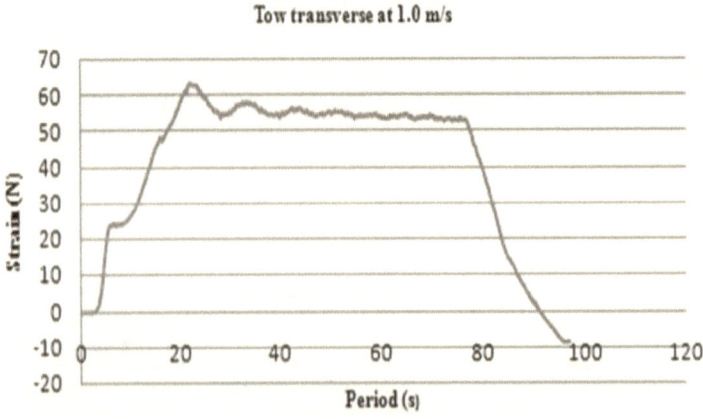

Figure 10. 12: Tow transverse at 1.0 m/s Current Speed

—

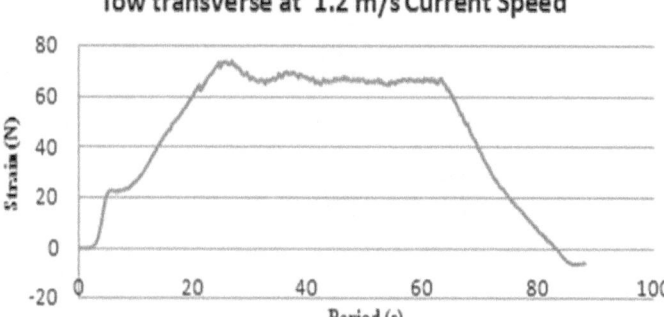

Figure 10. 13: Tow transverse 1.2 m/s Current Speed

Figure 10.11,10.12, and 10.13 shows that the strain values are nearly constant at the middle part of the graphs but the strains have different values for each graph which increase for increasing current speed. The increasing strain values presented in this result are mainly used to validate the accuracy of theoretical calculation which has been developed for this study. Note that the initial pretension in the strain (40N) was zeroed for measurement purposes. The value has to be added to the results in the figures to obtain the total tension in the lines.

For smaller towing speeds, it is possible to combine several towing speeds in a single test run to reduce waiting time. Figure 10.14 and 10.15 shows test results for such tests. The results show that strain increases with the increasing of current speed. The graph indicates that the second planting lines have a lower value compares to the first lines. This is due to the shielding effect which results from the wake deficit acting on second line resulting in the lower drag.

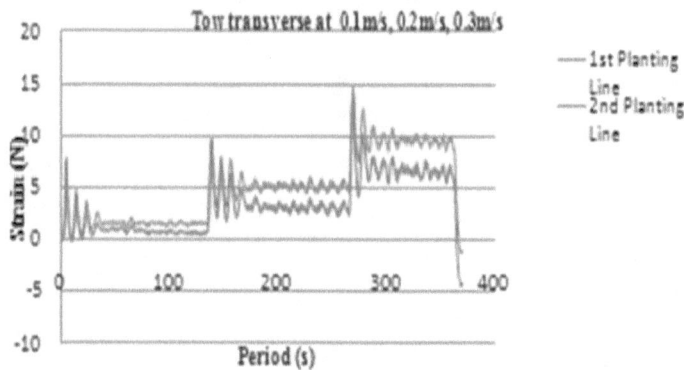

Figure 10. 14: Tow transverse at 0.1m/s, 0.2m/s, 0.3m/s (two lines)

–

341

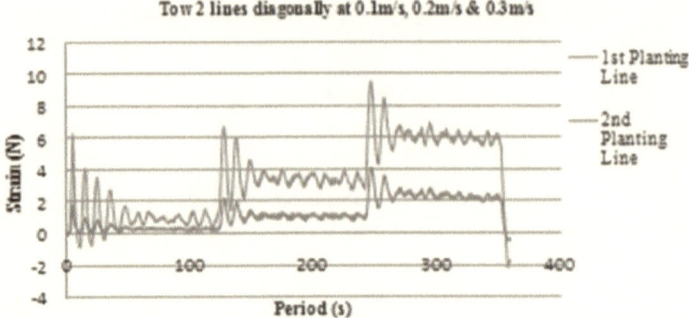

Figure 10. 15: Tow diagonally at 0.1m/s, 0.2m/s, 0.3m/s (two lines)

10.4.4 Mooring Line Dynamic

This part will discuss the mooring line analysis when experienced increasing current velocity to the dynamic of mooring line. Figure 10.16 shows the behaviour of mooring line profile at different current speed. In this analysis, three samples of current values are taken. The current speeds start with 0.5 m/s, 1.0 m/s and 2.0 m/s. Figure 10.16 shows the behaviour of mooring line profile when experiencing an increasing of current speed. From the graphs, obviously we can see the movement of mooring line is increased when the current speed increases. It means that, the horizontal excursion of mooring line distance is increasing simultaneously decreasing in line angle. Another logical behaviour from this analysis is that, the buoy point which attaches the mooring line to the structure keep on moving downwards towards the seabed when there is an increase of current speed. This means that, the mooring lines pull the platform downward.

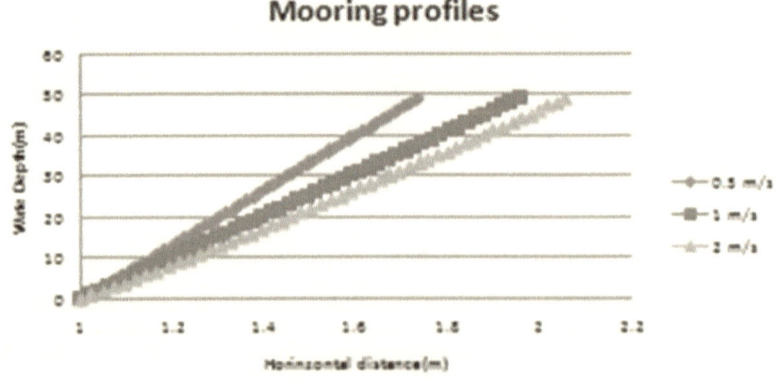

Figure 10. 16: Mooring profile

—

As for conclusion, the objectives stated in this report are achieved. This numerical model is developed by the formulas to determine the suitable mooring component to be equipped on the seaweed farming. Mooring behaviour and its design requirements have been determined for the feasibility of moving many similar floating structure used for offshore aquaculture to more exposed sites. In addition, modelling can serve as the basis for comparing and evaluating different mooring systems before purchasing and installation. Modelling will be an essential tool for choosing larger mooring systems for commercial-sized facilities. Performance of catenary mooring line such as tension and the suspended length required are obtained easily by entering the initial values into the input file data and will be calculated.

The initial results presented here are based on current drag load calculations only, which are expected to comprise the majority of loading on the aquaculture structure. As a next step, calculations are planned which will include the additional drag loading due to waves. Further studies including unsteady dynamic testing need to be carried out to adequately determine the additional loading components needed for completely robust design. It is hoped that this effort will lead to more full scale prototype systems being deployed from which much more is yet to be learned. Adequate feedback from actual systems to the design process will remain an essential part of an improved design process, leading to more robust and cost-effective aquaculture systems helping to meet the national goals of increased production.

References

1. Buck, B. H. And Buchholz, C. M. (2004). The offshore-ring: A new system design for the open ocean aquaculture of macroalgae. Journal of Applied Phycology, 16, 355-368.
2. North, W. J. (1987). Oceanic farming of Macrocystis, the problems and non-problems. InSeaweed Cultivation for Renewable Resources (ed. Benson, K. T. B. a. P. H.), pp. 3967. Elsevier, Amsterdam
3. S.Chakrabarti, 2005, Handbook of Offshore Engineering, Vol 1, Elsevier
4. John B. Herbich, Editor, 1992, Handbook of Coastal and Ocean Engineering, Vol 3, Gulf Publishing Company
5. Bergdahl, L.M. and Rask, I., 1987, "Dynamic vs. Quasi-Static Design of Catenary Mooring System", Proc. of the 19th Offshore Technology Conference, Houston, Texas, OTC 5530, Vol. 3, pp. 397-404.
6. Mavrakos, S.A., Papazoglou, V.J. and Triantafyllou, M.S. (1989), "An Investigation into the Feasibility of Deep Water Anchoring Systems", Proc. of the 8th Int. OMAE Conference, OMAE'89, The Hague, Vol. 1, pp. 683-689
7. J. P. Jones (2006).Fleet Moorings. Design of fleet mooring 110-135. Naval facilities Engineering Command
8. Goeller, J. and Laura, P. (1971). "Analytical and experimental study of the dynamic response of segmented cable systems." Journal of Sound and Vibration, Vol. 18, pp. 311-329.
9. A. Lloyd (1989). "SeakeepingBehavior in Rough Weather", Ellis Horwood Ltd.

CHAPTER 11

Sustainable Management of Navigation Channel for Inland Navigation Channel

"Do not worry about your difficulties in mathematics. I can
assure you, mine are still greater"
Albert Einstein

Summary

Maritime transportation industry is the cradle of all modes of
transportation where port is and ship is necessary to facilitate trading
through marine transportation. Human civilization and demand for
international trade has resulted to need for economic of large scale
for ships. Recently, there is continuous growth or need for larger and
sophisticated ship through increasing shipping activities and this has
lead to design and production of sophisticated state of art safety oriented
marine vehicle in term of size, speed and structure—albeit, this safety
based designed development is out of phase with conditions of navigation
channels. To create a balance for safe navigation in inland port and
navigation which are considered to be restricted water, this big ships will
ply, it is necessary to maintain the channel to keep accepting the target
largest vessel, and the channel should be maintain at frequency the ship
building are growing. This section presents the result of application of

—

best practice simplified method for channel maintenance against vessel design and reception requirement. Application of the model to predict channel depth for Port Tanjung Pelepas (PTP) in Johor, Malaysia shows a good agreement with actual channel depth. The chapter also discusses requirement for incorporation sustainability components in management of navigation channel.

11.1 Introduction

Marine transportation has recently witness continuous growth or need for larger and sophisticated ship through increasing shipping activities and demands, this has lead to design and building of sophisticated, state of the art, safety oriented marine vehicles in term of size, speed and structure, albeit, the design and production of vessels take little consideration in phasing them with navigation channel requirement of waterways. Management of navigation channel in inland waterway, require process to create a balance for safe navigation in restricted water these big ships will ply, It is important to maintain the channel according to and on proportion to the frequency the ships building are growing. Maintenance dredging is the activity that involve periodic removal of material which has been deposited in an area where capital dredging has been undertaken, management of the frequency of maintenance dredging varies from port to port, however, the objective remain to allow ships to enter and leave port at stated draft without delay and ensure efficiency of maintenance dredging, thus, the management require control step that must be taken during the process to minimize siltation and shoaling in channel.

11.2 Marine transportation growth trend

The ocean, ships and shipping remains a very important instrument for mobility of goods and people, if ships could no longer transit the waterways, the consumers will experience shortages of power, heat and food in days or weeks at the outside. Recent years have seen economic of scale due to improved trade, the significance of these trends is that larger ships will continue to use our waterways for the foreseeable future. But there are limits on size of ship that a channel can accommodate, and means of determining when special measures must be imposed on handling ships in order to ensure the continued safe, efficient, and environmentally friendly use of our channel.

To create a balance for safe navigation in restricted water and channel at a frequency the ship production are growing must be maintained. De

—

Jong et al. (1887), provided data on the explosion in the size of container ship that has occurred since the first "post Panama" vessel analysis shows that ships exceeding the panama canal limit (ship length, breath draught of 256mx32.2x11m) started to appear a few years ago. Other studies by transmarine also demonstrated that in recent time the vessel size is up to 18,000TEU. See figure 11.1 for growing ship sizes as presented by Transmarine.

RINA periodic recently report that Mearsk line has built 14,000 TEU ship that is ready for operations, however, safe operation of those big ships will operate remain is under deliberation. This shows the rapid the growing trends of container vessels and the need of suitable channel to match this growth. Recent projection at 18,000 TEU believed to be the technological capability for such target. As the ship sizes are increasing it is imperative to do periodic examinations of the requirement of the channel in regard to depth, width, squat, and alignment. Figure 11.2 shows the needs of the channel.

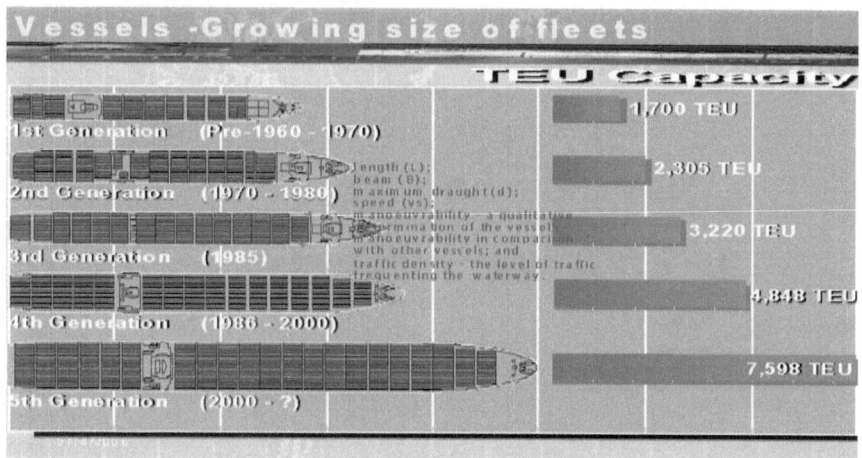

Figure 11. 1: Vessel size increase [Transmarine, 2005]

Mega- ship require specialized ports with high infrastructure investment

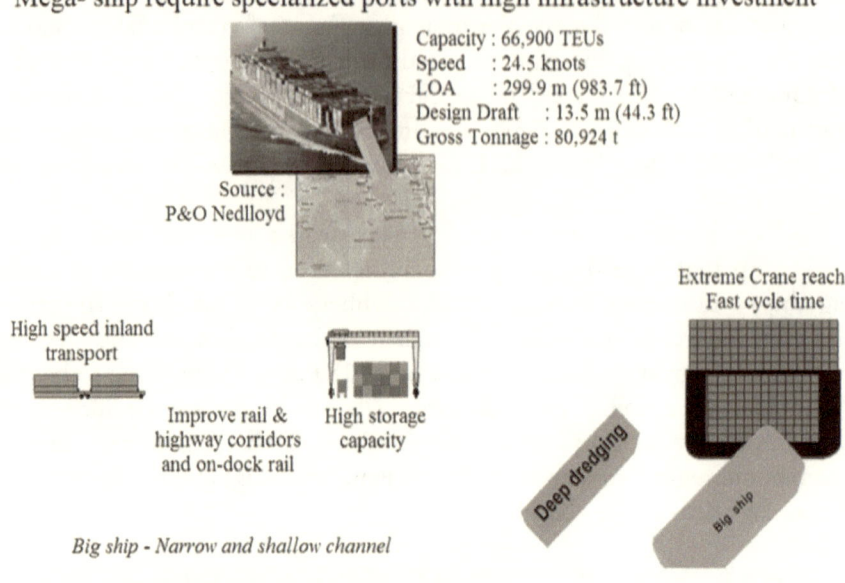

Capacity : 66,900 TEUs
Speed   : 24.5 knots
LOA   : 299.9 m (983.7 ft)
Design Draft   : 13.5 m (44.3 ft)
Gross Tonnage : 80,924 t

Source :
P&O Nedlloyd

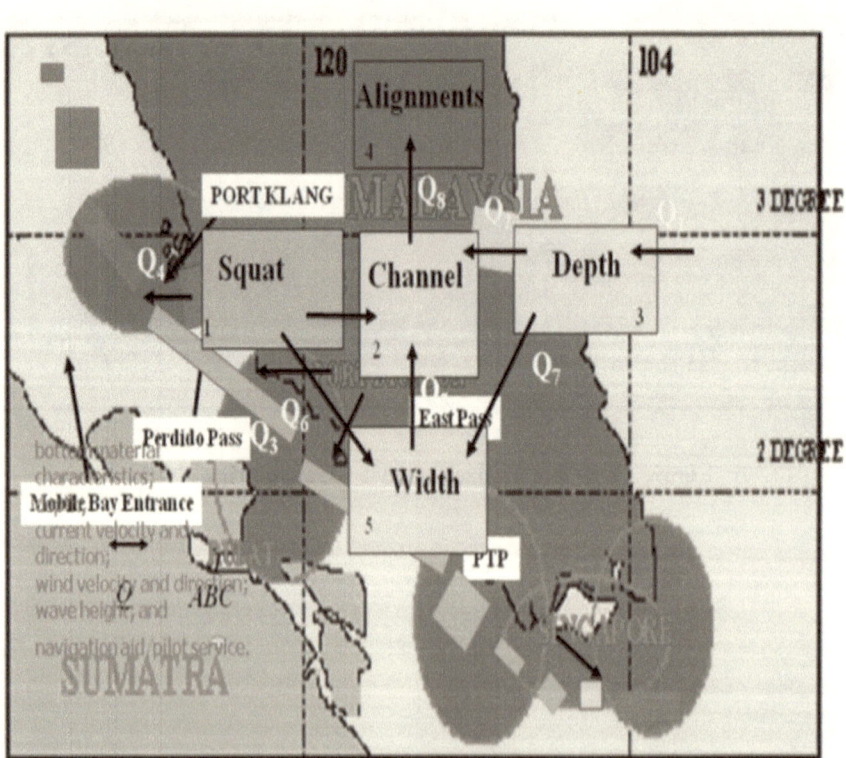

Figure 11. 2: Inland navigation channel requirement

—
348

Channel design and maintenance work fall among the works that require multivariable exercise that need model studies for good outcome. Shoaling remaining unavoidable part of most harbor and navigation channels and one method to preventing shoaling and associated siltation hurdles is using of maintenance dredging at economical frequency and sustainable manner. A study made by Mac Elverey, 1995 also stressed on the fact that the design of controllability of the ship is equally out of match with the size of the ship has been the main focus on design spiral is best design of basic requirement of speed, payload and endurance which focus is not placed neither on channel design nor its maintenance.

11.3 Present threat

Growing size of fleets and lack maintenance of channel has been going in human society for a long time, (INTERTANKO report in 1996 on the same issue regarding US water). Analysis drawn from marine departments revealed that disasters recorded at the Strait of Malacca collision where grounding takes the highest share of the risk. Det Norske Veritas studied on various navigation waterways in the Strait of Malacca as one of the high risk areas of the world. This issue considered very necessary and require diligent attention, especially in protected and restricted waters like the need to patronize the Strait of Malacca and its riparian which PTP is part of and where more than 800 ships pass through daily based on the causality figures by Malaysian Marine Department.

An Analysis made by the UNEP regarding region under coastal threat concluded that Asia coast is far more affected, because Asia have the largest river runs off to the sea than any other continents. According to UNEP report, maintenance of navigation channel remains one sensitive area of environmental degradation concern for environmental thematic problem especially dredging, its disposal problem to marine life.

Contemporary time is seeing flood waters taking over cities, serious enthrophication cut off transport, routes, communication power supplies, inundate destroy action of homes, crops and livelihoods, affect millions in rural and urban areas. Areas that are especially hard-hit have seen widespread devastation death from heaviest rainfall on record that has forced millions of people from their homes. Neighboring state has seen with unprecedented severe damaged ranging from agriculture and industrial units also widespread. Analysis drawn from marine departments in figures 11.2 show disasters record of the Strait of Malacca collision and grounding take the highest share of the risk. Also a risk assessment

–

studies carried out by Det Norske Veritas for various navigation waterways revealed similar risk situation for the Straits of Malacca.

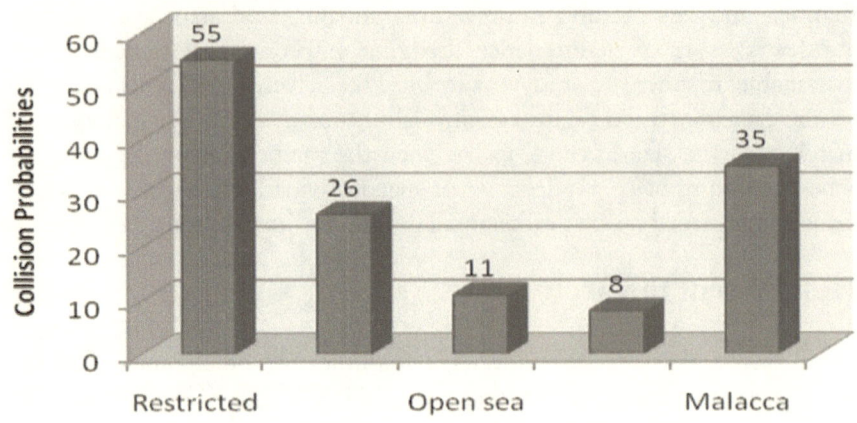

Figure 11. 3: World coast risk area (DNV, 2008)

A report by the UNEP regarding region under coastal treat indicated that Asia among all other region of the world will 6349 million tons per year, so far the highest (See Figure 11.3), this due to the so due to Asia having a lot of river runs off to the sea than any other continent.

11.4 Pollution source and Impacts to Port

The pollution found in the sediments that accumulate in harbors is one of the main causes of environmental impact of dredging operations. This is important feature since harbor waters and sediments are heavily polluted worldwide—containing high levels of a range of chemicals. The sources of pollution are multiples, in many cases they are linked to the harbor activity. TBT (tributyltin), is a compound used as antifouling that has been recognized as a harmful pollutant and whose use has been restricted at the international level. However, the problems caused by TBT will continue for many years, because TBT is kept stored in the harbor sediments. The main source of TBT to marine waters is the direct release from surfaces treated with antifouling paints containing TBT and other organizing compounds.

They have been used in order to prevent the attachment of aquatic organisms on the hull of ships or other devices that are immersed in the sea, such as the cages used for fish-farms (Evans, 2000). Paints incorporating TBT are regarded as the most effective anti foulants ever devised giving rise to important economic benefits. Their use reduces the fuel consumption of the vessels (and thus reduces CO_2 emissions), ships can go faster, and repainting costs are lower. Little is still known of the effects on marine organisms of dredging operations contaminated sediments, and the need of research on this issue is often claimed (Ten Hallers Tjabbes et al., 1994). Figure 11.4 revealed potential impact of poor management of Inland waterways channel.

To assess the potential impacts of a certain project is a difficult task. First, although the (a priori) main possible impacts of dredging can be identified, clearly it is not possible to review all the potential effects;, once the impacts of concern are selected, it might be costly, complex or impossible to assess the extent and the consequences of each of the effects caused by the activity assessed (Jensen and Mogensen (2000). Consequently, the dredging operation causes the re-suspension of sediments, solids in water to some extent produces an effect called turbidity, which is defined as an optical property of water related to attenuation of light (Peddicord and Dillon, 1997).

Other factors influence the level of turbidity (such as size distribution and shape of particles). Although, they are regarded as "short-term" impacts, the presence of large quantities of particles in the water can cause serious effects in areas where the system is not used to it, and particularly to sensitive species or areas (for example, coral reefs or aquaculture ponds), as well as reduction of oxygen in water, release of toxic components from suspended solid, covering of organisms, reducing food supply, etc (Bray, 2001). Moreover, lack of light may reduce photosynthesis, which might be relevant for sensitive species.

This make it incumbent for authorities concerned regarding waterways to evaluate and address the risks associated with ships that are plying them and find way and information sharing avenue systems for channel designers, naval architects, ship masters and pilots, and waterway managers that will help develop policy recommendations that will address the way channels are laid out, enlarged and how ships of various types using them should be designed and handled. And of course, ways to monitor existing and new ships operating at channel approach in order to guide ship designers to understand and review ships, pilotage, channel, current design and operational practices on how to incorporate needed improvements.

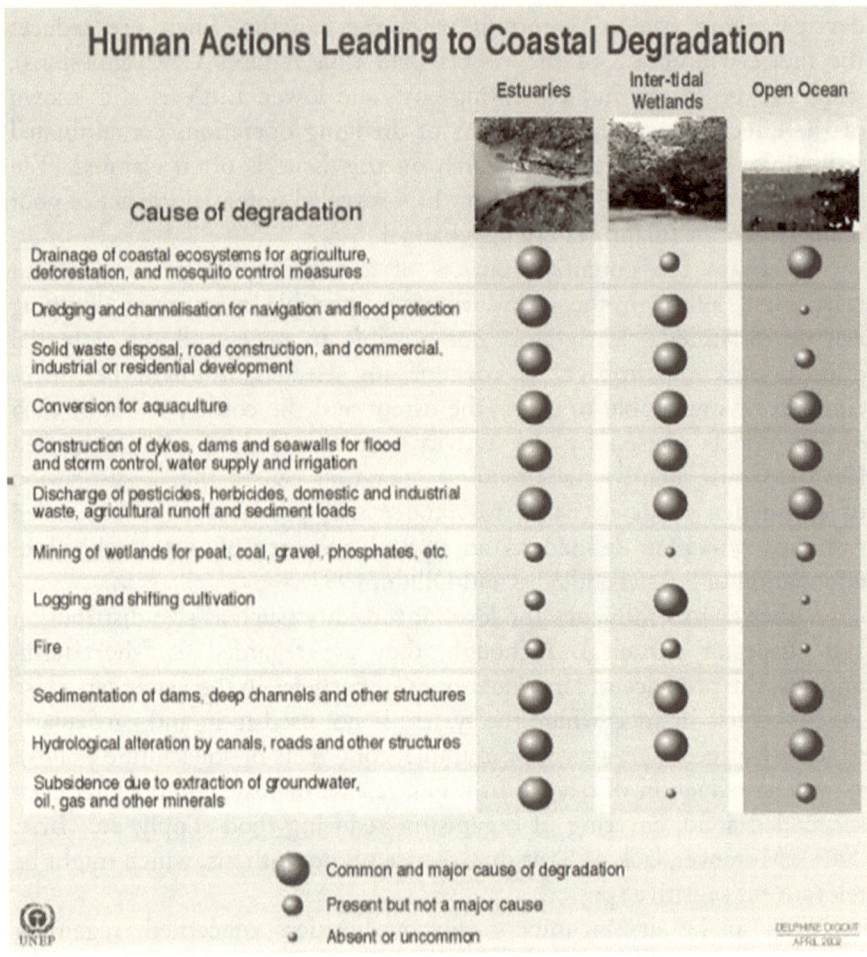

Figure 11. 4: Areas of Human action leading to environmental degradation (UNEP, 2008)

11.5 Channel maintenance process

Within the scope of this study, the main elements that method will analyse the following;

i. Channel dimension establishment through navigation requirements and Side slope tolerant.

ii. Hydrographic and dredge volume concept, calculation of yearly dredged quantity output.

iii. Dredge capacity and selection of dredge equipment.

352

iv. Disposal of dredged material, issue of transport distance and sustainability concept

v. Concept of uncertainty, risk cost and benefits assessment.

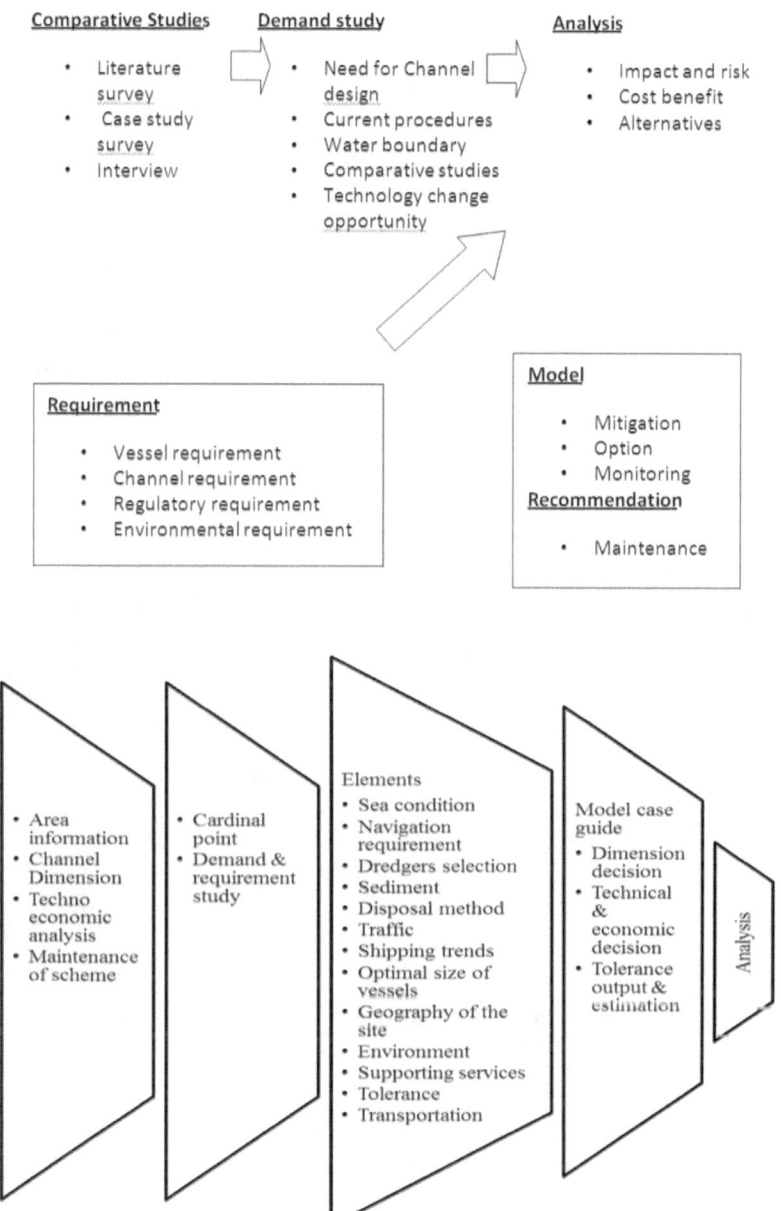

Figure 11. 5: Components of each stage

Navigation requirement: Having various vessels to channel equipment ratio, the size of the channel and the volume of dredged can be determined.

Maintenance dredging requirement: Analysis will go through iterative round of all the thematic part of the project:

i. Need for channel maintenance
ii. Old, Current and new practice
iii. Technological change and opportunity

Recommendation for channel deepening work requires the following (Figure 11.5):

i. Loaded vessel draft / squat, this involve the hydrodynamic sinking effect of lowering vessel keel relative to channel bottom with speed.
ii. Wave induced motions, safety clearance, and dredging tolerance.
iii. Advance maintenance dredging.

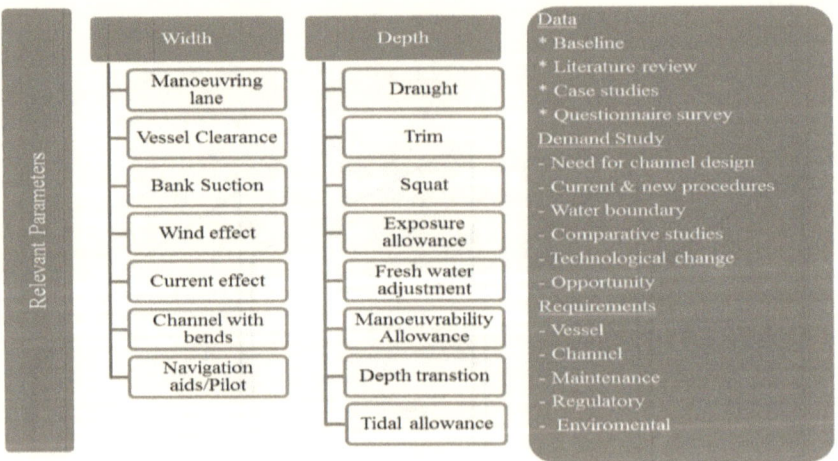

Figure 11. 6: Navigation parameters

To determine this iterative process with regulatory requirement, necessary projection base on the following data parameter as shown in the framework for depth calculation can be performed.

Table 11. 1: Navigation depth parameters requirement.

Depth parameters	Subparameters
Draught	vessel static draugh
Trim	Vessel depth
Squat	Vessel speed, draught, channel depth, block coefficient
Exposure allowance	Vessel size, traffic density, local wave climate
Fresh water allowance	Water salinity and vessel size
Maneuvering allowance	Channel bottom, operational characteristics, vessel speed and controllability
Overdepth allowance	Nature channel bottom, dredging tolerant and siltation
Depth transition allowance	Sudden change in channel depth
Tidal allowance	Reference datum, highest and lowest level tidal window

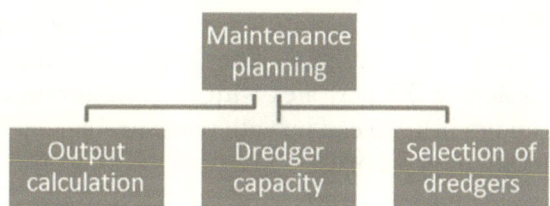

Figure 11. 7: Maintenance dredging components

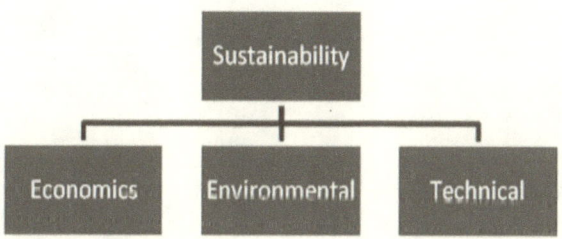

Figure 11. 8: Dredging and sustainability Process

This three studies have been done prior to establishment of PTP prior to the initial capital, however to maintain and capital dredging techniques for channel use similar process. Thus, methodology varies, and simplicity itself remains the beauty of design (Figure 11.7 and 11.8).

—

11.6 The case of Port of Tanjung Pelepas

The Port Tanjung Pelepas (PTP) of the Sungai Pulai located in Malay Peninsula's most southern tip in the State of Johor, close to the new Malaysia-Singapore Second Crossing, a new 1800-metre bridge linking Singapore with Malaysia's. The port development at PTP is one unique state of art capital project design work done on sensitivity and helps transform the river and mangrove area in 1998 into one of the world's most equipped container port. it remain one the significant implementation of Malaysian VISION 2020 plan, a 60 years concession 60-year concession for 800 ha port with Free Zone Status was made to Seaport Terminal (Johor), under operation of her subsidiary company, PTP by Malaysian government and syndicate of banks agreement of RM 2 billion loan. The port has stimulated rapid development in the region stretching from state capital Johor towards the west along the Johor Strait; it has changed the region to developed area with excellent infrastructure, housing facilities and new areas for industrial development.

What necessitated the port development initiative is again the demand and growth since the seventies, with forecast for potential critical capacity problem by the year 2000. The Johor Port Authority reached maximum expansion of the port area with the completion of Phase 4 of Pasir Gudang studies in 1990 end up with selecting Tanjung Pelepas as the most suitable location for Johor's.

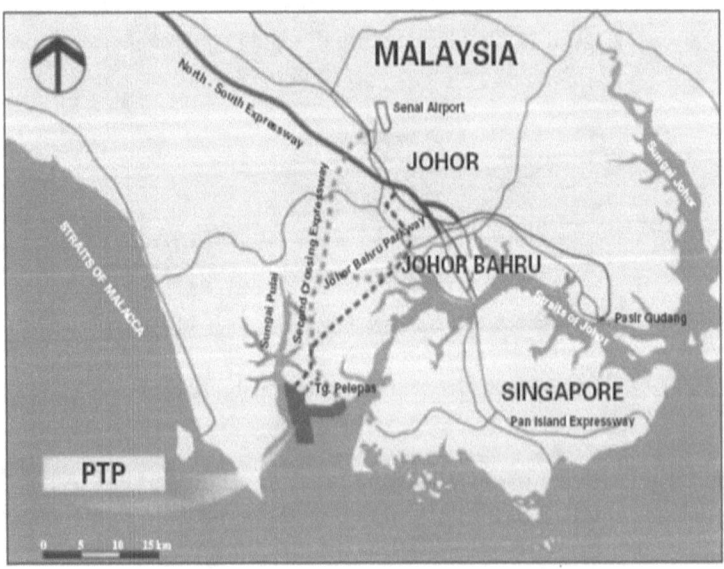

Figure 11. 9: Port of Tanjung Pelepas

—

The initial main dredging work done on Phase 1 of PTP development towards complementing a fully operational container terminal by end of 1999 or the dredging and reclamation scoped for the removal by dredging of existing soft materials to provide an approach channel, turning basin and bund foundation area and construction of the wharf bund and filling of the terminal and infrastructure areas to provide a stable platform for the container area. Summary of work done is as follows:

i.   Year 1997; contact cost—US$ 158 million
ii. 200 hectares of site clearance, mangrove and bush clearing; additional site investigation;
iii. Dredging of the 9-km long approach channel and turning basin, approximate volume of 16,000,000 m³;
iv. Dredging to foundation level below wharf bund, approximately volume 5,500,000 m³;
v.   Constructions of the wharf bund, approx. volume of sand 4,000,000 m³;
vi. Installation of 20,000,000 meters of wick drains as ground treatment
vii. Reclamation and surcharge of phase 1 area, Terra et Aqua in September 2000 for the foundation area and construction of the wharf bund and filling of the terminal and infrastructure to provide a stable platform for the container area.

The construction, completion, maintenance of the dredging and reclamation works involved the dredging of 16,000,000 m³ of soft and stiff material to form a 12 km access channel and turning basin, together with the dredging of 5,500,000 m³ of soft to medium material from a trench to form the base for the new wharf structure. The initial activity concentrated on dredging an access of 12 meter deep, 100 meter wide and approx. 5000 meters long (pre-dredging depth only approximately 4 meter by low tide) to allow the jumbo hopper dredgers to reach the Site. Dumping ground—in the Malacca Straits northwest of Karimun and along the coast of Pontian up to 120 km from the sand was won from Karimun southeast. The most economical filling method for the bund would have been:

i. Filling up to −6 to −7 CD direct dumping from a hopper dredger;
ii. From −6 to −7 to +4 CD rain bowing

11.7 Baseline Data Analysis

The input parameters are used to develop the requirements and design considerations for channel width and depth, as demonstrated in the flow chart shown above which proves detail on the width and depth parameters. Input data is captured from baseline studies that are undertaken involving an analysis and evaluation of the following data supplied by PTP:

11.8 Pressure Need for navigation optimizing navigation channel, economic and fairway analysis

The optimum channel depth requires studies of estimated costs, benefits and risk of various plans and alternative designs considering safety, efficiency, and environmental impacts in order to determine the most economical, functional channel alignment and design depth. Channel deepening design is often one of the major cost determining parameters for navigation project and design of such depth is of various types that require adaptability of each design to future improvements for increased navigational capability as in table 2).

11.9 Action and Reaction of Channelization Work

Need for navigation channel optimization, economic and fairway analysis leads to navigation channel maintenance. The case presented here is limited to depth analysis. The optimum channel depth requires studies of estimated costs, benefits and risk of various plans and alternative designs considering safety, efficiency, and environmental impacts in order to determine the most economical, functional channel alignment and design depth. Channel deepening design is often one of the major cost-determining parameters for navigation project and design of such depth is of various types that require adaptability of each design to future improvements for increased navigational capability. Figure 11.10 and 11.11 shows container cargo projection and fairway analysis for port PTP.

Table 11. 2: Baseline data for the target

Length (L)	22O-430
Beam (B)	
Maximum draught (d)	8—15M
Speed (Vs)	4-8 Knot

—

Maneuverability and speed	A qualitative determination of the vessel's maneuverability in comparison with other vessels; major company calling the port are well established and top liner trade operator in the world, under risk rating as might be needed for PTP.
Traffic density	The level of traffic frequenting the waterway—as applied to PTP, being a dredged channel with 14 m under risk rating requirement 1 is assigned.
Block coefficient	
Target vessel and other deep-draught vessels using the waterway:	
Types of smaller vessels and congestion	Based on PTP operations, frequency is less PTP is a time container port size.
Cross traffic	no cross traffic inner channel, there is cross traffic at outside channel −risk rating of 0 is assigned dimensions (length, beam, draught).
Type of cargo handled	Bulk, passenger, bunker barge, dredger, STS,
Types of smaller vessels and congestion	Based on PTP operations, frequency is less PTP is a time container port size.
Cross traffic	no cross traffic inner channel, there is cross traffic at outside channel −risk rating of 0 is assigned dimensions (length, beam, draught).
Type of cargo handled	Bulk, passenger, bunker barge, dredger, STS,
Channel characteristics: The waterway parameters, or waterway characteristics, are determined from field programs or existing information	
Bottom material characteristics	Silt /Mud depth
Channel	16m
Phase 1 terminal /basin	14m
Phase 1 terminal /basin berth pocket	15 m
Phase 2 terminal berth pocket	15m
Phase 1 terminal /basin	17ₘ
Phase 1 terminal /basin berth pocket	19.5m
Phase 2 terminal berth pocket	15m
Current velocity and direction	2Knots-

Outer area cross current	2knots
Wind velocity and direction;	
Channel basin—north—southerly flow	1 knot-
Flood tide	Northerly, southerly
Wave height	1.2-1.5m
Tides	Temperature, currents, tidal and/or river (velocity, direction, and duration);
Sediment	Sediment sizes and area distribution, movement, and serious scour and shoal areas
Sea bed	Type of bed and bank (soft or hard); SOFT
Alignment and configuration;	015 degree 40 minute, width 420, length 12.6 km
Freshwater inflow	NA, range between 0m-0.4m

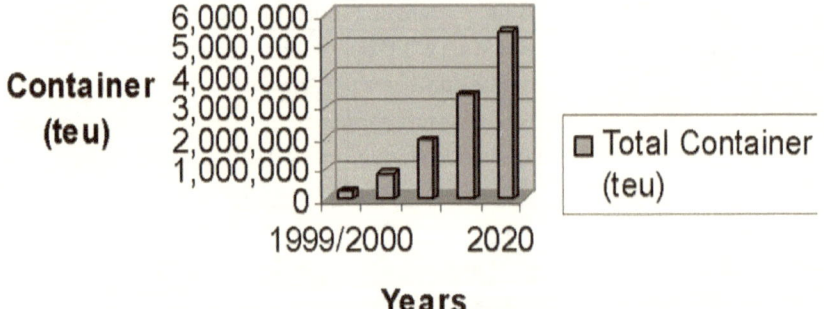

Figure 11. 10: Total container cargo (TEU) 2020 projection

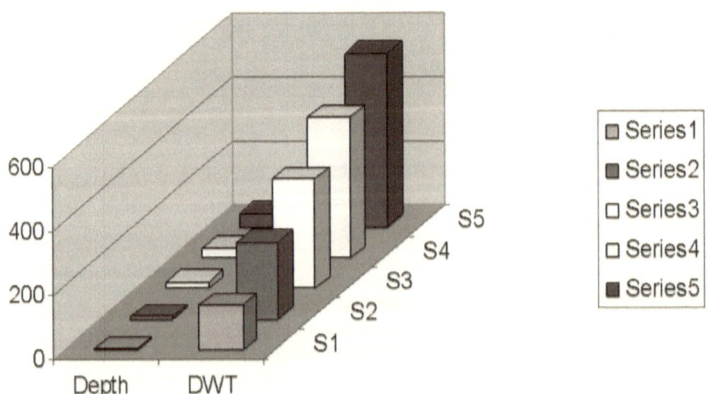

Figure 11. 11: Fairway analysis—Source PTP

The optimum economic channel is selected from a comparison of annual benefits and annual costs for each channel maintenance plans. Deeper channels will permit the use of larger ships, which are more economical to operate.

11.10 Cost, benefit and depth increase

In respect to PTP and channel maintenance work the following economic analysis based on need and projection data analysis represent the demands stage of this work for the fairway. Hourly cost of vessel is given as follows (Figure 11.12):

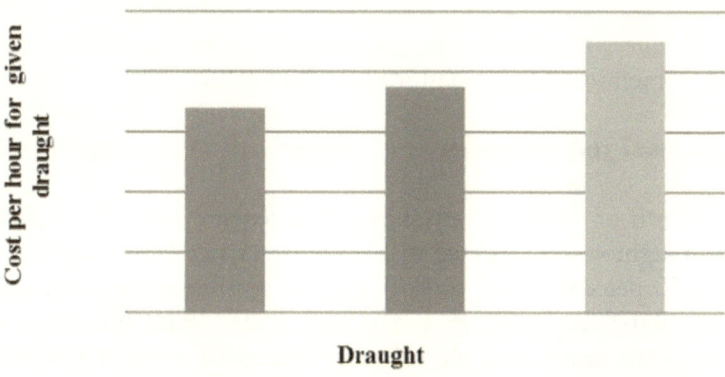

Figure 11. 12: Hourly cost of vessel (source UNDP, 2008)

Time saving in hours is a consequence of deepening of river Scheldt by 4 feet is as follows in Figure 11.13.

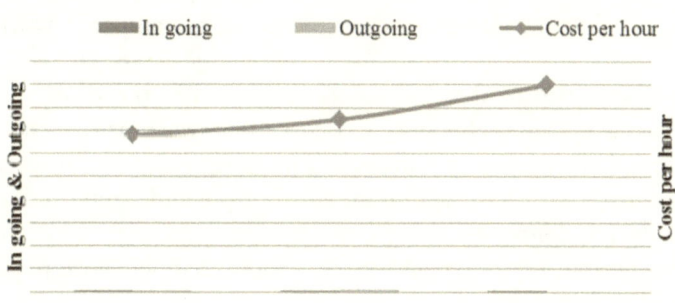

Figure 11. 13: Estimated time due to deepening

Total benefit can be calculated from the number of vessel, time saving per hour and cost of one hour (Table 3).

Table 11. 3 : Depth improvement advantage

Depth / Draught (m)	Cost per hour for given draught	Benefit
34-38	US$680	**US$100**
38-40	US$750	**US$ 150**
>50	US$900	**US$ 80**

11.11 Channel dimensioning: navigation vessel and channel requirement

This involves the input variables required, to determine the minimum waterway dimension requirement for safe navigation.

11.12 Vessel requirement

The depth of the waterway should be adequate to accommodate the deepest-draught vessel expected to use the waterway. However, this is not the case 100 percent of the time; it may be possible to schedule passage of the deepest-draught vessel during high water levels (i.e., high tide). Selection of the design draught should be based on an economic analysis of the cost of vessel delays, operation and light loading compared with construction and maintenance cost. Figure 11.14 shows vessel design parameter that should be considered in maintenance of channel. Normally, the channel depth is based on the development and parametric study of one or more design ships with an appropriately loaded or ballasted draft. Selection of the design ship and project design depth is determined jointly by an economic analysis of the expected benefits compared with project costs.

The design depth of the channel vary so that the design ship will be able to make a safe, efficient, and cost-effective transit of the channel under normal operational conditions. The design depths are considered, to be the authorized depths. This should not preclude minor adjustments in depth during continued design, construction, and operation delegated authorities permit. Figure 10 and Table 4 show channel depth requirement for channel maintenance.

—

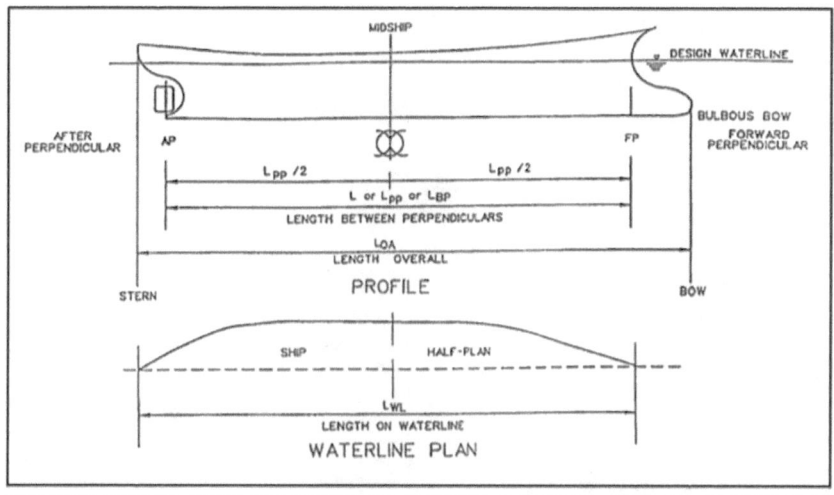

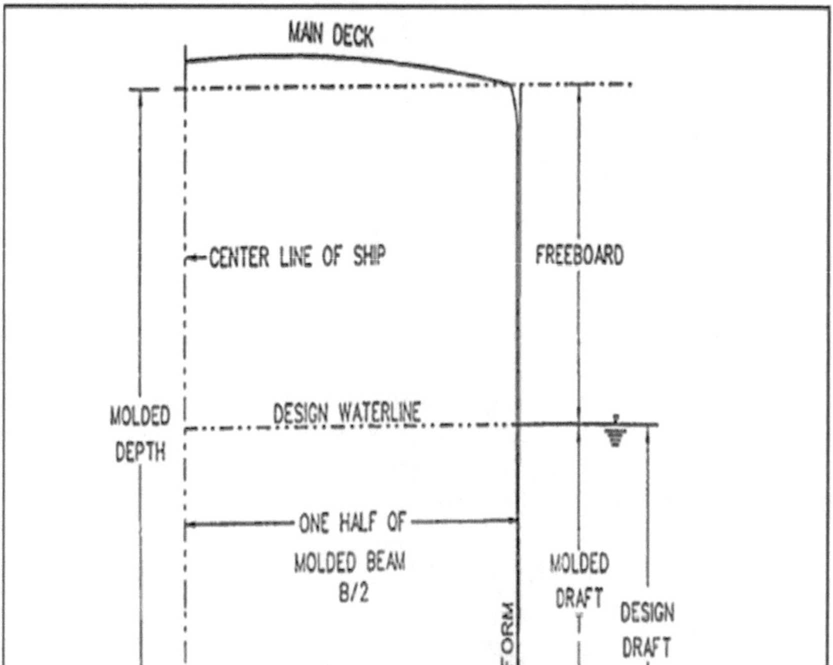

Figure 11. 14: Vessel requirement

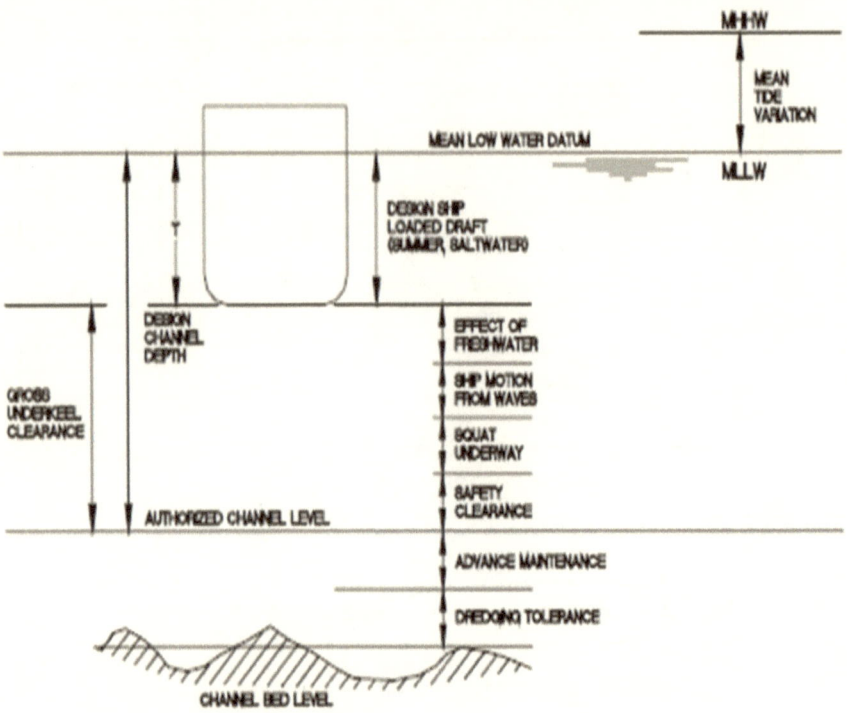

Figure 11. 15: Channel depth requirement

For channel to accept ship there must be corresponding depth required to maintain vessel maneuverability. Therefore, minimum value for water depth/draught ratio is necessary for assurance and reliability. In many parts of the world, a value of 1.10 has become acceptable, although a value of 1.15 is also often used. Example of standardized methods by the Japanese channel work and ship size is shown in Table 11.4.

Table 11. 4: Channel depth and ship size

Depth (m)	10	12	14	16	20	22
Ship size(DWT) (x10e3)	15	30	50	100	15	25

For channel to accept ships there must be corresponding depth required to maintain vessel maneuverability. Therefore minimum value for water depth/draught ratio is necessary to for assurance and reliability. See Table 5a and 5b for PTP channel parameters.

Table 11. 5: PTP channel parameters

Width	250
Maneuvering lane	1.6 to 2.0 times vessel beam
The ship clearance lanes	80% of the vessel beam
Bank clearance	80% of vessel beam is added to both sides of the channel
Depth	12.5m (Bellow MLLW) related to LAT
Under keel clearance	1m
Safety clearance	0.3
Advanced maintenance dredging Allowance	0.5m
Turning basin	600m
TEU estimate	6000 TEU
Berth	18m below LAT with containment dike for future dredging
Projection	350m wide by 20m with turning basin of 750m diameter
Side slope	1:8 to 1:6 vertical: horizontal) for silt and mud— assumption at 1:10, depending on material

Table 11. 6: Environmental criteria (PTP)

Pos	Parameter	Data/sizes/dimension	Sources
14.3.1	Max Wind speed during vessel berthing	22M/S	Master plan Volume 2
14.3.2	Max Wind at berth	27.7M/S	
14.3.3	Max significant wave height	Hmax=1.2m	
14.3.4	Max river current	1.0m/s	
	Wave period	Ts =4-5 sec	Sellhorn wind wave cal.

11.13 Result

Total Depth calculation: The design (authorized) depth will include the various allowances. Advance maintenance and dredging tolerance are provided in addition to the design depth. Minimum Waterway Depth for safe navigation is calculated from the sum of the draught of the design vessel as well as a number of allowances and requirements. The Canadian model the following recommended Canadian model formula is used.

—

Actual Waterway Depth = Target Vessel Static Draught + Trim + Squat + Exposure Allowance + Fresh Water Adjustment + Bottom Material Allowance + Over depth Allowance + Depth Transition—Tidal Allowance.

Channel (Advertised) Waterway Depth = Waterway Depth—Over depth Allowance

Table 6 shows estimate of volume that can be dredged to maintained PTP channel.

Table 11. 7: Estimate of dredge volume for 3 channel alignment

Channel	Depth (m LAT)	Volume (million Cubic meter)
Alignment 1	13	12.2
	14	14.0
	15	17.5
Alignment 2	13	13.0
	14	16.2
	15	20.1
Alignment 3	13	15.4
	14	18.2
	15	21.7

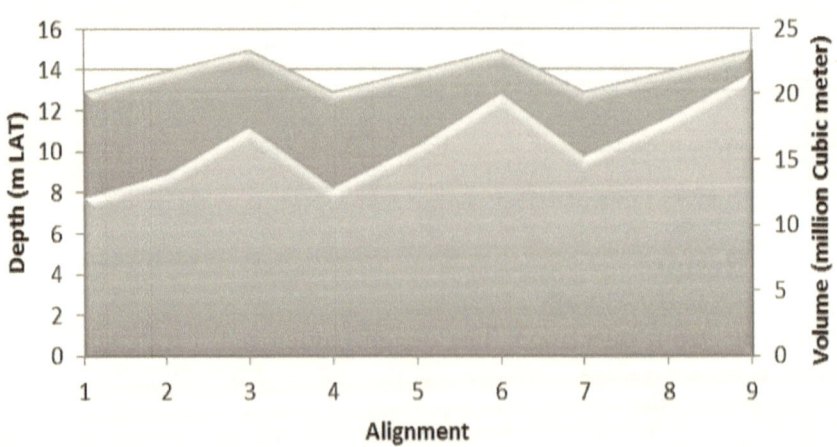

Figure 11. 16: Channel alignments

Alignment 1 sedimentation volume is 600,000—650,000 cubic meters per year, this is in line with the following channel characteristics (Figure 11.6):

i. Width: 250 m
ii. Maneuvering lane:1.6 to 2.0 times vessel beam
iii. The ship clearance lanes: 80% of the vessel beam
iv. Bank clearance—80% of vessel beam is added to both sides of the channel
v. Depth: 12.5m (bellow MLLW) related to LAT
vi. 1m under keel clearance
vii. Safety clearance: 0.3
viii. Advanced maintenance dredging allowance—0.5m
ix. Turning basin: 600m
x. TEU estimate: 6000 TEU
xi. Berth: 18m below LAT with containment dike for future dredging
xii. Projection 350m wide by 20m with turning basin of 750m diameter
xiii. Side slope: 1:8 to 1:6 vertical: horizontal for silt and mud—assumption at 1:10—depending on material

Table 7 shows environmental criteria required to be taken into consideration.

Table 11. 8: Environmental criteria's

Parameters	Data/size / dimension	Source
Max wind speed during vessel berthing	22m/s	PTP Master plan vol 2
Max significant wave high	24.7m/s	
Wave period	Hmax=1.2m	
Max. River current	1.0m/s	
Max wind at berth	Ts=4.5sec	Shell horn wind wave cal

Total Depth Prediction

The design (authorized) depth will include the various allowances as shown in Table 6.16. Advance maintenance and dredging tolerance are provided in addition to the design depth. Minimum Waterway Depth for safe navigation is calculated from the sum of the draught of the design vessel as well as a number of allowances and requirements. According the Canadian regulation Table 6.16 is used to model the depth of waterway.

—

Actual Waterway Depth = Target Vessel Static Draught + Trim + Squat + Exposure Allowance + Fresh Water Adjustment + Bottom Material Allowance + Over depth Allowance + Depth Transition—Tidal Allowance, (See Table 8) Components of Waterway Depth)

Table 11. 9: Model spreadsheet for channel depth calculation for PTP

Parameter Depth	
Target vessel static draught	9.0
Trim	1.0
Tidal window	0.0
Squat	1.2
depth allowance for exposure	0.0
Fresh water adjustment	0.0
Bottom material allowance	1.5
Maneuvering margin	2.0
Over depth allowance	0.5
Depth transition	0.0
Depth	15.2
Project (Advertised) Waterway Depth = Waterway Depth—Over depth Allowance	
Water way depth	16.5
Over depth allowance	0.5
Advertised depth	16.0

Alternative equation for validation by UNDP
H=D+Z+I+R+C+#

Squat Calculation

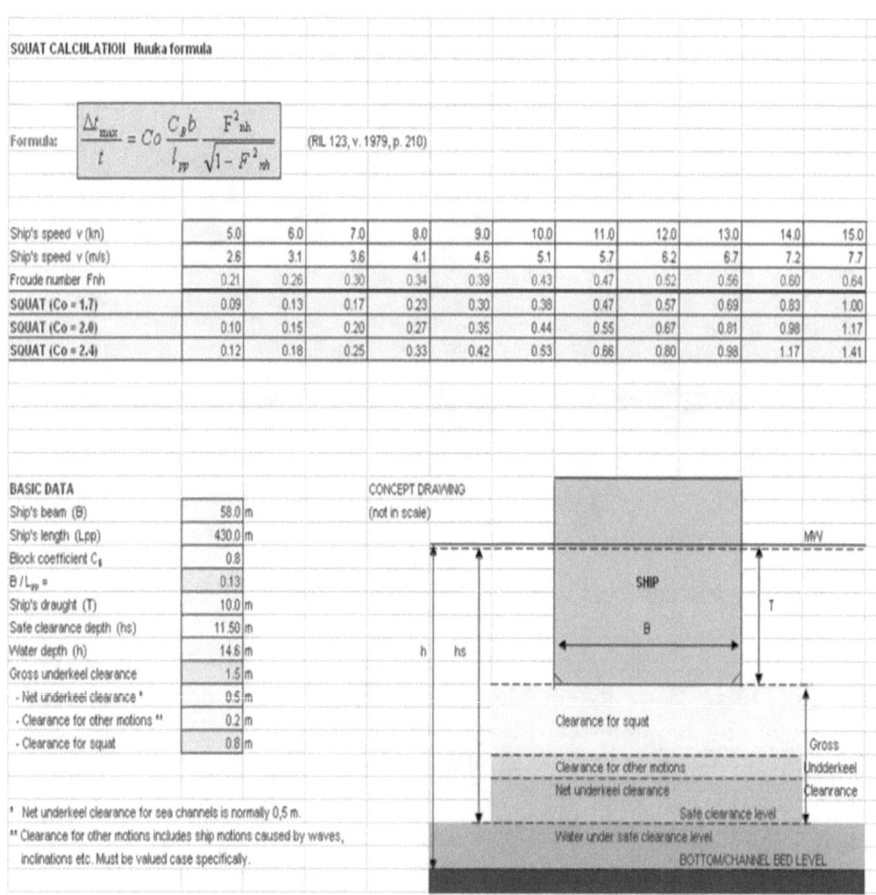

Figure 11. 17: Squat calculation

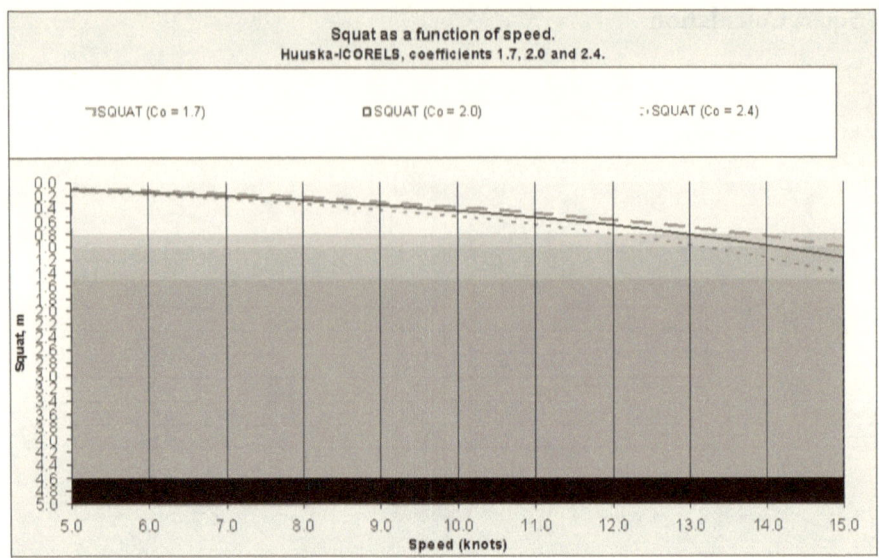

Figure 11. 18: Squat calculation

Maintenance dredging Capacity—sediments output and estimates

Maintenance dredging with objective to reduce channel delay, accept big ship to be done in environmental sustainable manner and optimal efficiency –in maintenance dredging quantifying the loss of depth pave wave for dredging requirement to be determined and this lead to optimal choice of dredger. Thus PTP is a new port with very big clearance to accept third generation ships, personal communication with the health, safety and environmental department there also confirmed regular survey for siltation towards planning to maintain balance which is put at 2—3 year for now(personal communication. Issue relating to investigation or communication about what size vessel will ply the channel in the 10 years is rarely discussed by channel workers. Generic calculation on data results from analysis of:

i. Vessel and channel requirement
ii. Channel dimension
iii. Hydrographic data
iv. Basic rate output of the dredger
v. Computation of volume
vi. Cycle time and Number of work day per year
vii. Working condition and Environmental discounting

—

Where;

output = *Number of cycle per day* x *Load Factor* x *Hoppercapacity* x *Number of working days*

Load Factor = $\frac{Volume}{Hopper\ Volume}$, *Number of working days* = *365 days*

For PTP:

Table 11. 10: Dredge output for PTP

Number of cycle	4—5 per day,
Hopper capacity =	2500—5000/6000
Number of days =	150,000/6000
Volume of maintenance dredging =	300,000-400,000 for 3 years
Load factor=	150,000/year
Output =	150,000/5000=30,000

Alternative equation for validation (USACE)—> V=0.5x (A1+A2) x (S2-S1)

Dredger selection

Hydraulic dredgers are based on the use of pumps to raise the material. The other main group is the mechanical dredgers. Dredger use for PTP dredging work was made using state of art combo slip hopper barge dredger; the uniqueness of this dredger has capacity to contain dredge material while in operation and transit operation until disposal location. Split hopper dredger is a modern hydraulic excavator, mounted on a platform fixed to the seabed. The material is excavated by the bucket of the excavator, kept and contained by the dredger. It is then raised above the water and transported directly to the disposal site. The soil at PTP is basically silt and mud, and the dredgers are well suited for this. Accuracy is only achieved if monitoring and control equipment are used.

Thus the containment has guaranteed no leakage during transportation. The type of dredger used may not be an important consideration for all dredging operations. For example when dredging in enclosed areas, such as docks or within locks, where there is little potential for any adverse effects on the wider marine environment or in highly turbid environments where any adverse effects due to sediment re-suspension are unlikely. Consideration should be given to the type of dredger used where adverse effects on marine animals and plants due

—

to suspended solids have been predicted which cannot be avoided by careful programming of the timing of the works. Assessments on the most suitable dredger to use must be made on a case by case basis, giving consideration to both practical and economic considerations. The type of dredger employed is often determined by the depth of water, scale of the maintenance operations, the type of material to be dredged, and can be a question of meeting supply and demand.

Protective silt curtains or screens can be used with certain dredging equipment (grab and backhoe dredgers) in order to reduce the amount of suspended sediment being transported outside the dredging area placed around sensitive marine features. The use of silt curtains is reported to considerably reduce the loss of suspended sediments from the dredged area, by up to 75% where current velocities are very low. However they are generally ineffective in areas with high wave action and current velocities which exceed 0.5 m/s. Over recent years, certain dredging methods have been used in ports and harbors that are not presently regulated under the licensing process, such as water injection dredging or sea bed leveling. These methods operate by moving material from one place to another along the seabed and as sediments are not raised from the surface of the water, and then strictly speaking no disposal takes place. Although the aim of these methods is to keep sediments in the vicinity of the seabed, there is potential for increased suspended sediments to occur possibly causing disturbance to marine animals and plants, especially where sediments are contaminated.

Agitation dredging, which encompasses a number of different techniques, is an example of type of operation, as its name implies. Agitation dredging aims to disturb sediments and raise them into suspension in order to move material through the water column. It is therefore inevitable that there will be greater increases in suspended solids and siltation levels. Subsequently, the magnitude and extent of impacts on the nature conservation interests of the site may possibly be greater, although they may remain within the range of natural variation, depending on the local conditions at the site. As with other types of dredging, where these dredging methods occur in systems with high background levels of suspended sediments there is likely to be little problem, however in other areas more caution may need to be applied particularly with regard to agitation dredging. Although, it should be noted that the amounts of material redistributed during agitation dredging may be no more than that occurs during natural phenomena, such as storm events.

When these dredging methods are proposed within the harbor area, either by the port themselves or a third party, consideration by the port

—

authority should be given to the potential effects of such an activity on safe navigation and the potential for effects on designated marine features. This should be based on information provided by those proposing to undertake the dredging, including answers to questions such as, where, when, over what area, how much material, and how often? When considering whether there are likely to be any effects on the communities of the designated features of the site, ports and harbors may consult with the country conservation agencies for advice.

Any identified effects of the proposed activities on designated features should be addressed and minimized by careful operation and by planning the dredge to avoid particularly sensitive times, as described relevant guidelines. The volume of soil is very important in dredger selection, a smaller and more economic and environmental sustainable dredger is preferred. Table 9 shows the guideline for dredger selection given by PIANC. Factors that are considered in selection of dredgers are;

i.   Kind of soil
ii.   Volume of soil and construction period
iii. Dredging depth and dredging thickness
iv.   Soil disposal methods
v.   Length and quality of the transport route to the disposal areas
vi.   Disturbance of the bottom sediments
vii. Meteorological, ocean logical, and geometric condition
viii. Tolerance of the excavation
ix.   Kind of dredger and auxiliary equipment
x.   Availability of the desired dredger

Considering PTP specification grab hopper is a good choice of dredger, for maintenance dredging. Considering PTP specification grab hopper is a good choice of dredger, for maintenance dredging. Traditionally, dredging quantities for purposes of design estimates and construction payment have been obtained from cross-sectional surveys of the project area. These surveys are normally run perpendicular to the general project alignment at a predetermined constant spacing. The elevation data are plotted in section view along with the design/required depth and/or allowable over depth templates. One or more reference or payment templates may be involved on a dredging project (e.g., zero tolerance, null ranges, etc.). Given sectional plots of both preconstruction and post construction (as-built) grades (or, in some cases, intermediate partial construction grades), the amount of excavated (cut) or placed (fill) area can be determined at each cross section. Figure shows the typical templates used to compute relative cut/fill quantities.

Table 11. 11: Dredger selection

Site conditions		Cutter Suction	Bucket Wheel	Standard Trailer	Grab Hopper	Bucket chain	Grab
Bed material	Loose silt	1	1	1	2	2	2
	Cohesive silt	1	1	1	1	1	1
	Fine sand	1	1	1	2	2	2
	Medium sand	1	1	1	2	2	2
	Coarse sand	1	1	1	2	2	2
Sea conditions	Enclosed water	1	1	3	1	2	2
	Shelter water	1	1	1	1	1	1
	Exposed water	3	3	1	3	3	N
Disposal to:	Shore	1	1	2	N	2	1
	Sea	N	N	1	1	1	1
Quantities	100,000 m^3	1	1	2	1	2	1
	250, 000 m^3	1	1	1	1	1	1
	500,000 m^3	1	1	1	2	1	1
	>500,000 m^3	1	1	1	3	1	3

1—Suitable 2—Acceptable 3—Marginal N—unsuitable

The model tested in this work from the records provide simple system for monitoring the channel and to draw the following conclusions, The depth of the channel is large; approximately 16m x 420 and taking ships in the order of approximately 350-420m LOA But there is tendency that the channel will soon get close to its limit. The rate of design ship to the channel state still exhibits non-linear behavior; bigger ship is coming there and the channel remains the same. The best that could be done is removing the shoal for enhanced management of the channel. Depth of 20m is projected against 2020, to meet the demand, however but. Critical study and employing a sustainable risk based methodology with good record of environmental change rationalization will be necessary in future. The phenomena of squat its effect are of major importance to the system, applying simplifies model tested in this work could help close monitoring towards reliability and confidence of the channel.

Approximately 600,000 million cubic yards of sediment will be dredged annually from the navigable water, and the condition environmental change of such sediment required. The contaminants of concern and their risk to the environment and to humans will vary widely depending upon site-specific factors ranging from ecological habitat to sediment particle size distribution. The soft alluvium in previous work allowed consolidate under high surface loads, resulting in settlement of finished ground surfaces, however continuous analysis on geo-technical engineering studies will always be required to complement sustainable planning work. Facing capacity difficulties are issue of concern everywhere today particularly the fate of the channels. Demand for ship has approached supply and the tradeoffs will be more and more carefully scrutinized by the resources available and environmental demands. The channel is very important for land and land use, and safety linked to environment required, as demands of the ship increase so does the need for system integration at local and international level. Under doctrine of sustainability, it has been widely accepted that new approach to design and maintaining system should focus on top down risk based, whose matrix will holistically cover all issues of concern including uncertainty. Uncertainty itself remains a big issue who's definitional and framing fall under complex circumstance. Future studies on this work then lies on issue of risk based assessment for channel design and decision support system, simple data inventory system, system integration, and extensive studies that cover type of uncertainty that exist. Waterways development need to have a strategy for the future of its marine structures program by examining the removal of non-core operations, and negotiate responsibilities for water depth forecasting with the Hydrographic Department. The study revealed reflections that could be helpful for understanding the meaning and relevance simplified monitoring procedure, understanding and the concept sustainability in practical cases of environmental management of inland navigation.

The case tests how to develop plans for simple waterways performance measurement information system, it is recommended that such method could be incorporate could tap existing information and communication to link reporting system of data to National Channel Inventory system. Such system could be implemented in line with the accountability framework), in order to provide improved monitoring and performance reporting capability on measures and indicators such as channel monitoring level of service compliance; subject matter expertise level of activities; timeliness of notification to mariners; and actual repairs to structures versus required.

References

1. Bridges, J. W. and Bridges, O. 2001. Hormones as growth promoters: the precautionary

2. Principle or a political risk assessment? In: Late lessons from early warnings: the precautionary principle 1896-2000. European Environmental Agency (ed.) Environmental Issue report No 22.

3. Burt, N., C. Fletcher, and E. Paipai. 1997b. Guide 2B. Conventions, Codes and Conditions:

4. Land Disposal. In: Environmental Aspects of Dredging. IADC and CEDA (eds.) The Netherlands.

5. Craye, M. 2004. Interactive uncertainty assessment through pedigree-based tools: what, why and how? European Commission-Joint Research Center. Ispra. Italy

6. Csiti, A. and Burt, N.1999. Guide 5. Reuse, recycle or relocate. In: Environmental Aspects of Dredging. IADC and CEDA (eds.). The Netherlands.

7. Proceedings of the International Symposium on Uncertainty and Precaution in Environmental Management. Technical University of Denmark. Copenhagen, June 2004.102

8. Evans, S. M., Birchenough, A. C. and Brancato, M. S. 2000. The TBT Ban: Out of the Frying Pan into the Fire? Marine Pollution Bulletin 40: 204-211.

9. Evans, S. M. 2000. Marine Antifoulants. In: Seas at The Millennium: An Environmental Evaluation. Sheppard, C. (ed.). Elsevier Science. pp. 247-256

10. Funtowicz, S. O. 2004. Models of Science & Policy: From Expert Demonstration to Post Normal Science. Proceedings of the International Symposium on Uncertainty and Precaution in Environmental Management. Technical University of Denmark. Copenhagen, June 2004.

11. Funtowicz, S. O. and Ravetz, J. R. 1990. Uncertainty and Quality in Science for Policy. Kluwer Academic Publishers.

12. Funtowicz, S. O. and Ravetz, J. R.. 1992. The Good, the true and the post-modern. Futures 24: 963-976. Funtowicz, S. O. and Ravetz, J. R. 1993. Science for the Post-Normal Age. Futures 25: 739-755.

13. Funtowicz, S. O. and Ravetz J. R. 1994. The worth of a songbird: ecological economics as a post-normal science. Ecological Economics 10: 197-207.

14. Gibbs, P. E., Bryan, G. W., Pascoe, P. L. and Burt, G. R. 1987. The use of the dogwhelk (Nucella lapillus) as an indicator of TBT contamination. Journal of the Marine Biological Association of the United Kingdom 67: 507-524.103.

CHAPTER 12

Power Integrity Requirement of New Generation of ROV for Deep Sea Operation

"I never think of the future—it comes soon enough"
Albert Einstein

Summary

Remotely operated vehicles (ROVs) system requires powerful vehicles to support the bollard thrust and tool power required for deepwater tasks. Evolving deeper waters, vehicle support for heavy-duty tasks demand, deepwater subsea construction, repair and maintenance require efficient ROV power pack to support these tasks. Typical work-class ROV systems provide maximum power levels ranging from 100 to 200 horsepower that produce impressive thrust in either vertical or horizontal directions. Problem associated with ROV power pack include inefficiencies in the power system designs that limit peak system performance thrust curves, inability of the hydraulic system to adjust to varying demands, environmental concern related to energy usage and ship husbandry. This section address the design and development of a variable pressure power delivery and propulsion system that significantly increases overall system efficiency to maximize use of available power.

–

378

12.1 Introduction

Environmental issue has been key driver to today technological decision. Deepwater marine operation has increased due to prohibitive nature of offshore activities in proximity to coastline. Deep water construction posed many challenges. This include the situation of water depth increases and subsequence requirement for surface vessels size increase in order to support the equipment needed to reach the seabed. This makes the use and demand of ROVs imperative. Consequentially, the source of energy that meets these demands is increasingly becoming important. Energy space, size, and economic energy efficiency is tackled through increase ROV functionality with larger onboard power systems that provide more available thrust to support higher variety of tasks. Subsea equipment and hardware improvement has target effective equipment handling and design of a variable-pressure power delivery and propulsion system for completion of ROVs mission.

All components of an ROV system should be rated to the maximum operating depth of the underwater environment anticipated, including safety factors. Pollution released from ROV devices have really been addressed, and the reality of environmental interaction makes it important for ROV system design to address ship husbandry problem. This chapter discusses the potential of using alternative energy hybrid to power ROV system with hope to reduce challenge of air prolusion released to the atmosphere.ROV deep water operation find application in the following areas: FPSO, diving support, research vessel, drillship (Klagesm. et al, 2002).

12.2 System failure and risk based design requirement for ROV

In order to improve reliability of system, a generalized version and analytical expression for this important principle have also been formulated for multiple failure modes. It is argued that the traditional approach based on a risk matrix is suitable only for single failure modes/scenarios. In the case of multiple failure modes (scenarios), the individual risks should be aggregated and compared with the maximum tolerable risk. Risk-based design is important in order to minimize the probability of system failure below a maximal acceptable level at a minimum total cost (the sum of the cost for building the system and the risk of failure).

Today, design shift towards knowledge intensive product, risk based design is believed to be key elements for enhancement of industrial competitiveness. The use of risk based design, operation and regulation open door to innovation and radical novel and inventive, and cost effective design solution. Risk based approach for ROV follow well established quantitative risk analysis used in offshore industries. The key to successful use of risk based design require advance tool to determine the risks involved and to quantify the effects of risk preventing/reducing measures as well as to develop (evaluation criteria to judge their cost effectiveness.

ROV operating capabilities requirement that can be investigated is under risk based design are:

i. Standardized intervention ports for all subsea with any available ROV.
ii. Visible mechanical indicator or redundant telemetry channel
iii. ROV testing requirements
iv. Electrical power requirement

General requirements—refer to SOLAS requirements, Part D, Chapter II-1—outlines requirements for Ship construction sub-division and stability, machinery and electrical installations

12.3 ROV System and Subsystem

The ROV system is one of the simplest robotic designs, where complex assignments can be accomplished with a variety of closed-loop aids to navigation. ROV system has its immovable locomotive part and counterparts that are capable to move under its own power. The power of locomotion has ability to navigate the robot, with levels of autonomy to achieve defined mission. Remote operated vehicle (ROV) are built with secondary control of the subsea blowout prevent or (BOP) stack, and most provide other tertiary control systems as well.

The ROV intervention capability is limited on some subsea BOP stacks while others have the ability to control multiple functions. ROV intervention capabilities for secondary control of all subsea BOP stacks, including the ability to close all shears and pipe rams, close the choke and kill valves. Deep water operation requires larger component wall thicknesses are required for the air-filled spaces (pressure-resistant housings) on the vehicle. This increased wall thickness results in an increased vehicle weight, which requires a larger floatation system to counter the additional weight. This causes an increase in drag due to a larger cross-section, which requires more power, hence large cable to become larger.

Today design culture is embracing the open source computer-based control models that allow users to design their own navigation and control matrix. This concept allows development of new techniques; define by the user's imagination. Open source platform take the control of the development of navigation capabilities including the mission from the hands of the design engineer (who may or may not understand the user's needs) into the hands of the end user (who does understand the needs). Cost efficient design of the systems with the user in mind is critical to the success of the ROV and the mission. Saving weight is also key cost-effective design and operation. Figure 12.1a sows components of ROV that must be incorporated in the design spiral of the electrical requirement (Michel J.-L. (1990). Figure 12.1b shows H-ROV.

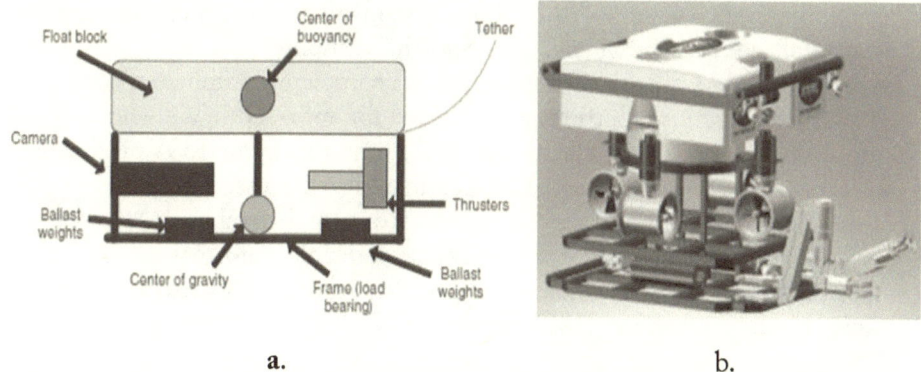

a.   b.

Figure 12. 1: ROV parts

The vehicle power system can be conveniently divided into transmission and distribution systems, which are described in sequence below. The transmission and distribution system prototyped mode is encouraged to designed, built, and tested before scale up and deployment. ROV subsystem includes (Renard V. et al (1993):

    i.   Lighting
    ii. Cameras
    iii. Sensors
    iv. Manipulators
    v. Electrical

Recent year have seen development of third generation ROV with Hybrid ROV that utilize hybrid design, one of such design is H-ROV which was developed in collaboration with Data Response Kongsberg by Sperre AS. H-ROV is built with an advanced propulsion system, auto-tracking, and an ingenious multiple control tool platforms for subsea DP and auto-traction operations. The redundancy system can benefit from robust electrical system design.

UMT ROV – STEALTH 2

The Stealth Remotely Operated Vehicle by Shark Marine Technologies Inc. is versatile ROVs on the market today. Small in size and portable with many features and capabilities. The Stealth ROV is packaged with plug and play ready for such options as scanning sonar, manipulator arm, sub-bottom profiler, and total positioning system. The size and weight (45kg) of this ROV system allows for operation from even small boats or inflatable. The Stealth2 computer controller with its daylight viewable, graphical interface allows completely automated control of the ROV functions. Settings are provided for auto-depth, auto-heading, auto altitude and vertical trim as well as for monitoring the ROVs internal environment. The computer controller may also be used for processing other Windows based options such as sonar or vehicle tracking. On-screen displays simplify navigation and provide valuable information during video playback as well as efficient high quality recordings of video, jpg and .mpeg. Figure 12.2 shows UMT Stealth ROV.

Figure 12. 2 :UMT Stealth ROV

The stealth can also fulfill other mission with manipulator arms, cutting arms, scaling lasers, various cameras; including zoom features or extreme low light, tracking systems, sonars; including multiple receiver units and sub-bottom profilers, gradiometers, magnetometers, recovery tools, cable reel systems and more. The stealth has application in different underwater operations from inspection services, to search and recovery, to environmental studies, to archaeological investigations. Vehicles are presently in use the world round by various navies, marine institutes, logging companies, underwater recovery units, commercial dive operations and more. Table 12.1 shows specification of the stealth.

Table 12. 1 : Specifications

Vehicle dimension	:	30"L x 22"W x 18.5 inc. handle
Vehicle weight	:	90 lbs. (40 kg)
Controller dimension	:	20"W x 18"D x 9"H
Controller weight	:	44 lb. (20kg) (Include hand control)
Hand control dimension	:	7.5"W x 7.5"D x 3"H
Hand control weight	:	41 lbs (1.8kg)
Hand control cable length	:	15 ft standard (longer optional)
Neutral umbilical description	:	Urethane Jacket with TPR Floatation jacket, 1000 lbs. minimum breaking load
Neutral umbilical size	:	0.53" Diameter (12.7mm)
Neutral Umbilical length	:	500 ft. Standard (up to 2000ft. optical)
Neutral Umbilical Weight	:	52 lbs per 500 ft. (20 kg per 150m), dry weight
Horizontal Thrusters	:	2 each, 1/3 Horsepower
Vertical Thrusters	:	2 each, 1/3 Horsepower
Lightning	:	2 each 150 watt quartz—variable control
Camera	:	180 degree viewable (pan optional)
Camera motion	:	High resolution colour 450TV line (Others optional)
On screen display	:	Depth, Heading date, time, title (Others optional)
Scanning Sonar	:	Pre-wired for plug and play(Sonar optional)
Depth Rating	:	1000 feet (300m)

12.4 New Generation of ROV for Deep Water Operation Challenge Electrical Power requirement

ROV power performance and efficiency depends on capability to effectively lifting heavy objects, pushing large equipment items into position, and acting as a supply for high-powered tooling at minimum cost, space and time. Increased input power of ROV system means

—

increased electrical current capacity requirements for the umbilical/tether system and increased motor, pump, and thruster sizes. As well as sub sequential system changes to support these primary size/capacity increases, use of more copper in the umbilical that requires more steel armor on the cable because weight of the conductors is entirely parasitic. The main components of the power system include (Fouquet Y., 2002) (See Figure):

i. Power source
ii. The tether
iii. Data
iv. The connectors

The ROV is simply a delivery platform for transporting the sensor package to the work location. The Human-Robot Interface (the intuitive interaction protocol between the human operator and the robotic vehicle) is still in its infancy; However, sensors are still outstretching the human's ability to interpret this data fast enough to react to the feedback. Beside this deep sea operation is imposing more requirements for the power design, rating and application of new generation of ROV. The majority of the company's assignments have involved the development of tailor-made solutions to solve specific problems in subsea operations for their customers.

12.4.1 Power Distribution System

To satisfy environmental problem, recent design also focus on minimized acoustic emissions, fiber optic telemetry system, and full integration of vehicle, navigation, and science sensor data streams. One of the evolving ROV technologies is the design pioneer by Mbari, where the ROV is designed to operate up to 4000 m depth rating, 100 kg payload with +/-35 kg variable buoyancy adjustment, precision 4 degree-of-freedom vehicle control. Operational features include a quick-change payload tools led, and extensive onboard fault detection and isolation capability. The ROV electrical power system to deliver and manage 15 kW of DC electrical power, primarily to meet the vehicle propulsion goals of 1.5 knot free speed and 0.75 knot full depth transit (i.e., with cable drag). The electrical load capability includes 3.7 kW (mechanical output) brushless DC permanent magnet motors. Distribution voltage selection is based on vehicle performance and personnel safety issues.

Traditionally ROV vehicle operate mostly at 120V, due to power requirement the industry is adopting 270 and 240 VDC full wave rectification of 120/208 three phase AC for manned submersible, this in line with aircraft power distribution, after apparent that the 5 kW demanded by the largest loads would require large and heavy switches, connectors, and wiring at 120 V. emerging practice for 270 VDC aircraft power distribution, and with. 48 VDC is presently the highest industry standard voltage that can be considered "low voltage" for safety purposes. However, due to deep sea operation environment future ROV will require all electric power operation with high voltage demand. Such system will require the use of SCADA and Distributed Computer System for the vehicle data management system.

The power distribution system include the DC busses, power switches, ground fault detection system, and motor regeneration control system. Mbarry system deisgn employ distribution and control system design where 15 kW of 240 VDC power and 2 kW of 48 VDC power on each of the A and B busses that have synchronization capability and leaves room for future upgrades to the transmission system as well. The ability to detect ground fault conditions on any circuit passing through seawater; the ability to switch off and fully isolate any faulted load circuit; and minimization of personnel exposure to 240 VDC circuits and wiring (See Figure 12.3a). This diagram shows the values for voltage, current, kVA, and power loss throughout the system, at no load and full load operating points. The system end-to-end load factor or ratio of power delivered to power lost in transmission. This value can be determined after a survey of load analysis requirements, as a tradeoff between voltage regulation and power delivery capability. Figure 12.3b. Typical uninterrupted power system for 480 volt system.

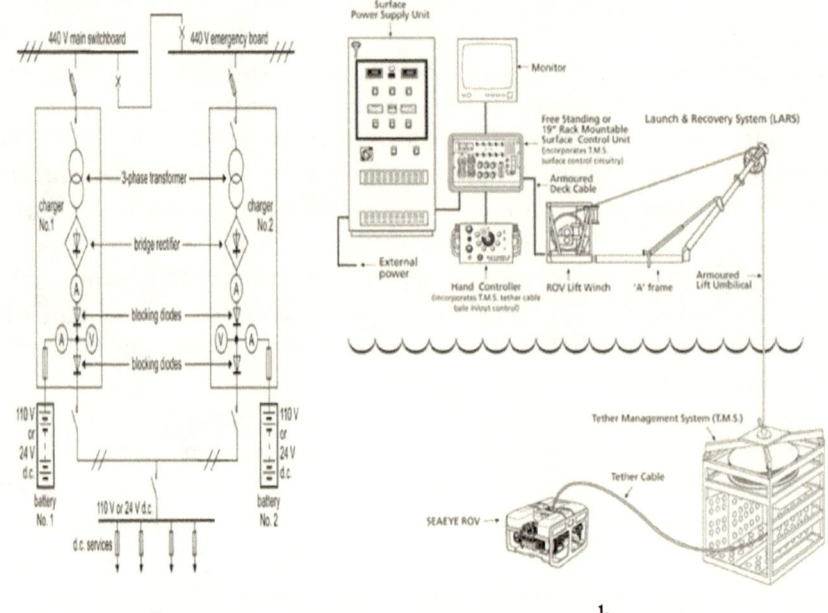

a.   b.

Figure 12. 3: One—line diagram of power transmission

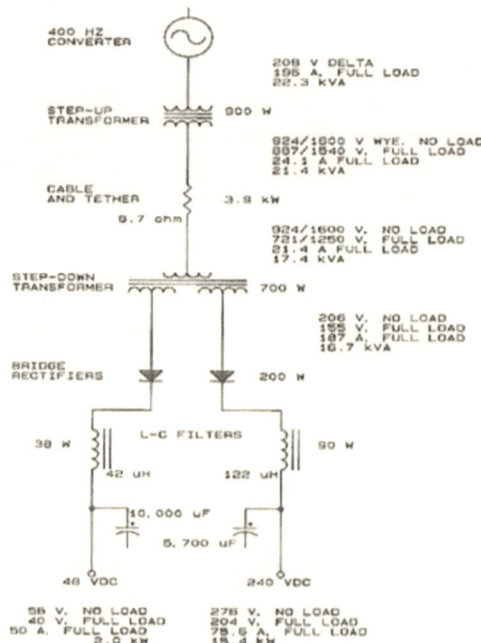

Figure 12. 4 : ROV with umbilical delivery system

The standard for work class ROVs is to use electrical power, from the umbilical, which is converted to hydraulic power. This requires an inefficient process that requires a lot of electric power. Electric thrusters could increase the reliability of an individual ROV (See Figure 12.4). The electric ROVs have fewer moving parts so they should be easier and cheaper to maintain over the long term. For ultra-deepwater operation efficient electric could provide more capability than current hydraulic ROVs cannot efficiently access. Traditionally, deepwater ROV designs were beefed-up versions of shallow water designs. What is needed now is change in technology that will generate all-electric ROVs with the power and versatility of the current fleet and the added ability to operate in ultra-deepwater (J. Newman et al, 1992). An all-electric remotely operated vehicle (ROV) is being popular for deep water operation. They have high reliability, layout flexibility, load diversity and economic part load running, easy control and low noise and vibration. Early ROV designs of every description relied on established electronic technology. In fact, the first ROV, the US Navy's CURV, used to recover a hydrogen bomb off the coast of Spain in the 1950s was all electric. One problem with the all-electric design was that as ROVs got larger, so is the thrusters. An electric-thruster ROV is more efficient.

Another primary reason all-electric ROVs will be used in ultra-deepwater has to do with the umbilical. The umbilical connects the ROV cage to the winch and control equipment on the surface. The umbilical provides power to the unit and communications back and forth between the operator and the ROV. The umbilical also hoists and lowers the ROV and its cage. To handle this strain, and protect the power and communication lines inside, the umbilical is armored by a steel coating. This coating is protective, but also very heavy. The larger diameter of the umbilical, the heavier the armor. At a certain depth, the size umbilical needed to transmit power to a hydraulic work class ROV would require an umbilical that is too heavy to support its own weight. The steel would no longer do the job. That require lightweight alloy such as titanium, or to Kevlar. Titanium would work, but is prohibitively expensive, as is Kevlar. Figure shows a typical system for all electric system. The university of Alaska in collaboration with industry are developing a new ROV system capable of rapid accost effective scientific response to dynamic underwater events such as hydrothermaldiking, catastrophic shelf slumping, phytosplanktonblooms and other transient phenomena. The general schematic includes (See Figure 12.5):

i.   Surface control console with pilot monitors and control

—

   ii.   Remote science and monitoring stations, and deck cabin
   iii. Winch, CTD cable and depressor weight
   iv. Vehicle tether and vehicle
   v.   Scientific payload

Control

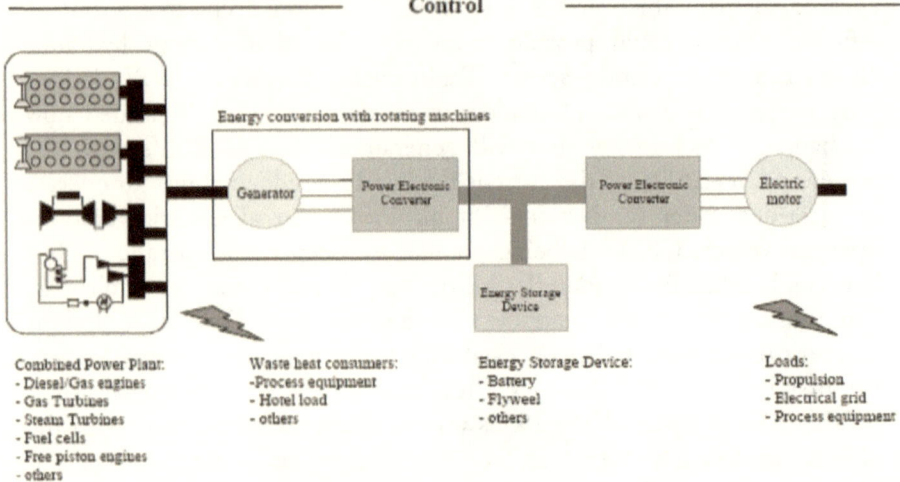

Figure 12. 5: All electric system

Safety for 240 V circuits are restricted to high power loads that are not frequently opened, and the circuits appear in only a limited number of wiring junction boxes. Both the 240 V and the 48 V systems is required to be fully isolated from frame ground, and ground fault monitor circuits to warn if the impedance to ground falls low enough to cause a hazardous condition. It is therefore essential that personnel are trained in safe working practices for these voltages. This will mean a considerable increase in the electrical content of all training.

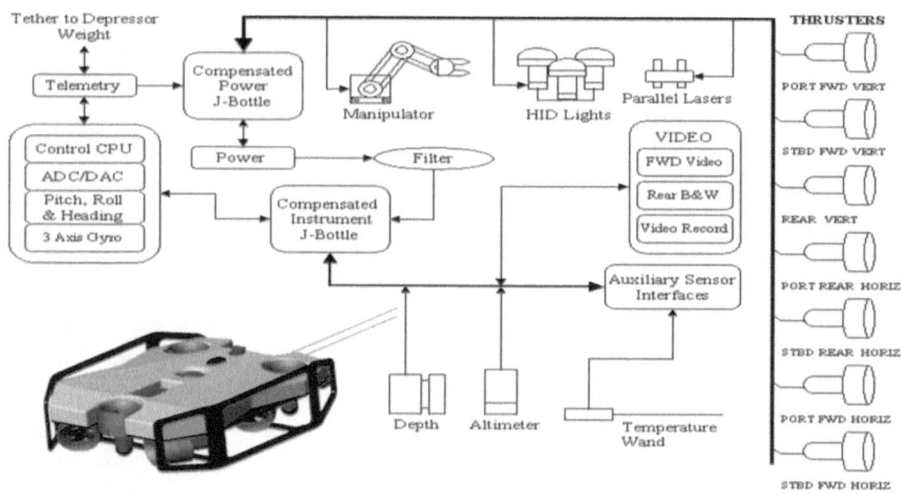

Figure 12. 6: ROV Distribution system

12.4.2 Power Source requirement

Electrical power transmission is an important factor in ROV system design due to their effect upon component weights, electrical noise propagation and safety considerations. The ROV power system design involves series of compromises and trade-off of cost, safety, and needed performance. The power system design reflects the overall vehicle. The design involve an iterative process that starts with goals for vehicle payload, operating depth, speed, support ship size, and vehicle and cable technologies. The payload, depth, and speed are derived from science requirements., the size is defined, and most technology choices are chosen based on common science and acceptable flexibility for required schedule and resource constraints. Payload and depth requirements and propulsion system are deduced from vehicle size and frontal area.

Consideration for choice between AC and DC is another challenge in the power design. Direct current (DC) allows for lower cost and weight of tether components; Since inductance noise is minimal, it allows for less shielding of conductors in close proximity to the power line as well as weight considerations for portability, and the expense of power transmission devices. Alternating current (AC) allows longer transmission distances than that available to DC while using smaller conductors as smaller systems use only DC as their power source. Submersible systems attempting to escape a hazardous bottom condition have been known to lose power at critical moments while the vessel is making power-draining

–

389

repositioning thrusts on its engines. This can cause entanglement of the vehicle. Submersible maneuvering power can be separately provided.

With the advent of the lightweight micro-generators for use with small ROVs, the portability of the ROV system is significantly enhanced. Battery/inverter combination for systems AC and DC power also contribute to light weight effort. Emergency system power source capable of uninterrupted power to the system at its maximum sustained current draw for the length of the anticipated operation is also a necessity for design requirement. On larger ROV systems, AC power is used for the umbilical due to its long power transmission distances, which are not seen by the smaller systems. AC power in close proximity to video conductors could cause electrical noise to propagate due to EMF (electromotive force) conditions. Larger work-class systems require the use AC power transmission from the surface down the umbilical to the cage (the umbilical normally uses fiber-optic transmission, lowering the EMF noise through the video) since the umbilical does not require neutral buoyancy. At the cage, the AC power is then rectified to DC to run the submersible through the neutrally buoyant tether that runs between the cage and vehicle. Uninterrupted power supply system is important to sustain power requirement of ROV and its recovery system. Potential energy source for ROV are:

i. Fuel Engines combustion engine could operate in form of:
    a. Internal combustion engines-Diesel engine
    b. External combustion engine-Braytoncycle (gas turbine) engines, Steam engine

ii. Batteries and Fuel Cells—Electrochemical processes at work
    a. − Canonical battery technologies
    b. − Fuel cell characteristics

iii. Others : Nuclear power sources, renewable energy,
iv. Emissions, green manufacturing, primary batteries, generators

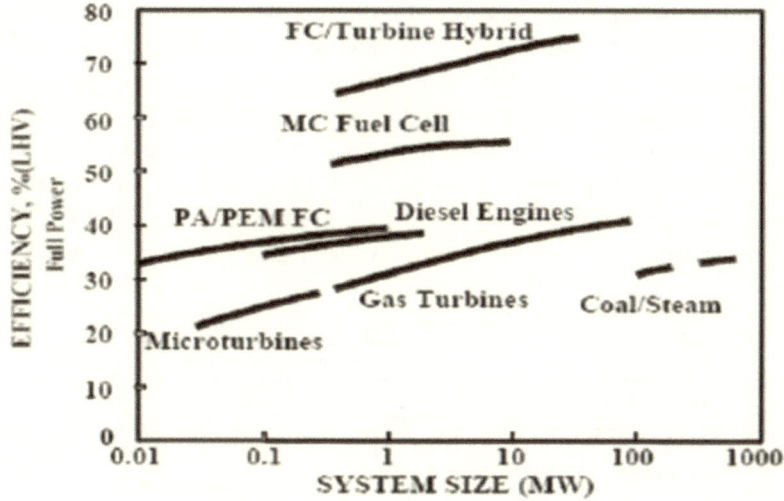

Figure 12. 7: Efficiency vs System size

Size and weight of power system matter in the design and estimation of resistance of marine vehicles, Figure 12.7 gives size standing information of power source option.

i. Requirement of power systems for marine applications include:
ii. Shows typical continuous UPS DC supported supply system
iii. Essential DC services supplied from 440V through charger 1— continuously in trickle charges
iv. During power loss, battery should be able to maintains transitional supply while emergency generator restores power to emergency board & charger 2
v. Either battery is available for few hours if both generators are unavailable
vi. Some critical emergency lights should have internal battery supported UPS i.e. battery charge continuously during non emergency conditions
vii. Main Supply of power energy source must be carried on board; has to last days, months, years.
viii. Weight and volume constraints may be significantly reduced compared to terrestrial and esp. aeronautical applications.
ix. Reliability and safety critical due to ocean environment.
x. Capital cost, operating costs, life cycle analysis, emissions are significant in design, due to large scale.

—

Understanding of the science of energy is also important requirement. Energy can be produced through electrochemical, combustion, electromagnetic, heat, mechanical system alternative or their combination. Electrochemical process involve engines convert chemical energy into heat energy or mechanical or kinetic energy where 1 MegaJoule is: 1 kN force applied over 1 km;1 Kelvin heating for 1000 kg air;1 Kelvin heating for 240 kg water; and 10 Amperes flowing for 1000 seconds at 100 Volts. Table 12.2 show various heating content for available energy option for ROV.

Table 12. 2: Energy source fuel heat content

Fuel	Heat content(MJ/KG)
Gasoline(C8H15)	45
Diesel(C13H23)	42
Propane(C3H8)	48
Hydrogen(H2)	130
Ethanol(C2H5OH)	28

$C_8H_{15}+47O_2->32CO_2+$other product

Gas turbines are preferable due to extremely high power density, and the high thermal energy content of traditional fuels. Li-based batteries now available at ~0.65MJ/kg (180kWh/kg); gold standard in consumer electronics and in autonomous marine vehicles. Fuel cells are still power—sparse and costly for most mobile applications, but continue to be developed. They are more suitable for power generation plants in remote locations. Example of specification of gas turbine engine that can be used for ROV is

LM2500 Specifications—
"Output: 33,600 shaft horsepower (shp)
Specific Fuel Consumption: 0.373 lbs/shp-hr
Thermal Efficiency: 37%
Heat Rate: 6,860 Btu/shp-hr
Exhaust Gas Flow: 155 lbs/sec
Exhaust Gas Temperature: 1,051°F
Weight: 10,300 lbs
Length: 6,52 meters (m)
Height: 2.04 m
Average performance, 60 hertz, 59°F, sea level, 60%
Relative humidity, no inlet/exhaust losses, liquid fuel,

LHV=18,400 Btu/lb"

http://www.geae.com/aboutgeae/presscenter/marine/marine_200351.ht

Energy storage technology remains a challenge for the use of alternative energy for ROV. An example of a simple battery would be one in which zinc and carbon are used as the electrodes, while a dilute acid, such as sulfuric acid (dilute), acts as the electrolyte. The acid dissolves the zinc and causes zinc ions to leave the electrode. Each zinc ion which enters the electrolyte leaves two electrons on the zinc plate. The carbon electrode also dissolves but at a slower rate. The result is a difference in potential between the two electrodes. The Dry cell is relatively inexpensive and quite portable. The anode consists of a Zinc is placed in contact with a moist paste of $ZnCl_2$ and NH_4Cl. A carbon rod surrounded by MnO_2 and filler is the cathode. The cell reaction varies with the rate of discharge. Lead acid cell are electrodes of lead and lead dioxide, dipping into concentrated sulfuric acid Nominal discharge rate C is capacity of battery in Ah, divided by one hour (typical). Lithium primary cells can reach 2.90 MJ/l. Table 12.3 and Figure 12.8 show performance battery.

$$P_b -> Pb^{2+} + 2e^- \text{ (Oxidized) or } P_b + S_{o4}^{e-} + 2e^- -> P_b S_{o4} + 2e^-$$

Gathering electron at the positive electrode

$$Pb^{4+} + 2e^- -> Pb^{2+} \text{ (Reduced) or } Pb^{2+} + So_4^{2-} + 4h^+ - 4h^+ -> 2Pbso_4 + 2h_2o$$

Total chemistry of the lead acid

$$Pb^{2+} + So_4^{2-} + 2so_4^{2+} + 4^+ \, 2\,Pbso_4 + 2h_2o$$

Table 12. 3: Comparison of Battery Performance

	Energy Density (MJ/Kg MJ/i)	Memory Effect	Maximum current	Recharge efficiency	Self-discharge %/min at 293 K
Lead–Acid	0.14, 0.36	No	20c	0.8-0.94	
Ni-Cd	0.29, 0.72	Yes	3c	0.7-0.85	25
NiM H	0.29, 1.08	Yes	0.6c	–	>20
Li-ion	0.43-0.72 1.03-1.37	No	2c	–	12

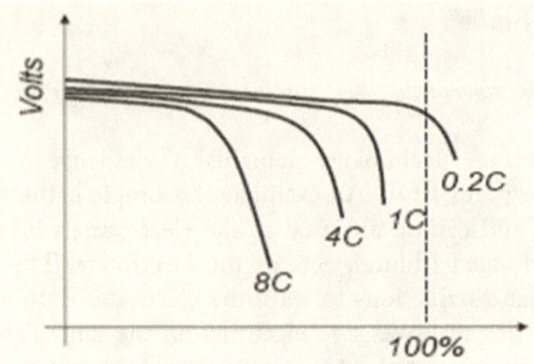

Figure 12. 8 : Battery performance

Typical Fuel cell employ electrochemical conversion work likes like a battery, but the fuel cell is defined as having a continuous supply of fuel.

At anode, electrons are released:

$$2h_2 - > 4h^+ + 2^{e-}$$

At cathode, electrons are absorbed:

$$O_2 -> 4e^- + 4h_2o$$

Fuel cell has high sensitivity to impurities: e.g., PEM FC is permanently poisoned by 1ppb sulfide. Weight cost of storage of H2 in metal hydrides is 66:1; as compressed gas: 16:1 while oxidant storage: as low as 0.25:1. Reformation of H2 from other fuels is complex and weight inefficient: e.g., Genesis 20L Reformer supplies H2 at ~ 0.05 kW/kg. Fuel cell also has characteristics to change load rapidly.

12.5 Power Transmission Conversion and Transformation Requirement

The power transmission system include the shipboard power source, step-up / step-down transformers, vehicle cable and tether, and power conversion equipment required to produce DC distribution power aboard the ROV. Once vehicle size, depth, and speed are determined, the main cable, power transmission system, and propulsion system co-designed can be taken through iterative process. AC and DC power distribution choice and routing is very important in the design of ROV. Power conversion for the system involves the use of solid state rectifier (diode, SCR).

These converters are also example of game changer in the decision analysis for use of AC/DC and hydraulic system. But they also require protection of large semiconductors, e.g. thyristors, which can additionally be destroyed by a fast rate-of-change of. Voltage and current caused by rapid switching. To suppress a rapid overvoltage rise (dv/dt) across a thyristor an R-C snubber circuit is used. Its action is based on the fact that voltage cannot change instantaneously across a capacitor. The series resistor limits the corresponding current surge through the capacitor while it is limiting the voltage across the thyristor. Significant heat will be produced by the resistor which, in some applications, is directly cooled by water jacket. An in-line inductive effect will limit the rate-of-change of current (di/dt) through the thyristor. (E. Mellinger, 1986). Special fast-acting Line fuses may be used as back-up over current protection for the thyristors. Circuit protection for the electric propulsion units (including excitation and harmonic filters) principally employs co-ordinate protective relays. The parallel of a conventional AC relay with solid state devices, in this case Insulated Gate Bipolar Transistors (IGBTs) provide arcless make and break for the DC current, while the relay contacts carry the steady state load with only a few watts loss. Logic on the card sequences, the switching events and responds to overloads, and a shunt resistor and A/D converter allow current to be sensed and reported (See Figure 12.9a and 12.9b show SCR system and protection (J. Schaeffer,1965).

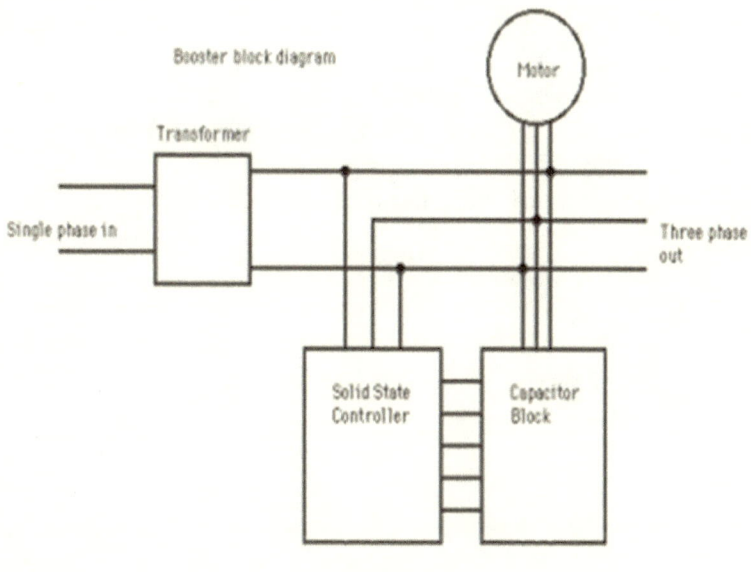

a.

–

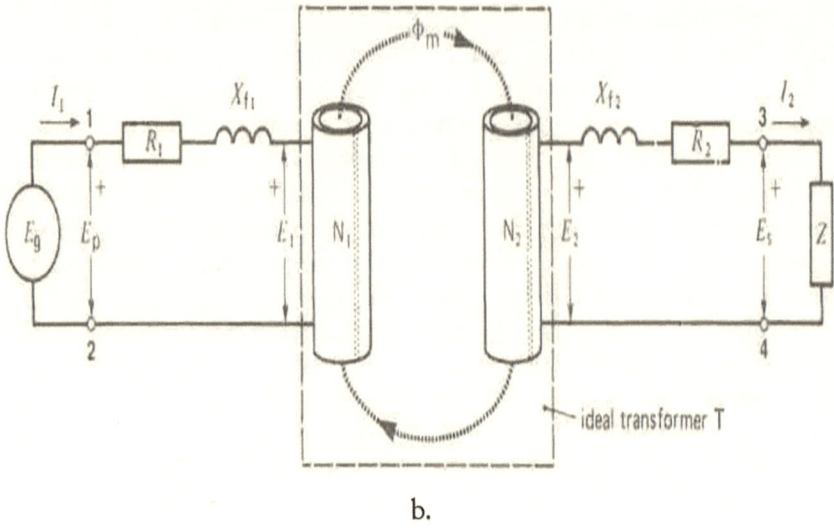

b.

Figure 12. 9 : a. SCR operation, b. SCR protection

Power transformation include the use of step-up and step-down transformers with use of material that target less losses—no load (iron) and full load (copper) losses. The transformation also depends on the connection (delta, wye, delta) arrangement of input, cable, and output circuits that can minimize the current waveform crest factor presented to the converter, so that each transformer has a delta winding for harmonic current control and a wye connection high voltage winding for minimum insulation stress. The vehicle step-down transformer contributes significantly to vehicle mass and volume budgets, and of scientific importance, to the vehicle acoustic signature as well. Figure 12.10a and 12.10b show power transformer and converter system.

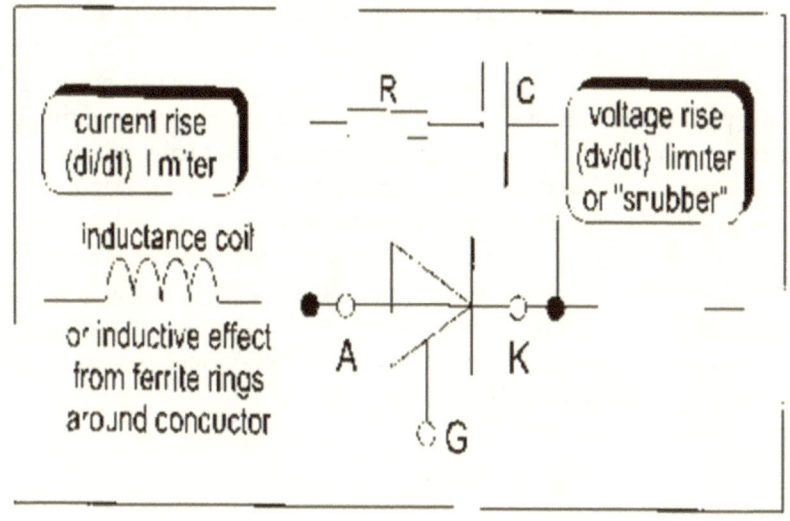

a.

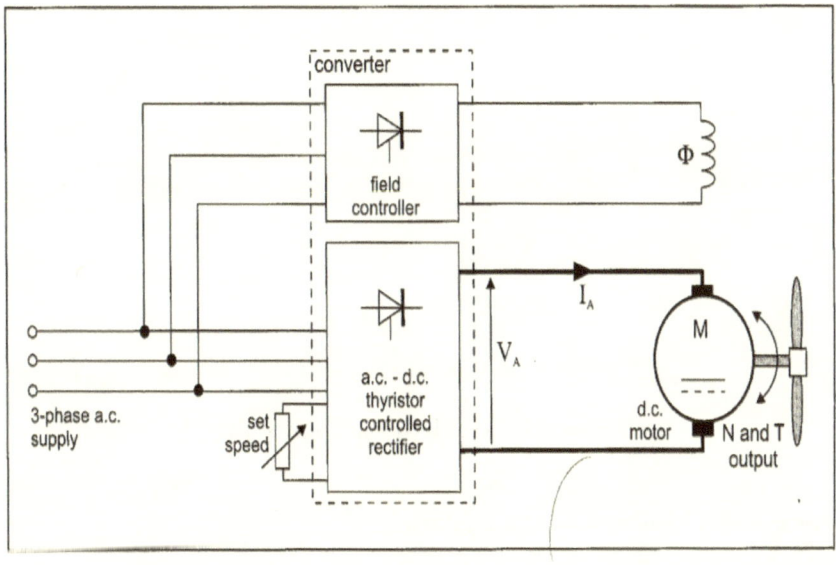

b.

Figure 12. 10: a. Power transformer b. Power converter

Transformer noise is largely due to core magnetostriction, and thus is present, and in fact maximum, when motors are off, loads are small, and input voltage is high, as during "quiet sub" operation. Reductions of transformer mass and volume are desirable, but these increase core flux level and winding current density, and thus increase both noise and thermal output. The keys to a small, light, quiet transformer thus became getting the waste heat out while keeping the noise in. This in turn meant breaking the acoustic path to seawater with an absorptive layer or a sharp discontinuity in acoustic impedance, while preserving high thermal conductivity.

By acoustically isolate the transformer using a gaseous vapor barrier, while using the vapor's latent heat of evaporation to carry the transformer's heat away. The choice of liquid is obviously critical since it must have high dielectric strength in both phases, high latent heat, and material compatibility, not to mention low toxicity, environmental correctness, and low cost. Figure 12.11 show the tether cross section and the cable sizing requirement (A. Kelley, 1992).

$$\frac{E_1}{E_2} = \frac{N_2}{N_1}$$

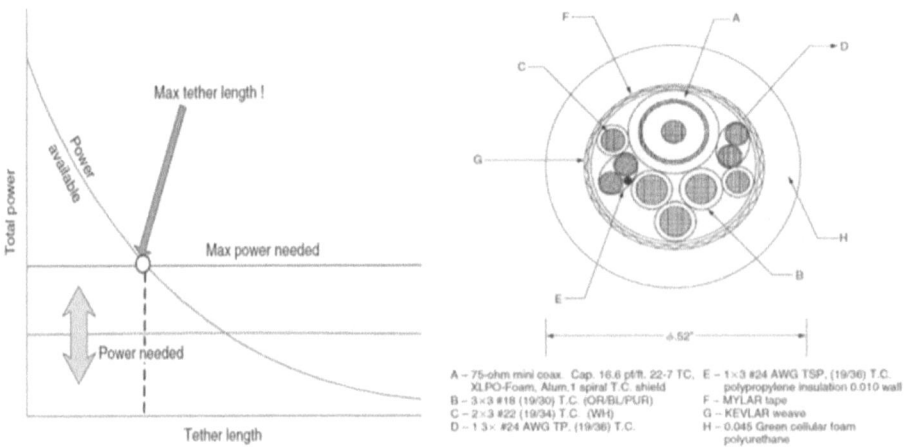

Figure 12. 11: a: Tether b. Cable sizing

Motor and Thruster Control System

The main are connected the propeller for horizontal and vertical thrust. Today, robust motor system comes which thyristor power management system that have control capability for maneuvering propulsion, trusting. On older analog systems, a simple rheostat controls the variable power to the electric motors, while newer digital controls and SCR are necessary for more advanced ROV movements. Figure 12 a and b show motor power requirement and torque speed characteristics.

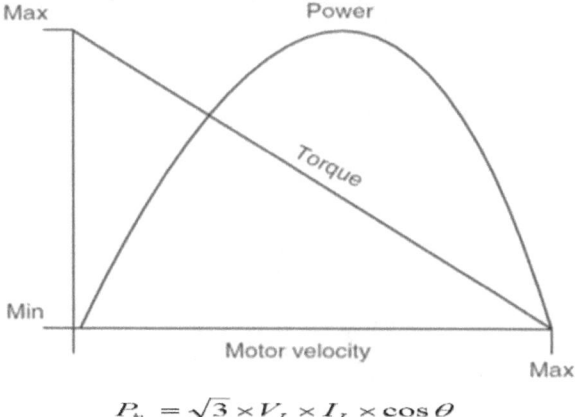

$$P_{in} = \sqrt{3} \times V_L \times I_L \times \cos\theta$$

Figure 12. 12: a:Motor Power

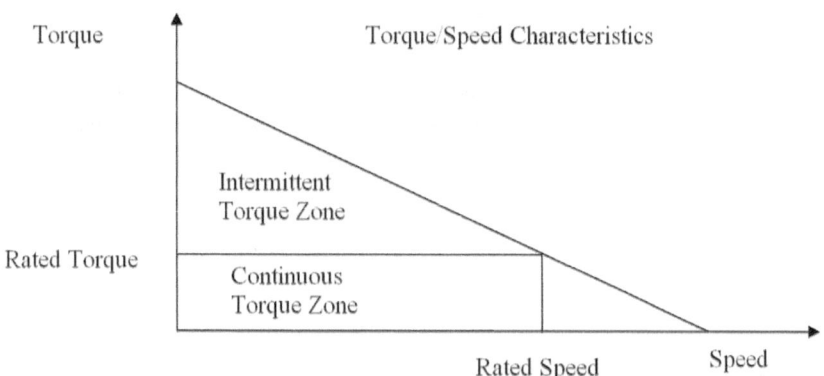

Figure 12.12b: Torque speed characteristics

$$S_{input} = \sqrt{3}.V_{line}.I_{Line}.....VA$$

$$\eta - \frac{P_{shaft}}{P_{input}} = \frac{P_o}{P_{in}}$$

$$P_f = \frac{P_{input}}{S_{input}}$$

Regeneration control reflect behavior of motors like generators during braking, this lead too high frequency voltage. High bandwidth thrust control, necessary for precision vehicle control, is expected to require frequent and repetitive motor braking, in order to minimize thruster response time.

Power Connector (Cable and Tether)

Umbilical refer to the cable linking the surface to the cage or tether management system (TMS). Tether is the cable from the TMS to the submersible. Any combination of electrical junctions is possible in order to achieve power transmission and/or data relay. AC power may be transmitted from the surface through the umbilical to the cage, where it is then changed to DC to power the submersible's thrusters and electronics. Further, video and data may be transmitted from the surface to the cage via fiber-optics (to lessen the noise due to AC power transmission), then changed to copper for the portion from the cage to the submersible, thus eliminating the AC noise problem. The umbilical/tether also should have strength member allowing for higher tensile strength of cable structure and Protective outer jacket for tear and abrasion resistance. The tether length is critical in determining the power available for use at the vehicle following law of resistance and Ohm law. The power available to the vehicle must be sufficient to operate all of the electrical equipment on the submersible. The maximum tether length for a given power requirement is a function of the size of the conductor, the voltage, and the resistance (G. Wilkins, 19987).

$R = Ro \, l/ \, A$
$V = IR$

Table 12. 4: Standard copper wire gauge resistance over nominal lengths (Deep Sea Power and Light)

Wire gauge	Ohm/1000ft (approx)
20	10
18	6
16	4
14	2.5
12	1.5

Salt water is highly conductive, causing any exposed electrical component submerged in salt water to short to ground. The result is the 'Ubiquitous ground fault'. The purpose of an underwater connector is to conduct needed electrical currents through the connector while at the same time squeezing the water path and sealing the connection to lower the risk of electrical leakage to ground. The underwater connector is lined with synthetic rubber that blocks the ingress path of water while allowing a positive electrical connection. Connectors sometimes experience cathodic delamination, causing rubber peeling and flaking from the connector walls. Connector maintenance include (N. Forrester,1982):

i.   Use small amounts of silicone grease to lubricate the connector, thus allowing easier slide on and off. Using too much grease, a widespread problem, can interfere with sealing.
ii. Always pull the connector by its body instead of its tail (cable), since the wire splice is located in the connection. Pulling on the tail could part the solder joint and ruin the electrical continuity within the connector.
iii. Keep the connectors as clean as possible through regularly scheduled maintenance tasks that include cleaning the contacts and lubricating the rubber lining.
iv. Spray the connector body with silicone spray to keep the housing from drying out, which could result in flaking and rubber degradation.

The connector materials must be able to withstand the environmental conditions without degradation. The physical size of the connector, its weight, ease of use (and appropriateness for the application), durability, submergence (depth) rating, field reparability, etc. should all be assessed. Other important requirement for cables include insulation spacing and right-of-way, operating capacitance and charging current, transmitted power, reliability and installation costs. Design element of cable includes metallic covering, outer coverings and corrosion protection, losses and temperature factors.

Power safety Stabilization Requirement

Power safety and harmonic stabilization are very important part of high demand regime of ROV vehicles. For the typical distribution arrangement earlier mentioned, power stabilization can be provided by four rectifier bridges actually contain Silicon Controlled Rectifiers (SCRs) which are fired by zero-crossing circuits and operate in on/off mode as electronic circuit breakers for their associated power busses. Fast fuses at each rectifier input protect against SCR or other catastrophic failure. Each rectifier bridge is followed by an L-C filter that reduces output ripple voltage, and reduces harmonic currents drawn from the power transmission system. Positive Temperature Coefficient (PTC) thermistors are used as constant-power capacitor bleeders

Two design features that increase the operational availability of the vehicle power transmission system are redundancy and fault tolerance. Redundancy incorporates the use of dual power busses for each distribution voltage. Thrusters are arranged so that failure of one 240 V bus leaves one vertical plus two horizontal thrusters available (lateral or fore-aft), which allows yaw control, translation, and vertical motion. The critical loads such as the main computer draw power from both A and B busses through diode-OR circuits. Fault tolerance is achieved through coordinated overload protection plus the ability to selectively isolate loads using switches in the distribution system. Here fuse and circuit breaker current-time characteristics are selected so that the over current device closest to the faulted load trips first, allowing operation on the non-faulted part of the system to resume with minimal interruption. The circuit breakers also function as controlled switches, and are commanded to disconnect loads when a ground fault is sensed on the associated supply bus, again allowing operations to continue.

Grounding implies an intentional electrical connection to a reference conducting body, with specific array of interconnected electrical conductors. Grounding systems should be serviced as needed to ensure continued compliance with electrical and safety codes, and to maintain overall reliability of the facility electrical system. All vehicle electrical systems are fully isolated from frame (seawater) ground. The insulation resistance must be continuously monitored for reasons of safety, and also to provide early warning of seawater intrusion. Figure 12.12a and 12.13b show the floating ammeter and the preferred ground connection for marine system. The available grounding systems include insulated neutral, earthed Neutral and resistance earth Neutral System. The insulted neutral is favored for marine application because of:

i.   This system is totally insulated from the ship's hull
ii. This system maintains continuity of power supply to the equipment even in the event of single phasing fault.
iii. This ensure power supply to critical equipment
iv. The power supply to the equipment can disrupt only if two single phase faults occur simultaneously in two lines which is then equivalent to short circuiting faults
vi. But such fault occur very rare

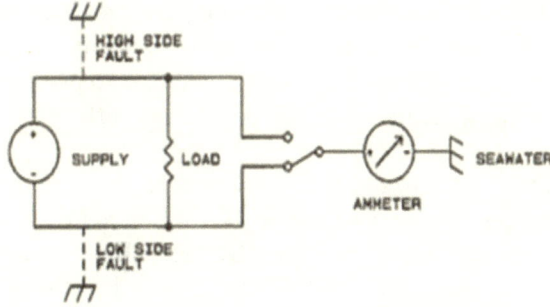

Figure 12. 13: Floating ammeter and insulated earthing

Each side of each supply voltage is alternately connected to frame ground through a current limited ammeter. If a ground fault exists on the opposite supply rail, current will flow through the meter. This approach can be extended to monitor several supplies of differing voltages with a shared common rail, at the expense of a more complex troubleshooting flowchart. Action must be initiated to continue to remove, or reduce to a minimum, the causes of recurrent problem areas. Personnel are encouraged to become familiar with Article 250 of the National Electrical Code (NEC), which deals with grounding requirements and practices. Factors which influence the choice of selecting system ground:

i.   voltage level of the power system
ii. transient over voltage possibilities
iii. types of equipment on the system
iv. cost of equipment
v.   required continuity of service
vi. quality of system operating personnel
vii. Safety consideration including fire hazards
viii. Distribution systems of ships are usually have their neutral points earthed to the ship's hull through a resistor
ix. The resistor in neutral line limits earth faults currents and protects equipment

Power Switching, Telemetry and Control

Power switches were required for each load, or group of loads, on the vehicle for power tolerant. High power DC switching is more difficult, due to two practical issues. Mechanical switching elements require elaborate arc suppression measures (vacuum or arc blowout), since unlike

AC current, DC has no naturally occurring zero crossings that allow the arc plasma to dissipate. Solid state switching elements inevitably have a few volts of "on" state voltage drop, and generate dozens of watts of waste heat. Both problems make compact packaging difficult (M. Chaffey, 1993).

It is important for ground fault isolation of the load to have switch control and telemetry as part of the ROV distributed data system. Some could have Instrument Bus Computer (IBC) switches are rated in amapere and voltage, mostly power by MOSFETs, driven directly by photovoltaic optoisolators. Shunt resistors allow current to be sensed by an onboard A/D converter and reported over the backplane..Beside the switch other power interlock devices that can be employed for switch board system are circuit breaker. Circuit breaker comes in form of air circuit breaks, oil circuit breakers, air-ballast circuit breakers, gas (sf6—sulphur hexafluoride) circuit breakers and vacuum breaker.

Air circuit breakers are used for low voltage where arc chutes and arc contacts are incorporated. Air blast circuit breakers is a different type that are use for high voltage line, they can handle high pressure at about $30kg/cm^2$ air blown during the operation of circuit breaker, thus the operation is too noisy. Oil circuit breaker normally use Napthenic base petroleum $[(CH_2)n]$ which have been carefully refined to avoid sludge or corrosion. the are expected to excellent dielectric strength high thermal conductivity and prone fire prone to fire hazard, leakage/contamination.SF_6 circuit breaker is most accepted circuit breaker, it is made of chemically very stable, non flammable, non corrosive, non poisonous, colorless and odorless gas with Limits the sonic velocity (1/3 of air). It has excellent dielectric strength, about twice of air. it can be used for high voltage and it has low GWP (global warming potential is high) and Lifetime 3200 years. Vacuum circuit breaker can also handle high voltage. The arc remains in the diffused column mode.

The control system controls the different functions of the ROV, this include the propulsion system, switching of the light(s), video camera(s), relay, digital fiber optics, digital, computer and subsystem control interface. The control system has to manage the input from the operator at the surface and convert it into actions subsea. The data required by the operator on the surface to accurately determine the position in the water is collected by sensors (sonar and acoustic positioning) and transmitted to the operator. Control systems are program to maintain required sequence and feedback operation. Today most control system utilizes PLC (Programmable Logic Computer). This is used in numerous manufacturing processes since it consists of easily assembled modular building blocks of switches, analog in/outputs, and digital in/outputs. Control stations vary from large containers, with their spacious enclosed working area for work class systems, to simple PC gaming joysticks. Figure

With the rise of robotics as a sub-discipline within electronics, further focus highlighted the need to control robotic systems based upon intuitive interaction through emulation of human sensory inputs. Digital control systems arose, more complex control matrices could be implemented much more easily through allowing the circuit to proportionally control a thruster based upon the simple position of a joystick control coupled with programmable logic circuits interface. The more sensors available to the 'human' that allow intuitive interaction with the 'robot', the easier it is for the operator to figuratively operate the vehicle from the vehicle's point of view.

Most ROV have spare twisted pair of conductors for hard-wire communication of sensors from the vehicle to the surface. This make sensor system to not need engineering support from the ROV manufacturer in order to design these sensor interfaces. The weakness is incompatibility of the transmission protocol to share the single data line; only one instrument may use the line at a time. Available industry standard protocols for transmissions is TCP/IP, RS-485, and RS-232, while useful and seemingly ubiquitous in the computer industry, is distance limited through conductors, thus causing transmission problems over longer lengths of tether. The move toward open source PC-based sensor data processing has led to the production of data protocol converters for use in ROV sensor interpretation. Most small ROV sensor manufacturers transmit data with the RS-485 protocol, requiring a converter at the surface to both isolate the signal and to convert it to USB (or RS-232) protocol for easy processing with a standard laptop computer. Standards for these protocol converters are slow in evolving (due to the size of the customer base).

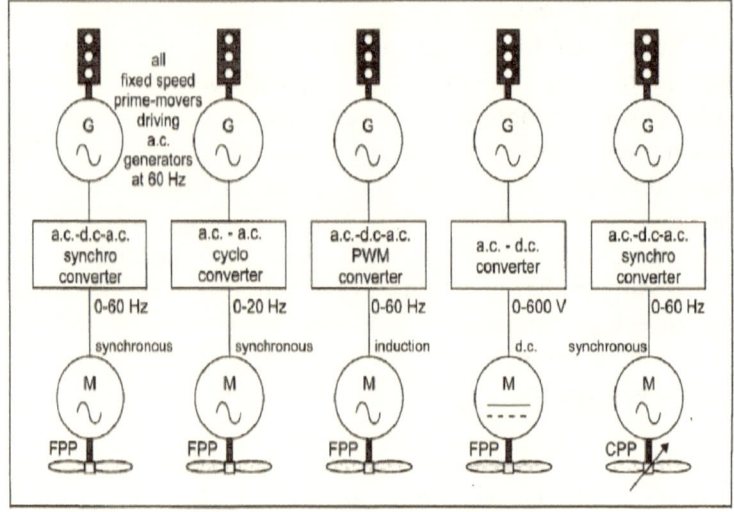

a.

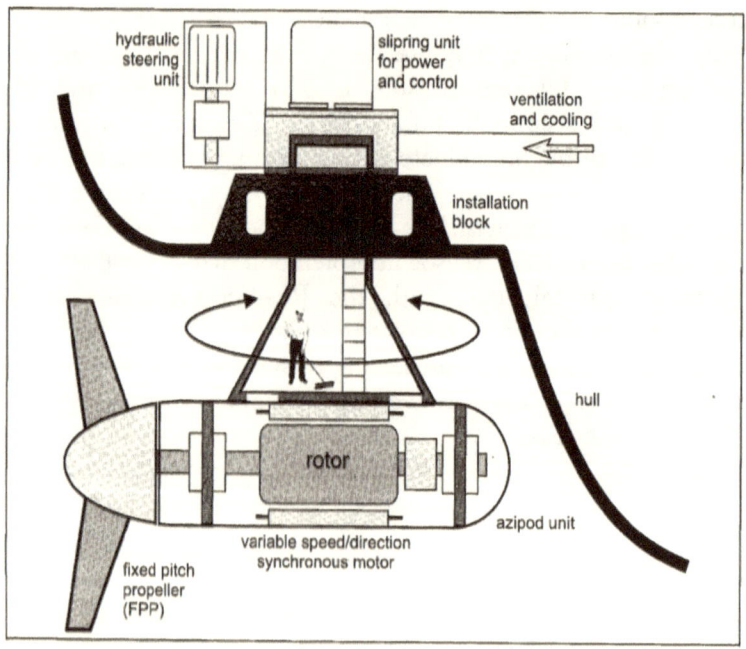

b.

Figure 12. 14: a: Motor frequency control, b. Azipod system

Data Transmission and Protocol

The challenges of proactive culture towards accident occurrence near population and prevention of environmental consequence of accident evolved requirement for maritime activities to operate deep water. The importance of ROV in development of new technology to meet this challenges is highlighted, this include, data collection, installation and monitoring. Likewise, the need for more power is highlighted and system requirement to meet power requirement ROV for deep water Operation is discussed. ROV system integrator must become familiar with the wiring and pin arrangement for these converters that will be instrumental to HVDC to ROV system as well as to assure data transmission from the sensor, through the vehicle and tether to the software at the surface, is achieved. Power sensor and data throughput reliability promise greater the ability for deepwater to deliver to the operator the necessary job-specific data as well as sensory feedback needed to properly propel,. Maneuver and control ROV for deepwater operation.

References

1. Klages M., Mesnil B., Soltwedel T, Christophe A (2002); The "AWI" expedition of RV "L'Atalante" in 2001. *Reports on Polar and Marine Research* 422: 65 pp.
2. Michel J.-L., Drogou J.-F., Floury L. (1990); "Subsea Work
3. Environment for Submersibles"; DA. Ardus and MA. Champ (eds)*Ocean Ressources*, Vol. II, 31-39. 1990; Kluwer Academic Publishers—Printed in the Netherlands.
4. Renard V., Sichler B., Masson D., Dias JMA, Herrouin G., Michel J.L.(1993); "AUVs Mission Analysis for Deep Sea Surveys"; *First ISR Workshop on AUV's*, Porto, 1-3 september 1993, Portugal.Sarradin P.-M., Olu Leroy K., Ondréas H., Sibuet M., Klages M.,
5. Fouquet Y., Savoye B., Drogou J.-F., Michel J.-L.(2002); "Evaluation of the 1st year of scientific use of the French ROV VICTOR 6000"; *Underwater Technology 2002*, pp. 11-16, Tokyo, Japan.
6. ANSI/IEEE Std 80-1986, IEEE Guide for safety in substation Grounding. Polar Engineering Conference, Honolulu, Hawaii, USA, May 25-30, 2003
7. J. Newman and B. Robison, "Development of a dedicated ROV for ocean science," MTS Journal, Vol. 26, No. 4, 1992, pp. 46-53.
8. G. Wilkins, "Fiber optics in the "optimum" undersea electro-optical cable," ASME Energy Sources Technology Conf., Dallas, TX, Feb. 15-18, 1987.
9. N. Forrester, "Power transformer design for tethered underwater vehicles," IEEE Oceans '92, Newport, RI, Oct. 1992.
10. Staff, Dept. Elec. Engr., Massachusetts Institute of Technology, Magnetic Circuits and Transformers, New York: J. Wiley and Sons, 1943.
11. J. Schaeffer, Rectifier Circuits: Theory and Design, New York: J. Wiley and Sons, 1965.
12. A. Kelley and W. Yadusky, "Rectifier Design for minimum line-current harmonics and maximum power factor," IEEE Trans. on Power Electronics, vol. 7, no. 2, pp. 332-341, April 1992.
13. M. Chaffey, A. Pearce, R. Herlien, "Distributed data and computing system on an ROV designed for ocean science," IEEE Oceans '93, Victoria, BC, in press.
14. W.L. Weeks, Transmission and Distribution of Electrical Energy, New York, NY: Harper and Row, 1981, p. 171.

15. E. Mellinger, K. Prada, R. Koehler, and K. Doherty, Instrument Bus, An Electronic System Architecture for Oceanographic Instrumentation, Woods Hole, MA: Woods Hole Oceanographic Institution, 1986, Technical Report 8630.

16. *"IMCA M 141, Guidelines on the Use of DGPS as a Position Reference in DP Control Systems". http://www.imca-int.com/divisions/marine/publications/141.html."IMO MSC/Circ.645, Guidelines for vessels with dynamic positioning systems". http://www.imo.org/includes/blastDataOnly.asp/data_id%3D10015/MSCcirc645.pdf.*

CHAPTER 13

Risk Aversion Requirement towards Best Practice Potential Waste Recycled—Based Bioenergy for Ships

"The most beautiful experience we can have is the mysterious"
Albert Einstein

Summary

Human status today can best be define as an age of knowledge, efficiency and sustainable developments towards fulfilling significant part of human existence in this beautiful planets we all share.—Previous time in human history has been dominated with various experimentation, knowledge acquisition which has resulted to new discovery and new philosophy of doing things in efficient, facilities, sensitive and cooperative and above all sustainable manner (maintaining quarto bottom balance (economic, technical, environmental, social between man techno sphere and environ sphere world in order to sustain continuous healthy existence of our planet and the right of future generation . new knowledge and technology have emerged, since there is no drain in this planet, the greatest challenge for humanity lies in recycling our waste to the lowest level of usage. This chapter will discussed the need to choose waste derived biofuel above all other food sources. The section describe risk based environmental philosophy procedure to described risk and risk abatement required to come up with a potential best sustainable bioenergy generation system.

13.1 Introduction

This 21st century has become an age of recycling where a lots of emphasize is placed on reuse of material to curb current environmental problems and maximize use of depleting natural resources and energy conservation. Modern day sustainable use and management of resource recommend need to incorporate recycling culture in our ways of life including technological process. Biomass is not left behind in this; the use of biomass energy resource derived from the carbonaceous waste of various natural and human activities to produce electricity is becoming popular. Biomass is considered as one of the clean, more—efficient and more-stable means of power generation. And it has become imperative for marine industry to tap this new evolving power generation mode especially the use of micro generation approach considering the mobile nature of ships.

Biofuels exist in solid, liquid or gas form thereby potentially affecting three of our core markets. Solid biofuels or biomass tend to be used in external combustion, however its use in the shipping industry has been limited to liquid biofuel due to lack of appropriate information economics forecasts, Sources of biomass include by-products from the timber industry, agricultural crops, raw material from the forest, major parts of household waste, and demolition wood, all things being equal using pure biomass that do not affect human and ecological chain make it suitable energy source. Biomass has low sulfur content means biomass combustion therefore considered much less acidifying than with coal, for example. Also, the ashes from biomass consumption, which are very low in heavy metals, can be recycled.

One advantage of biomass compared to other renewable-based systems that require costly advanced technology (such as solar photovoltaics) is that biomass can generate electricity with the same type of equipment and power plants that now burn fossil fuels, Many innovations in power generation with other fossil fuels may also be adaptable to the use of biomass fuels. Various factors have hindered the growth of the renewable energy resource, however. Most biomass power plants operating today are characterized by low boiler and thermal-plant efficiencies; both the fuel's characteristics and the small size of most facilities contribute to these efficiencies. In addition, such plants are costly to build.

Biomass remains potential renewable energy contributor to net reduction in greenhouse gas emissions by offsetting CO_2 from fossil

—

generation. The current method generating biomass power is biomass fired boilers and Rankine steam turbines. Recent research work in developing sustainable, and economic biomass focus on high-pressure supercritical steam cycles, use of feedstock supply system, and conversion of biomass to a low or medium Btu mass of gas that can be fired in combustion turbine cycles, resulting in efficiencies one-and-a-half times that of a simple steam turbine. biofuels has potential to influence marine industry, and it as become importance for designers and ship owners to accept their influence on the world fleet of the future especially the micro generation concept with co generation for cargo and fuel for ships.

The chapter discuss conceptual work, trend, sociopolitical driver, economic, development, and future of biomass with hope to bring awareness to local, national and multinational bodies making biofuels policies as well as maritime multidisciplinary expertise in regulation, economics, engineering, and vessel design and operation. The chapter also discusses how the shipping industry can take advantage of growing tide to tap benefit promised by waste use power generation system.

13.2 Biomass developmental trend

The concept of use of Biofuels for energy generation has has been existing concept, and in the face of challenges posed by environmental need, its growth is likely to dominate renewable energy market. Following the advent of peanut oil diesel engines developed by Rudolf Diesel in 1911 the production and use of biofuels worldwide has grown significantly in recent years. The current world biofuels market is focused on: Bioethanol blended into fossil motor gasoline (petrol) or used directly and biodiesel or Fatty Acid Methyl Ester diesel blended into fossil diesel. However the use of The Fischer-Tropsch model that involve catalyzed chemical reaction to produce a synthetic petroleum substitute, typically from coal, natural gas or biomass, for use as synthetic lubrication oil or as a synthetic fuel seem promising and negate risk posed by food based biomass. This synthetic fuel runs only in diesel engines and some aircraft engines. Oil, product and chemical tankers being constructed now are likely to benefit much more from use of biomas. However use on gasoline engines ignites the vapors at much higher temperatures, which pose limitation to inland water craft.

Biomass generation and growing trend can be classified into 3 generation types:

  i.   First generation' biofuels relate to biofuels made from sugar or
       starch, producing bioethanol, and vegetable oil or animal fats

—

producing biodiesel. First generation biofuels provoke increasing criticism through their dependence on food crops and issues over biodiversity, land use and human rights. Hybrid technology for percentage blending is being employed to mitigate food production impact.

ii. Second generation biofuels mitigate problem posed by the first generation biofuels. They do not affect food crops because they are made from waste biomass from agricultural and forestry, fast-growing grasses and trees specially grown as so-called "energy crops". With technology, sustainability and cost issues to overcome, second-generation biofuels are still several years away from commercial viability and many second generation mass produced biofuels are still under development including the biomass to liquid. Fischer-Tropsch production technique.

iii. Third generation biofuels are green fuels like algae biofuel made from energy and biomass crops that have been designed in such a way that their structure or properties conform to the requirements of a particular bioconversion process. They are made from such as sewage, and grown on ponds.

Just like tanker revolution influence on ship type, demand for biomass will bring, will bring capacity, bio—material or completed product from source to production area and then to the point of use, will bring technological, environmental change will require ships of different configuration, size and tank coating type. As well as impact on the tonne mile demand will change accordingly. Effect on shipping is likely to follow shipping large scale growth on exports and seaborne trade from key exporting regions, particularly South America. Brazil has a key role. Brazil has already been branded to be producing en-mass ethanol from sugar cane since the 1970s with a cost per unit reportedly the lowest in the world. And it is currently exploring ethanol

Table 13. 1: World ethanol consumption 2007

Consumption	
World ethanol consumption—	51 million tones, 2007
Us and brazil	68%
EU and China –	17%—surplus of 0.1 million tones
US deficit—	1.7mt
EU deficit—	1.3 mt
World—deficit	1mt

—

413

Table 13. 2: Biofuel growth

Vegetable oil	33 mt in 2000 to 59 mt in 2008	
Palm oil	13 mt in 2000 to 32 mt forecast in 2008.	a 7.5% p.a growth rate
Soya bean	7 mt to some 11.5 mt in 2008,	
EU	imports—5.7 mt in 2001 to an expected 10.3 mt for 2008	8.9%.
	3.1 mt in 2001 to 5.2 mt forecast for 2008	39%
	Production capacity—1.9 mt in 2002 to 11 mt in 2007, with 2007.	50% of total capacity.

Recent year is also witnessing emerging trade on biofuel product between the US, EU, and Asia and whilst Brazil exports the most ethanol globally at about 2.9 million tonnes per year, the top importers of the US, EU,Japan and Korea have increasing demand that will have to be satisfied by increased shipping capacity. Seaborne vegetable oil supply is increasingly growing.Recently biofuel is driving a new technology, Worldwide; the use of biofuels for cars and public vehicles has grown significantly. With excess capacity waiting for source material it seems inevitable that shipping demand will increase.

13.3 Inter industry best practice

13.3.1 Land based use

i. UK pumps mandate at least 2.5% biofuels. This target will rise to 5% by 2010. Also in the UK, the first train to run on biodiesel went into service in June 2007 for a six month trial period. The train uses a blended fuel, which is 20% biodiesel and the operator, Virgin Trains, is confident the mix can be increased to at least a 50% mix with the further possibility to run trains on fuels entirely from non-carbon sources.

ii. On January 15, 2006—Central Ohio Transit Authority (COTA lunch a program to test a 20% blend of biodiesel (B20) in its buses. In two months they used approximately 45,000 gallons of B20. As a result of the test, in April 2006 they began using biodiesel fleet-wide. In addition to using B20 in the winter months, COTA has committed to using 50-90% biodiesel blends (B50—B90) during the summer months. This is projected to decrease regular diesel fuel consumption by over one million gallons per year.

iii. 26th of October 2007. buses in the UK running on B100 was launched on In a pilot project. Argent Energy (UK) Limited is working together with Stagecoach to supply biodiesel made by recycling and processing animal fat and used cooking oil.

iv. For power stations, B&W have orders in the EU for 45 MW of two-stroke biofuel engines with a thermal efficiency of 51-52%. Specifically, these operate on palm oil of varying quality, and in the future, it is expected that more engines, whether stationary or marine, will be developed to run on biofuels.

v. US DOE has funded five new advanced biomass gasification research and development projects beginning in 2001(Vermont project.

vi. 2008—Ford announced a £1 billion research project to convert more of its vehicles to new biofuel sources. The first trial oft, Last year. BP Australia has now sold over 100 million liters of 10% ethanol content fuel to Australian motorists, and Brazil sells both 22% ethanol petrol nationwide and 100% ethanol to over 4 million cars, It is a trend that is gathering momentum.

vii. In a program initiated by the Swedish National Board for Industrial and Technical Development in Stockholm, several Swedish universities, companies, and utilities are collaborating to accelerate the demonstration of the advanced EVGT for natural-gas firing, especially in small-scale units. A natural-gas-fired EVGT pilot plant (0.6 megawatts of power output for a simple gas-turbine cycle) should start operation in Lund, Sweden, in 1998.

viii. AES Corporation is a leading company in biomass conversion internationally. At AES Kilroot in Northern Ireland, the team recently completed a successful trial to convert the plant to burn a mixture of coal and biomass. With further investment in the technology, nearly half of Northern Ireland's 2012 renewable target could be met from AES Kilroot alone.

13.3.2 Aero industry

Virgin Atlantic—Air transport is receiving increasing attention because of environmental concerns linked to CO_2 emissions, air quality and noise. Virgin Atlantic in collaboration with Boeing and General Electric aircraft alternative fuels project for aircraft. A successful test flight from London to Amsterdam flight took place on 24th February

—

of this year, running one of the four jumbo jet engines on a mixture of 20% coconut oil and babassu nut oil, with 80% conventional jet fuel. This fuel was specifically chosen due to its performance at low operating temperatures. The test was successful, with no noticeable difference in performance. Except that; imitation that biofuel mix used was in no way sustainable in the quantities required by the demands of the aviation industry. In a way to mitigate this Virgin is looking to us use of Algae based fuels as it is predicted that they may be suitable for use at low temperature.

13.3.3 Maritime industry

The use of land based transportation, is growing, however the use for sea based transportation need to be explored. Biofuels for ship will be advantageous. In recent UK pilot project where Buses are run on B100 Argent Energy (UK) Limited is working together with Stagecoach to supply biodiesel made by recycling and processing animal fat and used cooking oil. Marine engines with their inherent lower speed and more tolerant to burning alternative fuels than smaller, higher speed engines tolerance will allow them to run on lower grade and cheaper biofuels. Royal Caribbean Cruise Lines (RCCL) unveiled a palm oil-based biodiesel since 2005.Optimistic outcome of the trial made RCCL confident enough to sign a contract in August 2007 for delivery of a minimum 15 million gallons and for the four years after, a minimum of 18 millions gallons of biodiesel for its cruise ships fleet. The contract marked the single largest long-term biodiesel sales contract in the United States. In early 2007, United States Coast Guard indicated that their fleet will augment increase use of biofuels by 15% over the next four years. In the marine industry, beside energy substitute advantage, biolubricants and biodegradable oil are particularly advantageous from an environmental and pollution perspective. Bio lubrication also offer higher viscosity, flash point and better technical properties such as increased sealing and lower machine operating temperature advantageous use in ship operation.

Time has gone when maritime industry could afford natty gritty in adopting technology, other industry are already on a fast track preparing themselves technically for evitable changes driven by environmental problem, Global energy demands and political debate add further pressures to find alternative energy especially bio energy because of hybridization of old and new system advantage it offer. The implication is that shipping could be caught ill prepared for any rapid change in demand or supply of biofuel. Thus this technology is in the early stages

—

of development but the shipping industry need top be prepared for the impacts of its breakthrough because Shipping will eventually required be at the centre of this supply and demand logistics chain again. Table 13.3 shows the projection for the main present players.

Table 13. 3: Projection

Region	Growth (1990-1994)	Projection (2020)
United states	7%	15%
Europe	2%	15%

13.4 Sources of biomass

North American Electric Reliability Council (NERC) region. Supply has classified biofuel into the following four type's vizs: agricultural residues, energy crops, forestry residues, and urban wood waste/mill residues. A brief description of each type of biomass is provided below:

i.   Agricultural residues from the remaining stalks and biomass material left on the ground can be collected and used for energy generation purposes this include residues of wheat straw and corn stover.

ii. Energy crops are produced solely or primarily for use as feed stocks in energy generation processes. Energy crops includes hybrid poplar, and switch grass, grown on idled, or in pasture, and in the Conservation Reserve Program (CRP).

iii. Forestry residues are composed of logging residues, rough rotten salvageable dead wood, and excess small pole trees.

iv. Urban wood waste/mill residues are waste woods from manufacturing operations that would otherwise be landfilled. The urban wood waste/mill residue category includes primary mill residues and urban wood such as pallets, construction waste, and demolition debris, which are not otherwise used.

The most important agricultural commodity crops being planted in the United States are listed in *Table 13.4*. Corn, wheat, and soybeans represent about 70 percent of total cropland harvested.

Table 13. 4: U.S Agricultural Commodity Crops with the Largest Acreage, 2000 **(Million Acres)**

—

Crop	Acreage	Crop	Acreage	Crop	Acreage
Corn	79.54	Hops	36.12	Oats	4.48
Wheat	62.53	Catton	15.54	Rice	3.06
Soybeans	74.5	Grain Sorghum	9.19	Canola	1.51
Hay	59.85	Barley	5.84	Rye	1.34

Source: U.S Department of Agriculture, Agricultural Statistic
2001, web site www.usda.gov/nass/agro01.htm

Table 13. 5: Representative characteristics for different subcategories
of urban wood waste and mill residues.

Residue type	(%)	(Btu per Pound)		(Dollars per Wet Ton)
	Moisture Content	Heating Value, Wet	Heating Value, Dry	Collection and Processing Cost
Bark Residue (Primary Mill)	40	4.697	8.629	4
Wood Residue (Primary Mill)	40	4.661	8.568	4
Wood Yard Trimmings	25	6.15	8.6	12
Construction residues	15	7.103	8.568	12
Demolition Residues	15	7.103	8.568	12
Other Waste Wood	15	7.103	8.568	12

Source: Antares Group. Inc Biomass Residue Supply Curves
United State (Update), Report for the U.S Department of Energy
and the National Renewable Energy Laboratory (June 1999)

13.5 Risk and Uncertainties

Although a significant amount of effort has gone into estimating the available quantities of biomass supply, the following uncertainties that need to be incorporated into design and decision work on biodiesel use are:

i.   Risk to land use—Our planet only have 295 land, for example Brazil has some 200 million acres of farmland available, more than the 46 million acres of land, required to grow the sugarcane needed to satisfy the projected 2022

ii. Evolving competing uses of biomass materials, the large market consumption, pricing and growing need.

iii. In agricultural waste, the impact of biomass removal on soil quality pose treat to agricultural residues that need to be left on

—

the soil to maintain soil quality could result in significant losses of biomass for electric power generation purposes.

iv. Impact of changes in forest fire prevention policies on biomass availability could cause vegetation in forests to minimize the potential for forest fires could significantly increase the quantity of forestry residues available.

v. Potential attempt to recycle more of the municipal solid waste stream might translate into less available biomass for electricity generation.

vi. Impact on the food production industry as witness in recent food scarcity crisis

13.5.1 Regulatory impact

The EU has stated that by 2020 a target of 20% of community wide energy will be renewable. Further to this, all member states are to achieve a mandatory 10% minimum target for the share of biofuels in transport petrol and diesel consumption by 2020.. The legislation provides a phase-in for biofuel blends, including availability of high percentage biofuel blends at filling stations. The United States Congress passed the Renewable Fuels Standards (RFS) in February 2008, which will require 35 billion gallons of renewable and alternative fuels in 2022. In parallel to this, work is continuing to reduce emissions further in vehicles. Political drivers in Asia vary according to region. In Southeast Asia, the centre of world production for palm oil, coconut oil, and other tropical oils, political support for farming is the key driver. The issue affecting shipping is whether to refine and use biodiesel locally, or export the unrefined oil for product production elsewhere. In the short term the economics have favored the exports of unrefined oil—which is good news for us. Over the next ten years, with the cost of oil rising, and strict emission reductions in place, the need for increased biofuel production is likely to increase. As well as creating a net positive balance fuel. According to the IEA, world biofuels demand for transport could increase to about 3% of overall world oil demand in 2015 and double by 2030 over the 2008 figure. This does not sound so significant but as we show later it has a significant impact on the specialist fleet capacity demand.

As we said before, predicting the trade pattern of biofuels adds a layer of complexity to the overall energy supply picture and our oil distribution system. We also believe that this forecast will be the minimum seen as the political pressures will cause the level to rise beyond 3%. To put the scale in context, the current oil tanker fleet of vessels 10,000 dwt or larger comprises of some 4,600 vessels amounting to 386 million dwt. These include about 2,560 Handysize tankers. Additionally, there are some 4,400 more small tankers from 1,000 to 10,000 dwt accounting for 16 million dwt. Our projections show a significant role for seaborne transport, even using conservative bases with high proportions of locally supplied biofuels. This is a significant fleet segment that poses technical and regulatory challenges. As we have discussed, the requirements cannot be fully defined because many market factors remain uncertain, but ship owners who are building new vessels or operating existing vessels should consider this future trade through flexible design options that we will introduce later.

13.5.2 Potential Impacts to Shipping

The key political drivers for biofuels are environmental concerns, energy security and agricultural policy. The ton mile demand for future tankers will be greatly affected by national, regional or global policy and political decision making in these areas. There is a greater flexibility in the sourcing of biofuels than there is in hydrocarbon energy sources and this may be attractive to particular governments. Once the regulatory framework is clear, economics will determine how the regulations will best be met and seaborne trade will be at the center of the outcome. In many parts of the world, environmental concerns are the leading political driver for biofuels. Reflecting these concerns, the global Kyoto Protocol, was negotiated in 1997, and this further provides a driver for the use of biofuels.

13.5.3 Shipping Routes and Economics Impacts

The above trend analysis discussed indicate potential capacity requirement from shipping, so far North America, Europe and South East Asia are the key importing regions where this growth is concentrated. This includes the Latin American counties of Brazil, Argentina, Bolivia, and Paraguay and Southeast Asia's Indonesia and Malaysia will remain key suppliers for the palm oil, Philippines and Papua New Guinea have potentials for vegetable oil and agricultural while Thailand has potential for sugarcane.

This trade potential will determine future trade route from Malacca Straits to Europe, ballast to Argentina, to load soybean oil to China, and then make a short ballast voyage to the Malacca Straits, where the pattern begins again, a typical complicated front haul / backhaul combinations that can initiate, economies of scale need to reduce freight costs and subsequent push for bigger ship production and short sea services like recent experience of today's tankers. According to plateau case study the following regional impact can be deduced for shipping.

Table 13. 6: Bio fuel Demand by region

	Biofuel	Demand
North America	ethanol	33 million tons
Europe	ethanol and biodiesel.: 50:50	30 million tons
Asia	ethanol and biodiesel.: 50:50	18 million tons

North America demand—policy work support biofuel use in the US and 32 Handysize equivalent tankers will be needed to meet US demand in 2015. With technological breakthrough there will be need for 125 vessels 2030.

European demand—Due to environmental requirement and energy security believed to be politically acceptable in the EU but economics may drive a different outcome.80 Handysizes with some due to the growth in trade and longer voyage distance. With technological breakthrough for 2nd and 3rd generation biofuel growth will need growing to 145 in 2030 Aframax vessels if the technical issues can be overcome.

Asia demand—In plateau case 50 Handysize equivalents are required in 2015 and 2030 with forecast vessel sizes being Handysizes with some Panamax vessels 162 vessels total in the three regions.

By adding up all the regions, with biofuels as only 3% of world transport demand, we are looking at a fleet of about 400 Handysize vessels to accommodate the demand and supply drivers by 2030 and 162 by 2015. The total vessel forecast for 2030 could means 2,560 vessels of 81 million deadweight tons. As regions identify these growth markets and recognize the economies of $/tonne scale that can be achieved, as shown here, with bigger tonnage, we are seeing natural investment occurring. New port developments in concerned trade rout will be required to accommodate large Panamax vessel and parcel size for palm oil exports. on the long haul routes.

13.5.4 Biomass Ship Technologics Impacts

Generation

A variety of methods could turn an age-old natural resource into a new and efficient means of generating electricity. Biomass in large amounts is available in many areas, and is being considered as a fuel source for future generation of electricity. Biomass is by its nature both bulky and widely distributed and electricity from conventional, centralized power plants requires an extensive distribution network. Traditionally power is generated through centralized, conventional power plant, where biomass is transported to the central plant, typically a steam or gas turbine power plant, and the electricity is then distributed through the grid to the end users. Costs include fuel and transportation, power plant construction, maintenance, and operation, and distribution of the electric power, including losses in transmission.

Table 13. 7: Biomass and Coal

	Electrical efficiency	Capacity
Biomass	thermal efficiency—40 %	$2,000 per kilowat
Coal	45 %	$1,500 per kilowatt,

However, micro-biomass power generators located at the site of end-use seem to offer a path for new solution for energy. Recent development in towards use of micro biomass will equally offer best practice adaptation for marine power. Biomass is used at or near the site of end-use, with heat from external combustion converted directly to electricity by a biomass fired free-piston genset. Costs include fuel and acquisition and maintenance of the genset and burner. Since the electricity is used on site, both transmission losses and distribution costs are minimal.

Thus, in areas without existing infrastructure to transmit power, there are no additional costs. In this case it is also possible to cogenerate using the rejected heat for space or hot water heating, or absorption cooling. Previously, option two has not been feasible, since there have been no small (less than ~50 kW) devices for directly and efficiently converting biomass energy to electricity. Micro-biomass power generation is a more cost-effective means of providing power than central biomass power generation. In particular, areas where there is a need for both power and heat—domestic hot water and space heat and absorption chilling—are attractive for cogeneration configurations of this machine. Biomass can be generated using single or ganged free-piston Stirling engines gensets. These micro-biomass generators offer a number of advantages over centralized biomass fueled power plants. They can be placed at the end-user location taking advantage of local fuel prices and do not require a distribution grid. They can directly provide electrical output with integral linear alternators, or where power requirements are larger they can be ganged and drive a conventional rotary turbine. They are hermetically sealed and offer long lives through their non-contact operation.

Biomass for electricity generation is treated in four ways in NEMS: (1) new dedicated biomass or biomass gasification, (2) existing and new plants that co-fire biomass with coal, (3) existing plants that combust biomass directly in an open-loop process,[18] and (4) biomass use in industrial cogeneration applications. Existing biomass plants are accounted for using information such as on-line years, efficiencies, heat rates, and retirement dates, obtained through EIA surveys of the electricity generation sector.

Emissions offsets and waste reduction could help enhance the appeal of biomass to utilities An important consideration for the future use of biomass-fired power plants is the treatment of biomass flue gases. Biomass-combustion flue gases have high moisture content. When the flue gas is cooled to a temperature below the dew point, water vapor starts to condense. By using flue-gas condensation, sensible and latent heat can be recovered for district heating or other heat-consuming processes; this increases the heat generation from a cogeneration plant by more than 30 percent. Flue-gas condensation not only recovers heat but also captures dust and hazardous pollutants from flue gases at the same time. Most dioxins, chlorine, mercury, and dust are removed, and sulfur oxides are separated out to some extent. Another feature of flue gas condensation is water recovery, which helps solve the problem of water consumption in evaporative gas turbines.

Biomass open door for another way rather than competing with fossil fuel plants a substantial opportunity exists to generate micro-biomass electric power, at power levels from fractions of a kilowatts through to tens or hundreds of kilowatts, at the point of en d use. At these power levels neither small internal combustion engines, which cannot use biomass directly, nor reciprocating steam engines, with low efficiency and limited life, can offer the end user economic electric power. Free-piston Stirling micro biomass engine engines are an economic alternative. Stirling offers the following advantages over significantly larger systems:

i. Stirling machines have reasonable overall efficiencies at moderate heater head temperatures (~600fC) cogeneration is simple large amounts of capital do not have to be raised to build a single evaluation plant with its associated technical and economic risks

ii. A large fraction of the value of the engine alternator can be reused at the end of its life

iii. Stirling systems can be ganged with multiple units operating in parallel.

13.5.6 Technological Change Requirement for Ships

It is clear that biomass will fuel freight increase as well specialized new design of chemical tankers. biodiesel is an IMO 2 cargo, Its vegetable oil feedstocks are IMO cargoes with double hull IMO 3 vessel configuration required. Ethanol typically transports in chemical tankers due to its cargo requirement but technological change break through could bring potential regulatory design change. Flexibility for ship conversion and retrofitting system could upset initial cost problem.

Biodiesel marine energy generation –The use of biofuels as a fuel has increased in most transport sectors, adopting this technology in marine industry is still slow despite flexibility offered by use of energy on ship compare to mass requirement for land based industry and ambient temperature performance for aviation industry. Cost is the driver and slow speed diesel engines can run on lower quality fuels so therefore replacing distillate marine oils with technical difficult hat May arise is that relating calorific energy value for main propulsion could result in a reduced service speed, range or larger bunker tanks.

Fuel quality and standards issue will also come in as a barrier as power generation will definitely depend on energy source and property of biofuel. Currently the fuel standard for marine applications, ISO 8217 relates solely to fossil fuels, and has no provision for biofuels. Thus land based standards has been developed which can be adapted as required for marine application, for example the European EN 14214 for automobile. The new MARPOL Annex VI, which was agreed earlier this year and proposed for adoption in October 2008, has in the definitions of "fuel oil" any fuel delivered to and intended for combustion purposes for propulsion or operation on board a ship. This leaves the door open for biofuels.

Another factors that need to be considered is the supply chain supply of biofuel to ship can be done through pre-mixed to the required blend, or the biofuel and diesel are supplied separately to the ship, and then mixed on board. The latter gives the operator the chance to dictate the exact blend of biofuels depending on conditions but that would require retrofitting or new technology to be installed on board together with additional complexity for the crew. The first option, where the biodiesel is blended prior to delivery to ship is also affected by the biodiesel shelf life. Fuel management will therefore become more complex in this new era because fuel aging and oxidation can lead to high acid number, high viscosity and the formation of gums and sediments.

13.5.7 Biodiesel Machinery Design Requirements

Temperature monitoring system—Technical problem that need to be further mitigated is the CFPP indication of low temperature operability of range between 0°C and 15°C for different types of biodiesels and can cause problems with filter clogging, this can only be overcome by carefully monitoring of the fuel tank temperatures. This could well affect ships operating in cold climates, where additional tank heating coils and heating may be required to avoid this from happening.

Corrosion control—Biodiesels are hygroscopic and maintain 1200-1500 ppm water, which can cause significant corrosion damage to fuel injection systems. Mitigation can be exercise through appropriate fuel conditioning prior to injection. biodiesels is injector fouling. biodiesel especially the blend type produce deposits due to presence of fatty acid and water in the fuel, this can apply to increased corrosion of the injector system but also in the presence of glycerol and viscous glycerides can contribute to further injector coking. Biodiesel due to its chemical properties degrades, softens or seeps through some gaskets and seals with prolonged exposure. Biodiesels are knows to be good solvents and therefore cause coating complexity. Reports of aggressiveness of biodiesel and bioethanol on tank coatings have been reported. In its pure form biodiesel, as a methyl ester, is less aggressive to epoxy coatings than ethanol, and therefore ethanol should be carried in tanks coated with dedicated tank coatings such a phenolic epoxy or zinc silicate tank coatings.

Fuel monitoring—It is also important to monitor the fuel acid number value to ensure that no rancid, acidic fuel is introduced to the injection system. A typical layout should involve separators to ensure water is removed from the fuel, as well as heaters at various stages to ensure the fuel is at the correct temperature for before the enter engine.

Lubrication—Biofuel rankcase lubricant may have impact on engine crankcase cleanliness and the potential consequences of fuel dilution. The droplet characteristics and lower volatility of biodiesel compared with conventional diesel, together with spray pattern and wall impingement in the modern diesel engines, can help noncombusted biodiesel past the piston rings, make contact with the cylinder liner and be scrapped down into the oil sump. The unburnt biodiesel tends to remain in the sump and the level of contamination may progressively build up over time. This can result in reduced lubricant viscosity and higher risk of component wear. A serious concern is the possibility that the unburnt biodiesel entering the oil sump may be oxidised, thus promoting oil thickening and requiring greater oil changes.

13.5.8 Port, Inland waterways and coastal vessel

There is potential for us of biodiesels for small craft that operate within inland water because of air and water pollution sensitivity associated with inland water transportation. The port facilities in Malaysia and Indonesia are already being improved to handle Handysize and Panamax tankers. There is also potential requirement for transshipment and supply vessel, supply chain for short sea service.

13.5.9 Cargo

3rd generation biofuels will require to be processed from solid cargoes to liquid cargoes. Of the wood currently harvested, 30% is waste. We propose that this is not going to be a waste in the future and will be converted by a Fischer Tropsch biomass to liquid processing plant. For coastal shipping to handle this trade, there will be need for new generation of 5,000 tonne deadweight dry cargo vessels. It is expected that these voyages will be regulated under the new Dry Bulk Cargo Code (BC Code), which is due to become mandatory in 2011. Larger vessels will develop if the trade develops but longer journeys will require design attention to the effects of condensation, ventilation and firefighting aspects In addition to wood chips and wood pellets, the other main type of common solid biofuel is palm seed cake, which is the residue of the palm oil production process. This is used as a cargo in co fired power stations as it is easy to light. As a cargo, seed cake comes under the BC Code as a fire hazard with the necessary changes in design required to accommodate also bulk liquid carriage will be another issue. Ethanol is listed in chapter 18 of the IBC code as a mild pollutant and not a safety hazard but it has a low flash point requiring explosion proof equipment. One complication of ethanol however is that it absorbs water from atmosphere, and to stop this occurring there is a current provision for a nitrogen blanket in the tank. This is usually supplied from shore before the cargo is loaded in the tanks, and is kept topped up by nitrogen bottles on the vessel.

13.5.10 Restricted water requirement

Panamax vessels—The Panama Canal is a crucial gateway for transport by sea and future vessels to support the biofuel trade will undoubtedly be required to consider the constraints of the current, or opportunities of the future, Panama Canal dimensions.Chemical tanker conversion and upgrade offer a good advantage Could lead to need for economies of scale to build large tankers which require midship section to have double bottom height of 2.15 M and side protection to MARPOL Annex I of 2 meters. If a vessel has to carry higher proportions of vegetable oil/ethanol to petroleum oil of more than 15%, it may require a Chemical Tanker notation with additional requirements both structural and equipment wise. The pre-mixing of biofuel on the vessel, or in the refinery/terminal, has also been a subject that has been discussed at IMO. At the last Bulk Liquid and Gas meeting in February 2008, IMO stated that at present it does not come under any international requirements, but may be developed further by the IMO in the future.

Aframax vessels—Brazil is looking to increase its ethanol export capacity and is therefore investigating carrying ethanol in even larger than Panamax size, possibly Aframax size. For an oil/ethanol Aframax Tanker, we would need to have a Noxious Liquid Substances Certificate that need an Approved Procedures and Arrangements Manual in accordance with MARPOL 73/78 Annex II, a Shipboard Marine Pollution Emergency Plan for Noxious Liquid Substance, a stripping test, and an initial survey. The stripping test requirement will be 75 litres if built after 1st January 2007, or best possible extent if built before 1st January 2007. Achieving these figures on a new Aframax size vessel may be difficult.

The main challenge to using biomass for power generation, therefore, is to develop more-efficient, lower-cost systems. Advanced biomass-based systems for power generation require fuel upgrading, combustion and cycle improvement, and better flue-gas treatment. Future biomass-based power-generation technologies have to provide superior environmental protection at lower cost by combining sophisticated biomass preparation, combustion, and conversion processes with post combustion cleanup. Such systems include fluidized-bed combustion, biomass-integrated gasification, and biomass externally fired gas turbines. Ships life cycle is around 20—25, for owners to make the most of the upcoming markets, it is necessary to be prepared for the new cargoes. Current ship designs may not be suited for biofuel ships. Therefore there is potential for pressure on organizations to adopt new standards to accommodate the demand driven by governmental legislation This in itself has some risk involved also the trade routes could create economy of large scale leading to larger ship production and sub sequential requirement from designer .other evolving challenges to secure energy and environment are Fuel cell technology, nuclear, natural gas and fuels made from waste plastics.

References

1. Research, sponsored by U.S. EPA and Acurex Environmental Corporation, Washington, DC, June 27-2 9, 1995.
2. Craig K. R. and Mann M. K. (1996). Cost and Performance Analysis of Three Integrated Biomass Gasification Combined Cycle Power Systems. http://www.eren.doe.gov/biopower/ snowpapr.html, August 1996.
3. Fuldner, A. H. (1997). Upgrading Transmission Capacity for Wholesale Electric Power Trade. http://www.eia.doe.gov/ fuelelectric.html, April 1997
4. Lane, N. W. and Beale, W.T. (1996a). Stirling Engines for Micro-Cogen and Cooling. Proc Strategic Gas Forum, Detroit, MI, June 19-20, 1996.
5. Sustainable Energy Ireland, CO_2 and Other Environmental Emissions Data, Table 7D, web site www.irish-energy.ie/ publications/ index.html.
6. M. K. Mann and P. L. Spath, "A Comparison of the Environmental Consequences of Power from Biomass, Coal, and Natural Gas," Presentation at the Energy Analysis Forum (Golden, CO, May 29-30, 2002), web site www.nrel.gov/analysis/ pdfs/m_mann.pdf.
7. Further information on the projects can be obtained at web site www.eren.doe.gov/biopower/projects/index.htm.
8. G. Wiltsee, Lessons Learned from Existing Biomass Power Plants, NREL/SR-570-26946 (Golden, CO: National Renewable Energy Laboratory, February 2000), web site www.nrel.gov/docs/ fy00osti/26946.pdf.
9. G. Wiltsee, Lessons Learned from Existing Biomass Power Plants, NREL/SR-570-26946 (Golden, CO: National Renewable Energy Laboratory, February 2000), web site www.nrel.gov/docs/ fy00osti/26946.pdf.
10. Switchgrass is a perennial warm season grass. Its native range includes the eastern United States (east of the Rocky Mountains). It can be planted, managed, and harvested like traditional hay crops with existing agricultural equipment. This description from M. Walsh et al., "The Economic Impacts of Bioenergy Crop Production on U.S. Agriculture"
11. (Oak Ridge, TN: Oak Ridge National Laboratory, May 2000), web site http://bioenergy.ornl.gov/papers/wagin/index.html.

–

12. U.S. Department of Agriculture, *Agricultural Statistics 2001*, web site www.usda.gov/nass/pubs/agr01/agr01.htm.
13. POLYSYS can be used to simulate other scenarios, such as those assuming high wildlife diversity.
14. U.S. Department of Agriculture, Economic Research Service, *Economic Indicators of the Farm Sector: Costs of Production—Major Field Crops, 1995* (Washington, DC, 1996).
15. R.L. Graham, B.C. English, and C.E. Noon, "A Geographic Information System-based Modeling System for Evaluating the Cost of Delivered Energy Crop Feedstocks," *Biomass and Bioenergy*, Vol. 18 (2000), pp. 309-329.
16. U.S. Department of Commerce, *Statistical Abstract of the United States: 1996*, P1-R96-STAB-00-NTH (Washington, DC, November 1996). 1.

CHAPTER 14

Utilization of Simulation for Training Enhancement

"The most beautiful thing we can have is the mysterious. It is the source of all true art and science"

Albert Einstein

Summary

Engineering system design, operation and maintenance has been handled for a long time through mathematical and real time models. The advent of computers, multimedia age and improvement in visualization has further proved the reality of fact that picture speaks more than words; also research in education and training has proven that visualization has great effect in improving learning. The complexity of real world situation of engineering education has obvious limitation of instructional presentation and training. Simulation gives result from theoretical representation of complex phenomena when hardware for the task is lacking, or in situations when enough time is not available for explanation.

This chapter will discuss, opportunities brought about by simulator as a tool in the training and certification of Malaysian Maritime Academy cadets training program. The usefulness of simulators in continuous education program to amplify and enhance competency based education and instructional training to meet goals of safety, cleaner ocean and protection of marine environment will be highlighted.. The chapter will also present the potential of simulators as training tool in other field of knowledge for enhanced outcome and competency based education.

14.1 Introduction

The world of man and the quest for knowledge to facilitate human activities including developing things that surround us has gone through various phases of development. The early man, used memorization as a tool, and wrote information on leaves, trees and mountains to store knowledge which was to be passed to the next generation. The main tools for everything related to learning has likewise gone through various phases of change and the most significant of these changes has been the emergence of ICT in the last one decade. Today, the developments in ICT have greatly accelerated the pace of knowledge delivery and the Simulation-Based studies and training is one typical example of such an evolution.

Simulation refers to the application of computational models for the study and prediction of physical events or the behavior of engineered systems. The development of computer simulation has drawn resources from a deep pool of scientific, mathematical, computational, engineering knowledge and methodologies. from the depth of its intellectual development and wide range of applications, computer simulation has emerged as a powerful tool, one that promises to revolutionize the way research in engineering and science are conducted in the twenty-first century. Simulation has long been identifies in several areas of knowledge and it is playing a remarkable role in promoting developments vital to the health, security, and technological competitiveness of the nation. Engineering and scientific communities have become increasingly aware that computer simulation is an indispensable tool for resolving a multitude of technological problems.

Basically, computer simulation represents an extension of theoretical science in that it is based on mathematical models. Such models attempt to characterize the physical predictions or consequences of scientific theories. With simulation engineers are better able to predict and optimize systems affecting almost all aspects of our lives and work, including our environment, our security and safety, and the products we use and export. The use of computer simulations in engineering science began over half a century ago, but only in the past decade or so has simulation theory and technology made a dramatic impact across the whole engineering fields. That remarkable change has come about mainly because of developments in the computational and computer sciences and the rapid advances

—

in computing equipment and systems. Clearly, the use of simulation is quickly becoming indispensable for goal based engineering education.

Simulation is an important feature in engineering systems or any system that involves many processes. Most engineering simulations entail mathematical modeling and computer assisted investigation. Mathematical model used to dominate simulation world however mathematical modeling is not reliable and the incorporation of physical model often helped to improve today complex system simulation. The development and use of such frameworks require the support of inter-disciplinary teams of researchers, including scientists, engineers, applied mathematicians, and computer scientists.

Maritime industry due to its nature and need for safety to maintain Cleaner Ocean has institutionalized and incorporated opportunity offered by simulation to training of marine personnel to fulfill objective of having competent personnel to man the ships that sail the ocean of the world. Encouraged by belief that knowledge, understanding, application and integration which are requirement for outcome and competency based education could enhance traditional instruction delivery method, through incorporating audio visual and multimedia tools, led the IMO to adopt resolution to use simulation as part of STCW requirement. Simulation is thus becoming central to advancement in maritime competency based training and education as well as educational training in biomedicine, nanomanufacturing, microelectronics, energy and environmental sciences, advanced materials, and product development. And there is ample evidence that developments in these new disciplines could significantly impact virtually every aspect of human experience.

This chapter explores potentials and prospects of incorporating simulation in engineering and science education structure. Good practice and experience enjoined by maritime industry including ALAM will be discussed. The authors will also discuss the core issues of simulation, the major obstacles to its development and the impact of simulation on training, educational and research.

14.2 Maritime simulation and simulators

Marine simulators bears similarity to flight simulators, which are used to train pilots on the ground in extreme hazard situations such as landing with no engine, pilot to crash aircraft without being hurt or complete electrical or hydraulic failures. The most advanced simulators have high-fidelity visual systems and hydraulic motion systems. However, marine simulators train ships' personnel, Simulators like these are mostly

—

used to simulate large vessels. They consist of a replication of a ships' bridge, with operating desk, and a number of screens on which the virtual surroundings are projected. The complexity of shipping activities from design to operations training and maintenance remain one of the factors that have made IMO to enact strong regulations to ensure safety at all times.

Due to new issues of imbalance in human activities and environmental behaviors, ship and its operating areas in closest proximity to ocean that cover two third of the world has further put maritime work a target by public and land based environmental agencies whose pressure has given IMO further challenges of protection of environment that has called for new way of doing things based on risk and proactive manner. Simulator is obviously one of the tools that fit in such proactive measures to prevent accidents as its consequence leads to serious environmental problems.

While International laws are best implemented and enforced through local authorities law, the performance and control are best achieved through third eyes. DNV is one such third eyes under classification society in maritime industry which has laid down some guidelines marine simulators. Certification of the simulators by DNV ensures that simulator systems have qualified personnel giving realistic and high quality simulator training conforming to SCTW requirement. Table 14.1 and 14.2 show approved STCW courses and DNV certified simulation institutions.

Table 14. 2: STCW Courses

Course Provider	Certificate no.	Course Title	Expiry date	Country
Seagull AS	04/014	Assessor Training CBT	03/12/2007	Norway
Nordic Crisis Management AS	091/061214	ECDIS Training	14/12/2009	Norway
	005/050304	CBT STCW Crowd Management Course	02/03/2008	
Furuno INS Training center	047/051122	ECDIS Training Course	22/11/2008	Denmark
Malaysian Maritime Academy (ALAM)	055/060306	Nautical Studies and marine Engineering Courses	06/03/2011	Malaysia
NYK Ship Management PTE-LTD	102/070320	BRM courses	20/05/2010	Singapore

Table 14. 3: Approved simulators (manufacturers)—Source DNV classification

Manufacturer	Certificate no	Expiry date	Simulator type	Class	Country
Kongsberg Maritime AS	011/060606	06/12/2007	Machinery Operation	A	Norway
	012/060606	06/12/2007	Cargo Handling		
	A-9616	25/05/2010	Bridge Operation		
Transan Ltd.	A-9641	20/06/2010			Ireland
	A-9642	23/06/2010			
L-3 Communications MPRI	A-9335	30/12/2009	Machinery Operation		USA

However the use and the corresponding training program has been developed on ad hoc basis of individual training center working with many shipping companies where simulation is often inserted into existing program rather incorporating simulation as part of the objectives. Neither has there been any standardization for simulation. The fact is that simulation itself does not train but its benefit to training comes from the way it is used, it become imperative to make part of training aim and objective that comes with education requirement. STCW was built on conventional approach that focus on knowledge to determine competency with oblivion to job task and performance in training that arise from reality of mismatch between training course and corresponding needs. The ability of simulation fidelity, producing real life task in a safe environment and provide mitigation option to this where simulation come in as fundamental tool to bridge gap between theory and application.

14.3 Marine simulation curriculum and training process

In maritime industry, IMO classified simulation under STCW amendment into the following groups:

> Category 1—Full mission capable of simulating full visual navigation including maneuvering.
> Category 2—Multitask—full mission capable of simulating full visual navigation excluding maneuvering ie radar simulator.
> Category 3—Limited task—capable of simulating environmental for limited extreme conditions.

Category 4—Special task—capable of simulating maneuvering with operator outside the environment.

Simulation in STCW code is made of two parts, where:

> Part A is mandatory and includes the minimum standards of competency for seagoing personnel, simulator used in both training and assessment. And the requirement for ARPA (Automatic Radar Plotting Aid—thus simulator equipment being used prior February 1, 2002 may be exempted at the discretion of parties involved).
> Part B deal with guidance to those involved in education, training or assessing the competence under STCW provision concerning application of various safety and environmental regulation and conformity.

Maritime training is divided into two groups:

i. License—This include experience trainee that undergo additional training designed toward improving their existing skill, performance and awareness. Such training includes: Watch keeping, ARPA, Control, Ship resources management, Ship team management, Emergency procedure, Ship handling, Vessel traffic management, Search and rescue, Area familiarization
ii. Unlicensed—this involved cadets working towards first certificate of competency under standard structure program. Simulation application for this training includes: Watch keeping, knowledge of international regulation, Communication, Radar, electronic navigation, main and auxiliary machineries.

Most of these courses are not actually mandatory, how good it will have been if they shipping company could incorporate them in their program. Thus recent amendment of STCW recommendation for use of simulation is a significant stepping stone in this direction. The training should be designed by considering the following:

i. Training need through identification of gaps between training, required knowledge, skills abilities and actual knowledge, skills and ability.
ii. Specific training objectives which should include performance measurement

iii. Training method including assessment weather simulation is required to achieve the objective
iv. Total training program
v. Experience of trainee
vi. Level and type of simulation
vii. Instructor qualification and experience
viii. Effectiveness and benefit of the training

Once this is determined a scheme of work which includes: Aim and general objective, trainees and numbers, structure and schedule.

i. Individual simulation exercise that include specific objective, planning, debriefing session and instruction to staff on methods of debriefing
ii. Method of assessment

The instructional process itself required the following consideration: Instructional process, Scenario design, Use of control and monitoring station, and Debriefing technique.

Rational for simulation can be stressed out from the fact that traditional class room method of teaching that use tool like chalk, overhead projector, occasional use of video material to amplify training objective give unbalanced advantage to instructor to have direct control that may not entertain training participation. Adding simulation to curriculum and allow the instructor to fill gap between theory and practical could change this equation. Simulation is expected to have program where and when it will be effective and useful in operation. Apart from concentration on fidelity of simulation—the degree of realism of simulators and simulation learning process includes the following factors:

i. The progression from easy to more complex and difficult task and operation
ii. The involvement of more than one sense
iii. The need to concentrate interest on single problems
iv. The trainees control of his own activities and possible mistakes

Effectiveness of simulation based depends on the following attributes of simulation that add up to the advantage simulation has for the future of man and knowledge.

i. Exceptional Bandwidth: The conceptual basis of materials modeling and simulation encompasses all of the physical sciences.

ii. Elimination of Empiricism: A virtue of multiscale modeling is that the results from both modeling and simulation are conceptually and operationally quantifiable.

iii. Visualization of Phenomena: The numerical outputs from a simulation are generally data on the degrees of freedom characterizing the model.

iv. Prediction of the consequences of threats and Countermeasure responses—Extreme engineering processes such as structural responses, fluid transport of contaminants, power distribution, and transportation systems, as well as the response of the human population.

14.4 ALAM's simulation and simulator experience

ALAM training towards a world class one stop maritime institution aims to deal with challenges of today and future through "beyond competency" partnership with DnV seaskill in order to add value of soft and hard skill required of marine personal and industry to conform with international local statutory requirements and often multi-national operations. ALAM simulation adds value to competence for ships staffs to be able to plan, define, develop and improve the competence of employees according to external requirements and established business goals to meet targets of safer, efficient operations, Cleaner Ocean and protection of marine environment.

Figure 14. 1: ALAM's simulator

—

DNV SeaSkill Standard for Competence under STCW focus on providing necessary tools and expertise to evaluate the competence of individuals, through test questions and practical assignments, in relation to specific jobs or positions wwith competence standards developed with the industry and outlining competence requirements for given positions; individuals may both be assessed and certified. The program also helps ALAM to recruit suitable mariners at suitable position good degree of reliability.

ALAM has undertaken a major effort in cooperation with DnV to improve the professional competency importance of maritime vocational and qualification by basing them on standard of proficiency required in employment. Towards this end, ALAM simulation is used as required as part of competency assessment on the ability of trainee to perform on board ship according to standards. Thus it is very important to incorporate cost comparison for necessary differentiation. ALAM invest on what to be achieved through simulation training involve task analysis and performance criteria developed to meet trainee and employers needs, IMO and classification society requirements whose aim concentrate on competency training. However ALAM invest on what to be achieve

ALAM simulator was built by Transas Marine USA, a simulator manufacturer known for building of state of the art simulators consisting of Transas Navi-Trainer Professional 3000 full mission ship simulator system with integrated GMDSS communications simulation capabilities, as well as ARPA/Radar simulator. Transas Marine's unique combination of simulation software, dedicated hardware (real ship controls) and commercial-off-the-shelf components, the simulator is an ideal tool in the training and certification of Maritime education programs. The ship simulator, IMO STCW training standards and the latest Det Norske Veritas 'Standards for Certification of Maritime Simulator Systems' to Class A as well as meet USCG. DnV has certification for simulators that meet the need of the marine industry quality training solutions.

Transas simulators are based on mathematical model that allows processes to be accelerated without detriment to physical realism at considerable reduced time. The simulations can be form under the following areas:

i.   Full mission Ship—Handling simulation
ii. Engine room simulation
iii. Cargo operations simulation

The simulator has diverse cargo types databases that facilitate selection and on almost all types of cargo for operation training. Cargo operation simulations include:

i.   Large Crude Oil Carrier (LCC);
ii. Liquefied Petroleum Gas (LPG) Carrier;
iii. Chemical tanker (CHT);
iv. Liquefied Natural Gas (LNG) Carrier.

14.4.1 Full Mission Ship-handling Simulator

The Full Mission Ship-handling Simulator (FMSHS) with capacity to simulate extensive exercise scenarios is certified by DNV as Class A Standard. Consisting of a single main bridge with nine high resolution projectors, 270° field of view visual scene (with a panning and tilting facilities to provide rear and over-the-side view), an extensive bridge mockup complete with a full complement of bridge equipment, environmental effects (consisting of wind, water current, depth, and bank forces), and high-fidelity own ship and passing ship hydrodynamic effects, the system realistically presents the total marine scene.

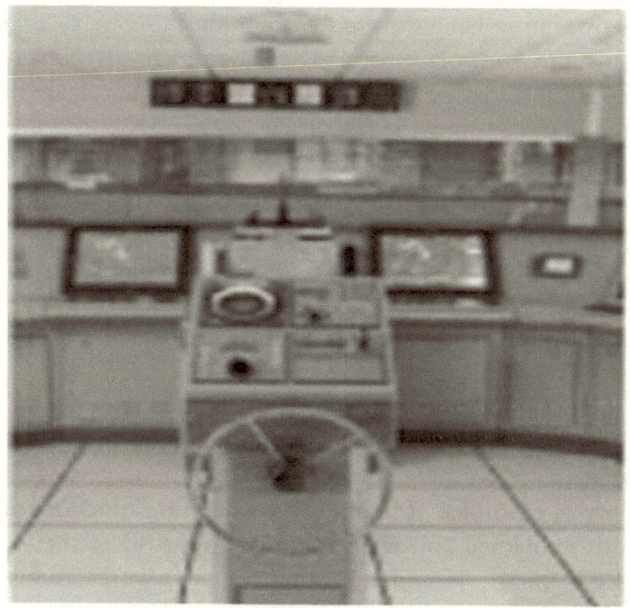

Figure 14. 2: ALAM"S Full Mission Simulator

—

Three additional cubicle bridges with 120° field of view are similarly equipped to provide interconnected operation and total ship-handling interaction between the simulators. In addition, one of the cubicle bridges is equipped with a dynamic positioning system. With a library of more than a hundred geographic databases, 79 ship models and a facility to generate a new geographic database.

14.4.2 Engine room simulator

Built to DNV Class A specification, the Engine Room Simulator (ERS) consist of the machinery space, engine control room and a computer workstation laboratory. The machinery space is equipped with local operation stations to provide appropriate indicators and controls for local power plant control. Four 42" plasma screens provide an innovative 3D virtual reality of the engine room compartments, the machinery layout, and the physical realism of the ship environment. The engine control room is equipped with main engine control console, diesel generator control console and main switchboard console to allow trainees to operate valves and machinery throughout the engine room. Realistic engine room sounds and alarms are also simulated in the engine control room to provide aural cues. Equipped with twelve (12) student workstations, the computer workstation laboratory provides trainees access to a wide variety of equipment and controls associated with the various power plant and auxiliary system and may be used for both individual and team instruction.

Five engine models provide trainees with comprehensive knowledge of major engine makers like MAN B&W, SEMT Pielstick, Wartsila Sulzer and Kawasaki Heavy Industry. . When interconnected, the FMSHS will response in accordance with the ERS models, which have been accurately modeled and validated.

14.4.3 Liquid Cargo Operation Simulator

These are designed to conform to latest liquid cargo operation system found on modern ship, the Liquid Cargo Operation Simulator (LICOS) is equipped with twelve (12) student workstations that provide access to a wide variety of examples of liquid cargo operation functions and may be used for both individual and team training. Eight (8) of the workstations have two LCD monitors and the other four (4) workstations are fitted with four LCD monitors which may be configured as an oil terminal when integrated with the VLCC model for team training.

—

The calculations of tanks, hull strength and ship loading in the simulator are carried out by the Load Calculator System, which is a real on-board Load Calculator. The Graphics User Interface (GUI) is optimized for familiarization with the entire system operating principles and for acquiring practical skills in equipment handling. The main tanker units are implemented as 3D objects, showing cross sections of individual assemblies. Computer animation is used to display the current processes.

Since operations by the bridge team on the main FMSHS will impact team operations of the ERS, and vice versa, the main Full Mission Ship-handling Simulator will be fully integrated with the Engine Room Simulator (ERS) to provide team training. When interconnected, the FMSHS response is in accordance with the ERS models, which have been accurately modeled and validated—making the ALAM Simulation Center a world-class simulator-based learning environment.

Figure 14. 3: ALAM's Liquid Cargo Operation Simulator

14.5 Benefits of simulation

Benefit offer by incorporating simulation in education are further amplified by the following:

i. Simulation allows us to explore natural events and engineered systems that have long defied analysis, measurement, and experimental methodologies. In effect, empirical assumptions will be replaced by science-based computational models.

ii. Simulation also has applications across technologies—from microprocessors to the infrastructure of cities.

iii. Simulation methods will lay the groundwork for entire technologies that are only now emerging as possibilities.

iv. Simulation will enable us to design and manufacture materials and products on a more scientific basis with less trial and error and Shorter design cycles.

v. Simulation improves our ability to predict outcomes and optimize solutions before committing resources to specific designs and decisions.

vi. Simulation will expand our ability to cope with problems that have been too complex for traditional methods. Such problems, for example, are those involving multiple scales of length and time, multiple physical processes, and unknown levels of uncertainties.

vii. Simulation will introduce tools and methods that apply across all engineering disciplines—electrical, computer, mechanical, civil, chemical, aerospace, nuclear, biomedical, and materials science. For instance, all engineering disciplines stand to benefit from advances in optimization, control, uncertainty quantification, verification and validation, design decision-making, and real-time response.

viii. Simulator certification benefits training institutions seeking assurance on heavy investments result in optimal training conditions and marketing of simulator training centers services

In addition to this simulation also offer advantage of:

i. Protection against Air Contaminants:
ii. Optimization of Infrastructures:
iii. Prediction of Long-Term Environmental Impacts
iv. Optimization of Emergency Responses
v. Optimization of Security Infrastructures for Urban Environments
vi. Planning of countermeasures
vii. Prediction of treat and countermeasure responses

14.5.1 Educational strategies of the future for Engineers and Scientists

Our time has seen significant dramatic expansion of the knowledge base required to advance modern simulation. The expansion ignores the traditional boundaries between academic disciplines, which have long been compartmentalized in the rigid organizational structures of today's universities. The old silo structure of educational institutions has become an antiquated liability. It discourages innovation, limits the critically important exchange of knowledge between core disciplines, and discourages the interdisciplinary research, study, and interaction critical to advances in simulation.

Today's demands nonetheless call for:

i. Citadel of learning to change their organizational structures to promote and reward collaborative research that invigorates and advances multidisciplinary science. It has also become a matter of need for universities to implement new multidisciplinary programs and organizations that provide rigorous, multifaceted education for the growing ranks of simulation trainers and researchers.

ii. Simulation need to be incorporated in our educational discipline as a engineering tool and proponent life-long learning opportunity

iii. Simulation requires a broad range of interdisciplinary knowledge that tomorrow's engineers and scientists with substantial depth of knowledge in computational and applied mathematics, as well as in their specific engineering or scientific disciplines. Participation in multidisciplinary research teams and industrial internships will give students the broad scientific and technical perspective, as well as the communication skills that are necessary for the effective development and deployment of simulation education.

iv. Integrating simulation into the educational system will broaden the curriculum for undergraduate students. Undergraduates, moreover, will have access to educational materials that demonstrate theories and practices that complement the traditional experimental and theoretical approaches to knowledge acquisition.

v. Simulation will also provide a rich new environment for undergraduate research, in which students from engineering and science can work together on interdisciplinary teams.

vi. As in any entrenched culture, change is hard to come by. To change the culture of separate disciplines in our universities will require well directed, persistent, and innovative government initiatives.

vii. The necessary changes in educational structure will come without strong directives from leaders from academia, industry, and government laboratories.

viii. And provide funding for multidisciplinary graduate education programs that offer students simulation integrated approach of team research and career development.

—

14.5.2 Challenges

There are challenges, that need to be faced regaining multiscale and multi-physics modeling, real-time integration of simulation methods with measurement systems, model validation and verification, handling large data, and visualization for discipline that want to incorporate it for the first time. It is only by these challenges involved in resolving open problems associated with simulation. Significantly, one of those challenges is education of the next generation of engineers and scientists in the theory and practices of simulation in every subject.

There is no doubt that a lot of money will be involved, and research will be required for specific feed where simulations need to be plugged into the program, But the risk worth taken research to exploit the considerable promise of Simulation in education.

Therefore it also necessary to provide new cyber—infrastructure that will allow teacher and scientists to pursue Competency and Objective Based Education and research in new ways and with new efficiency" by utilizing:

i. high-performance, global-scale networking,
ii. middleware,
iii. high-performance computing services,
iv. observation and measurement devices, and
v. improved of interfaces and visualization services.

Building simulation center require serious consideration of feasibility of developing a parallel programs in simulation that interfaces to multiple divisions of engineering education in concert with cyber infrastructure.

Progress in simulation will also require the creation of interdisciplinary education teams that work together on leading-edge simulation problems. A sweeping overhaul of our educational system towards simulation and initiative for change will not likely come from academia alone; it must be encouraged by the engineering and scientific leadership and throughout the organizational structure of our universities as well as simulation institution in industry. The payoffs for meeting these challenges are profound. We can expect dramatic advances on a broad frontier of knowledge and practice. The following list is a summary of its current limitations:

i. The development of models is very time consuming, particularly for geometries of complex engineering systems such as ships, automobiles, and aircraft. Moreover, the determination of material properties often requires extensive small-scale testing before simulation can be started, especially if statistical properties are needed. This testing also lengthens the time to obtain a simulation and hence the design cycle.

ii. Methods are needed for linking models at various scales and simulating multi-physics phenomena.

iii. Simulation is often separate from the design optimization process and cannot

iv. Simultaneously deal with factors such as manufacturability, cost, and environmental impact.

14.5.3 Overcoming the challenges of now

As design processes increasingly rely on computer simulation, validation and verification procedures will become increasingly important. Although some efforts have been made at providing validation benchmark problems for linear analysis, nonlinear simulation software has not been subjected to extensive validation procedures. Clearly, a basic understanding of verification and validation procedures is urgently needed. After all, to be useful, the simulation tools used by industry and defense agencies must provide reliable results. Furthermore, since many real-world phenomena are not deterministic, statistical methods that can quantify uncertainty will be needed. Design optimization is also in its infancy, and it too has many obstacles to overcome. The constraints on the optimization of a product design relate to manufacturability, robustness, and a variety of other factors. In order to be effective for engineering design, optimization methods must be closely coupled with simulation techniques. Overcoming the barriers behind simulation will require challenges and progress in our basic understanding and in the development of powerful new methods. Among these challenges are the following:

i. Multiscale methods that can deal with large ranges of time and spatial scales and link various types of physics.

ii. Methods for computing macroscopic phenomena, such as material properties and manufacturing processes, in terms of subscale behavior.

iii. Effective optimization methods that can deal with complex integrated systems, account for uncertainties, and provide robust designs.

iv. Frameworks for validation, verification, and uncertainty qualification.

v. Methods for rapidly generating models of complex geometries and material properties.

vi. Multiscale methods will provide extensive benefits.

The following are a few of the areas requiring development:

i. Quantitative models of the processes to be simulated must be developed. For many of those processes, models of some level of fidelity already exist, or they are being developed for narrower engineering purposes.

ii. A comprehensive simulation system is required that integrates detailed models of a wide range of scales. The comprehensiveness of the simulation system is a requirement if simulation in education applications is to simulate multiscale complex systems. Some of the issues are generic, but others are problem specific.

iii. New models of exceptional fidelity are required. The development and validation of such models entail the acquisition of data of extraordinary detail.

iv. A better understanding of the role of uncertainty is required. Some degree of uncertainty is inevitable in the ability of a model to reflect reality and in the data the model uses. We need to find ways to interpret uncertainty and to characterize its effects on assessments of the probable outcomes.

Generally, however, we still lack a fundamental understanding of what constitutes an optimal design and how to find it in a complex multi-criteria design environment. Once optimization methods are developed that can deal with these complexities, we can expect to see chemical plants, automobiles, laptop. Simulation has the potential to deliver, within a short design period, designs that are optimized for cost performance and total impact on the environment. The rewards of meeting those challenges are great: enhanced efficiency, security, safety, and convenience of life in the digital, infrastructure, city and ecosystem, a social infrastructure of unparalleled efficiency, rational responses to natural events and optimal interactions with natural environments.

The need for Simulation based education is at a crossroads in our global technological development. For almost half a century, developments in mathematical modeling, computational algorithms, and the technology of data intensive computing have led to remarkable improvements in the health, security, productivity, quality of life, and competitiveness of nations. We have now arrived at an historic moment where simulation is the key elements for achieving progress in engineering and science. The challenges of making progress, however, are as substantial as the benefits. We must, for example, find methods for linking phenomena in systems that span large ranges of time and spatial scales. We must be able to describe macroscopic events in terms of subscale behaviors. We need better optimization procedures for simulating complex systems, procedures that can account for uncertainties. We need to build frameworks for validation, verification, and uncertainty quantification.

In today's competitive world, in order to b at the frontier of knowledge it has become important to explore the possibility of incorporating in our engineering educational system to reflect the multidisciplinary nature of modern engineering and to help students acquire the necessary modeling and simulation skills. Thus simulation required good computer speed, funding and efficiency. However, this barrier can be solved by promoting interaction between multiple disciplines that fit naturally and strategically in parallel with or within the Cyber infrastructure framework.

Simulation definitely has the potential to deliver designs that are optimized for cost performance and their total impact on the environment (from production to disposal or recycling), all within a short design cycle. This achievement is not possible, however, simply by extending current research methods and taking small, incremental steps in simulation based education development. The barriers to the realization of simulation in education relate to our entire way of conducting research development and educating engineers. Other field of engineering can surely use experience of the shipping industry as a guide to incorporate simulation in education work.

Reference

1. Down, D.F., and R. Mercer. 1995. "Applying marine simulattion system to improve mariner proffesional development ". In the proceeding PIANC, Port '95, TAMPA Florida, March 1995
2. Roger D. Smith: *"Simulation: The Engine Behind the Virtual World"*, eMatter, December, 1999
3. Gynter, J.W.,TJ Hammell, J.A. grasso, and V.M. Pittsley. "Simulator for marne traningand licensing: guidelines for deck officer training systes.CAORF
4. Johnson, C., Moorhead, R., Munzner, T., Pfister, H., Rheingans, P., and Yoo, T. S., (Eds.); NIH-NSF Visualization Research Challenges Report; IEEE Press, 2006.
5. Hammel, T.J.) . the training devive is more than a simulator. Proceeding internatonal coference on marine simulation, 1981
6. Aldrich, C. (2003). Learning by Doing : A Comprehensive Guide to Simulations, Computer Games, and Pedagogy in e-Learning and Other Educational Experiences. San Francisco: Pfeifer—John Wiley & Sons
7. STCW 95:STYCW Convention, IMO London
8. Johnson, C.R., "Top Scientific Visualization Research Problems", IEEE Computer Graphics and Applications, pp. 2-6, July/August 2004
9. Hays, R.T. and Singer, MJ."Simulation fidelity in Training System Design—Bridging the gap between reality and Training "New York, 1989
10. Percival, F., Lodge, S., Saunders, D. The Simulation and Gaming Yearbook: Developing Transferable Skills in Education and Training. London: Kogan Page, 1993
11. Deutch, Christian "from work analysis tool to a training simulator: the realism of teaching tools.
12. Paramore, B and P King, study of task performance in report of Collision, Rammings, grounding in harbors and entrance, 1979
13. P. Humphreys, Extending Ourselves: Computational Science, Empiricism, and Scientific Method. Oxford: Oxford University Press, 2004
14. Wolfe, Joseph & Crookall, David, *Developing a scientific knowledge of simulation/gaming . Simulation & Gaming: An International Journal of Theory, Design and Research*, 1998

Appendix

Simulation Product Types

Simulator Centre	Certificate no	Expiry date	Simulator Type	Class	Country
Maririme Institute Willem Barnentsz	002/060222	2010.05.31	Bridge Operation	A	Netherlands
	003/060220	2010.05.31	Bridge Operation	A	
	004/060220	2010.05.31	Machinery Operation	A	
	005/060220	2010.05.31	Cargo handling	A	
STC B.V	006/060223	2010.05.10	Bridge Operation	A	
	007/060223	2010.05.10	Bridge Operation	A	
	008/060223	2010.05.10	Bridge Operation	A	
	009/060223	2010.05.10	Bridge Operation	A	
Malaysian Maritime Academy (ALAM)	001/060217	2011.02.17	Bridge Operation	A	Malaysia

	Certificate no	Expiry date	Simulator Type	Class	Country
	001/060523	2011.05.23	Bridge Operation	C	
	SSP-103	2006.06.30	Bridge Operation	A	
Sperry Marine Training Centre	SSP-104	2006.06.30	Bridge Operation	C	
	SSP-105	2006.06.30	Bridge Operation	C	Germany
	SSP-106	2006.06.30	Bridge Operation	C	
	SSP-202	2006.06.30	Machinery Operation	A	
Evergreen Seafarer Training centre	SSP-203	2006.06.30	Machinery Operation	X	
	SSP-301	2006.06.30	Radio Communication	B	
	SSP-204	2008.05.23	Machinery Operation	A	
National Kaoshiung Institute of Marine Technology	SSP-109	2006.07.25	Bridge Operation	X	
	SSP-115	2008.05.23	Bridge Operation	A	Taiwan
	SSP-118	2008.05.23	Bridge Operation	C	
	SSP-303	2007.10.11	Radio Communication	A	
	A-9467	2010.02.14	Cargo handling	A	

Simulator Centre	Certificate no	Expiry date	Simulator Type	Class	Country
	SSP-205	2008.05.23	Machinery Operation	A	
National Taiwan Ocean University	A-9368	2009.12.06	Cargo handling	A	Taiwan
	013/060825	2011.07.21	Bridge Operation	C	
	SSP-113	2007.10.10	Bridge Operation	A	
	SSP-116	2008.05.23	Bridge Operation	C	
Split Ship Management Ltd	SSP-110	2006.10.31	Bridge Operation	A	Croatia
	SSP-111	2007.06.28	Bridge Operation	A	

MPA Integrated Simulation Centre of singapore	SSP-112	2007.06.28	Bridge Operation	A	Singapore
Haeyong Maritime Service Co. Ltd	014/061026	2011.09.11	Bridge Operation	B	Korea
NYK Shipmanagment Pte Ltd Training Centre	015/061120	2011.11.20	Bridge Operation	A	Singapore
Mearsk Training Cantre AS	018/070323	2012.03.23	Cargo handling	A	Denmark
	016/061212	2011.11.15	Cargo handling	A	
Changi Naval Training Base FMSS Centre	016/061205	2011.12.05	Bridge Operation	A	Singapore
Warsash Maritime Academy	017/070228	2012.03.01	Cargo handling	A	United Kingdom

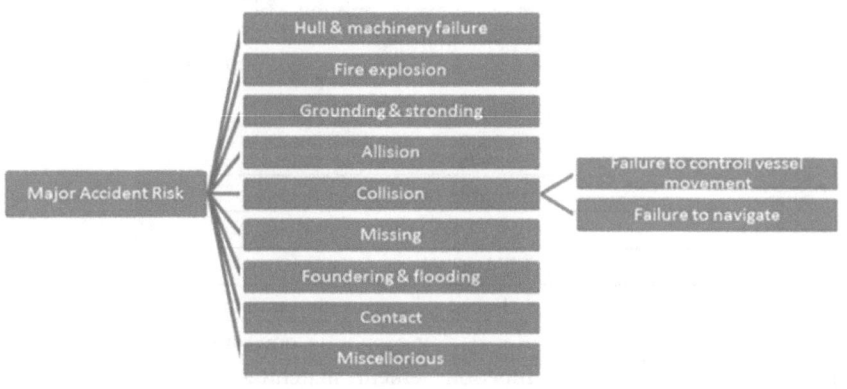

Accident scenario

Models used for design system based on risks

Model	Application	Drawback
Brown et al (1996)	Environmental Performance of Tankers	
(Sirkar et al (1997))	consequences of collisions and groundings	difficulties on quantifying consequence metrics

Brown and Amrozowicz	Hybrid use of risk assessment, probabilistic simulation and a spill consequence assessment model	Oil spill assessment limited to use of fault tre
Sirkar et al (1997)	Monte Carlo technique to estimate damage + spill cost analysis for environmental damage	Lack of cost data
IMO (IMO 13F 1995)	Pollution prevention index from probability distributions damage and oil spill.	Lack (Sirkar et al (1997)). rational
Research Council Committee(1999)	alternative rational approach to measuring impact of oil spills	Lack employment of stochastic probabilistic methods
Prince William Sound, Alaska, (PWS (1996)	the most complete risk assessment	Lack of logical risk assessment framework (NRC (1998))
((Volpe National Transportation Center (1997)).	Accident probabilities using statistics and expert opinion.	Lack employment of stochastic methods
Puget Sound Area (USCG (1999)).	simulation or on expert opinion for cost benefit analysis	Clean up cost and environmental damage omission

Process table

Process	Suitable techniques
HAZID	HAZOP, What if analysis, FMEA, FMECA
Risk analysis	FTA, ETA
Risk evaluation	Influence diagram, decision analysis
Risk control option	Regulatory, economic, environmental and function elements matching and iteration
Cost benefit analysis	ICAF, Net Benefit
Human reliability	Simulation/ probabilistic
Uncertainty	
Risk Monitoring	

Process table

Frequency classes	Quantification
very unlikely	once per 1000 year or more likely
Remote	once per 100- 100 year

—

452

Occasional	once per 10- 100 year
Probable	once per 1- 10 years
Frequent	more often than once per year

Process table

Quantification	Serenity	Occurrence	Detection	RPN
current failure that can apply to death failure, performance of mission	catastrophic	1	2	10
failure leading to degradation beyond accountable limit and causing hazard	critical	3	4	7
controllable failure leading to degradation beyond acceptable limit	major	4	6	5
nuisance failure that do not degrade system overall performance beyond acceptable limit	minor	7	8	2

Influence diagram

			Consequence Criteria				
			1 – Insignificant	2 – Minor	3 – Moderate	4 – Major	5 – Catastrophic
Likelihood	A	Consequence certain to occur	Medium (M)	High (H)	High (H)	Very High (VH)	Very High (VH)
	B	Consequence likely to occur	Medium (M)	Medium (M)	High (H)	High (H)	Very High (VH)
	C	Consequent possibly likely to occur some time	Low (L)	Medium (M)	High (H)	High (H)	High (H)
	D	consequence unlikely to occur but could happen	Low (L)	Low (L)	Medium (M)	Medium (M)	High (H)
	E	consequence may occur only in exceptional circumstances	Low (L)	Low (L)	Medium (M)	Medium (M)	High (H)

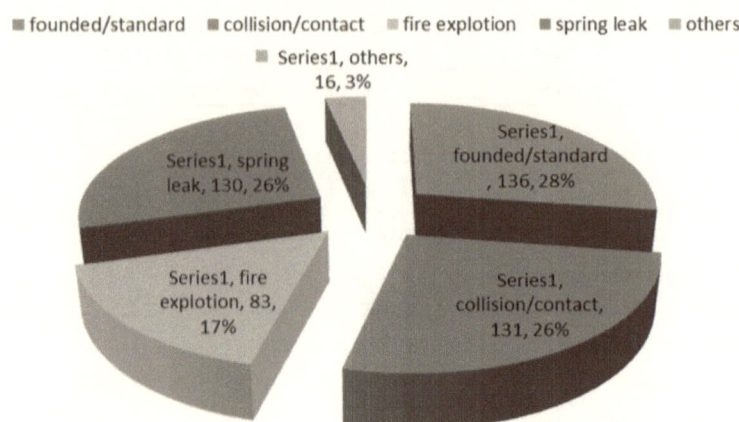

Waterways risk by accident categories

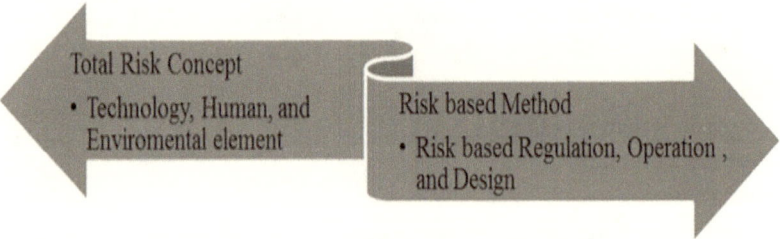

Risk modelling process

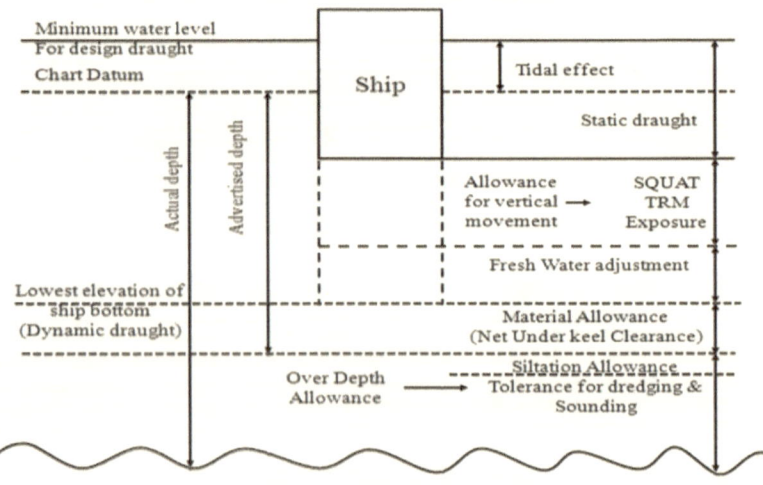

Channel and vessel dimension

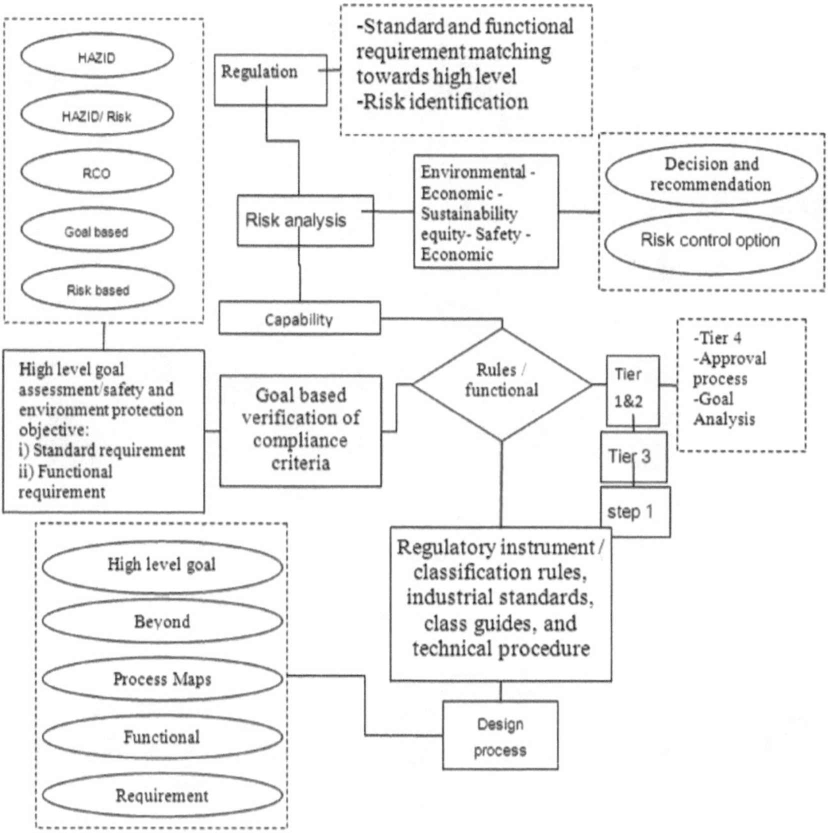

HAZID
HAZID/ Risk
RCO
Goal based
Risk based
Regulation
-Standard and functional requirement matching towards high level
-Risk identification
Risk analysis
Environmental - Economic - Sustainability equity- Safety - Economic
Decision and recommendation
Risk control option
Capability
High level goal assessment/safety and environment protection objective:
i) Standard requirement
ii) Functional requirement
Goal based verification of compliance criteria
Rules / functional
Tier 1&2
-Tier 4
-Approval process
-Goal Analysis
Tier 3
step 1
High level goal
Beyond
Process Maps
Functional
Requirement
Regulatory instrument / classification rules, industrial standards, class guides, and technical procedure
Design process

Holistic Risk analysis Process

COLLSION
C4
power transmital problem (Mode B- Gear- RPM, transmit power)
Event17
NAVIGATION FAILLURE
Gate19
VESSEL MOTION FAILLURE
Gate20

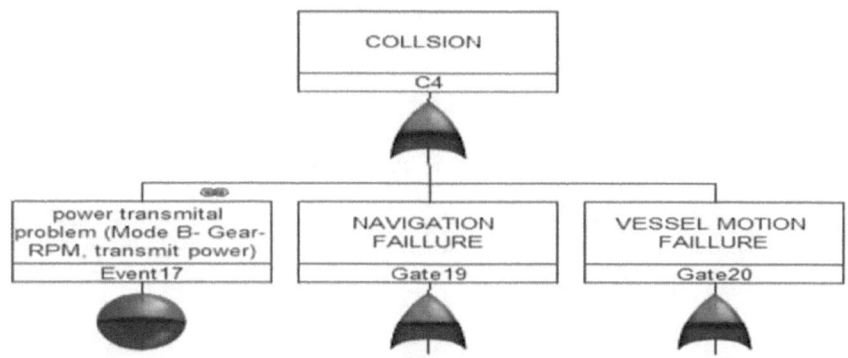

Collision contributing factors modelled from RELEX software

Risk assesment managment question

What can be done
• What option are available and what are their trade off in term of cost, benefit and risk?
• What are the impacts of current moment decision on future options?

Risk assesment Question
• What can go wrong?
• What is the likehood that it will go wrong
• What are the consequences?

Source of failure
• HArdware
• Software
• Organization
• Human

Total risk concept

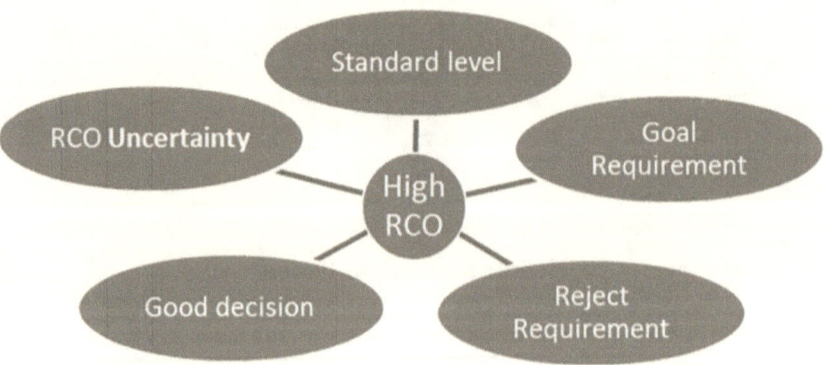

Cause of collision

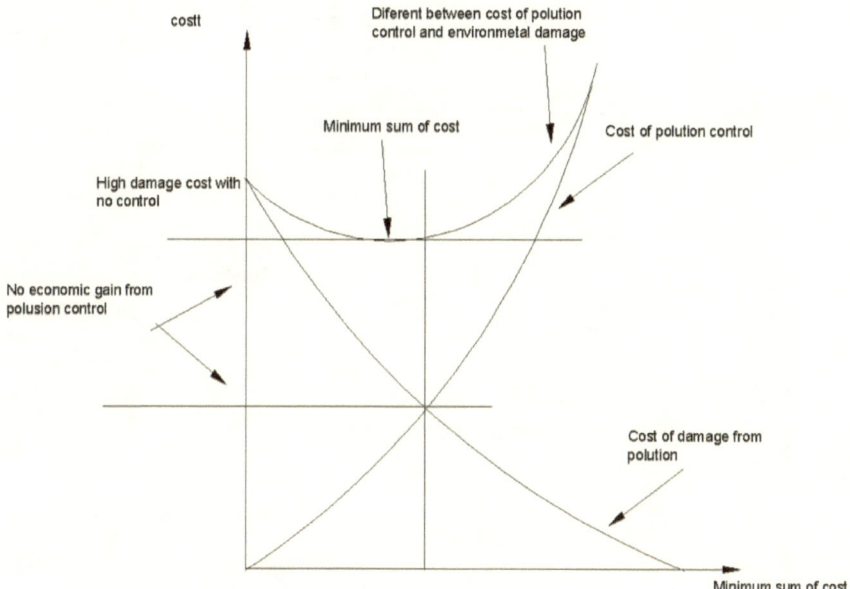

Sustainability analysis

Process outline

SIN	ACTIVITY	INPUT	INTERACTING SUB-PROCESS	CRITICAL ISSUES	CONTROLS	CONTROLING MEASUREMENT OUTPUT
1	Input from problem definition	Scope & detail from checklist	Making checklist	Scope of research , relevant w.r.t rule & regulation		
2	Stake holder	Profile		Contribution, availability		
3	Select recorder	Data recording		Ability to capture relevant inputs	Use of software, tape recording	Monitoring of records
4	Obtain necessary information , data &supporting documents Schedule & organize meting distribute material	Casualty statistic, data , expert input ALL. engagement	Root cause analysis of accident s and incidents			
5	Brainstorming to identify accident scenario using e.g DELPHI, what- if / checklist, FMEA, HAZOP, RCA, Task Analysis	Casualty statistic, data , expert input				
6	Comprehensive?	Accident scenario				
7	Estimate frequency: select appropriate technique e.g FMEA, FTA(with HRA)	Each accident scenarios				
8	Classify and rank hazards	Accident scenario(F &C)				

—

457

Regulation impacts

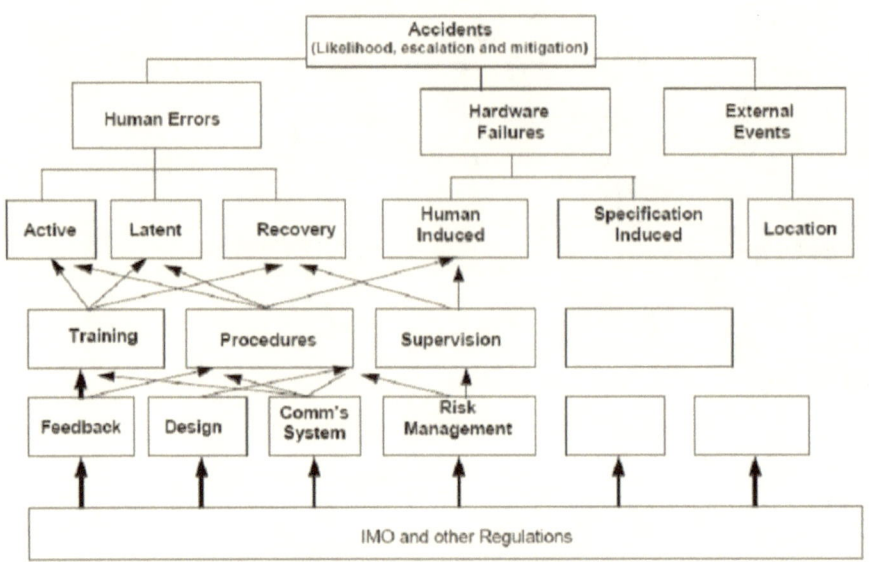

Preliminary hazard analysis

		triggering event1	hazardous condition	triggering event 2	potential accident	effect	corrective measures
kinetic energy	hazard element	loss of navigation control	ship1 sail on random course	another ship is on ship1 course	collision, rupture of cargo tanks	fatalities, environmental damage, damage to hull	Improving navigational standards
		loss of navigation control	ship1 sail on random course	stationary obstacle on ship 1 course	power grounding, rupture of cargo tank	fatalities, environmental damage, damage to hull	
		Obstacle on ship1 course	retardation (i.e. reverse)	movement of unfastened material on board vessel	crushed personnel, material damage	fatalities, environmental damage	

HAZOP

No	Guideword	Description	Causes	Safety measure
1	No Pitch	No rotational energy is transformed	operation, control mechanism, alignment failure	address by 2, 3, 4, 5
2	No blade	No rotational energy is transformed	Object in the water break the blade	implementation of propeller protection such as grating jet, sail in ice free water, +7& 9
3	No control bar	All blade on random pitch, loss of operational control	material weakness	improve design and construction
4	No crank wheel	On all blade have independent pitch		improve design and construction
5	NOT tough material strength	part of propeller breakdown	wrong design, corrosion or cavitations, alignment different pitch, extra load on bearing	validate propeller design, catholic protection, appropriate propeller material, test the propeller against cavitations, periodic alignment adjustment
6	MORE pitch than optimal	Too heavy load on propulsion system. Cavitations		surveillance, increase operator competency
7	LESS pitch than optimal	Too little load on propulsion system. Cavitations	operation failure	surveillance, increase operator competency
8	LESS draft than allowed	Propeller I not sufficiently submerged. Loss of Thrust		surveillance, increase operator competency
9	LESS depth than necessary	Propeller hit the ground and it is damaged		technical equipment, surveillance, increase operator competency

Nomenclature

Nm	Number of ship movement	Nm	Number of ship movement
V	Speed	Pi	Probability of impact
T	Draft of the ship	C_{head}	Head on Collision
B	Beam of the ship	Cov	Overtaking Collision
D	Depth of waterways	Ccr	Crossing Collision
W	Width of fairy	Can	Collision at specific angle
B1	Mean beam of meeting ship (m)	CCi	Collision at circular situation
B2	Mean beam of subject ship (m)	Pi	Impact probability
V1	Mean speed of meeting ship (knot)	E1	Accident drafting collision energy
V2	Mean speed of subject ship (knot)	E2	Accident Power collision energy
Theta	Angle	E	Consequence energy
L	Length of ship	Pa	Probability of accident
Ps	Traffic density of meeting ship (Ship/nm²)	Pc	Probability of losing control per passage of the fairway

www.ingramcontent.com/pod-product-compliance
Lightning Source LLC
Chambersburg PA
CBHW020720180526
45163CB00001B/45